AF785939

35 Springer Series in Solid-State Sciences

Edited by Manuel Cardona

Springer Series in Solid-State Sciences

Editors: M. Cardona P. Fulde H.-J. Queisser

Volume 40 **Semiconductor Physics** - An Introduction By K. Seeger

Volume 41 **The LMTO Method** By H.L. Skriver

Volume 42 **Crystal Optics with Spatial Dispersion and the Theory of Excitations** By V.M. Agranovich and V.L. Ginzburg

Volume 43 **Resonant Nonlinear Interactions of Light with Matter** By V.S. Butylkin, A.E. Kaplan, Yu.G. Khronopulo, and E.I. Yakubovich

Volume 44 **Elastic Media with Microstructure II** Three-Dimensional Models By I.A. Kunin

Volume 45 **Electronic Properties of Doped Semiconductors** By B.I. Shklovsky and A.L. Efros

Volumes 1 - 39 are listed on the back inside cover

J. Bourgoin M. Lannoo

Point Defects in Semiconductors II

Experimental Aspects

With a Foreword by G. D. Watkins

With 116 Figures

Springer-Verlag Berlin Heidelberg New York 1983

Dr. *Jacques Bourgoin*

Université de Paris VII, Groupe de Physique des Solides
de l'Ecole Normale Supérieure, Place Jussieu,
F-75221 Paris Cedex, France

Dr. *Michel Lannoo*

Laboratoire d'Etude des Surfaces et Interfaces, Physique des Solides,
Institut Superieur d'Electronique du Nord, 3, Rue François Baës,
F-59046 Lille Cédex, France

Series Editors:

Professor Dr. Manuel Cardona
Professor Dr. Peter Fulde
Professor Dr. Hans-Joachim Queisser

Max-Planck-Institut für Festkörperforschung, Heisenbergstrasse 1
D-7000 Stuttgart 80, Fed. Rep. of Germany

ISBN-13:987-3-642-81834-9 e-ISBN-13:978-3-642-81832-5
DOI: 10.1007/978-3-642-81832-5

Library of Congress Cataloging in Publication Data. (Rev. for vol. 2) Lannoo, M. (Michel), 1942- Point defects in semiconductors. (Springer series in solid-state sciences : 22, 35,) Includes bibliographical references and index. Contents: v. 1. Theoretical aspects – v. 2. Experimental aspects. 1. Semiconductors–Defects. 2. Point defects. I. Bourgoin, J. (Jacques), 1938-. II. Title. III. Series: Springer series in solid-state sciences ; 22, etc. QC611.6.D4L36 537.6'22 81-5354

This work is subject to copyright. All rights are reserved, whether the whole or part of the material is concerned, specifically those of translation, reprinting, reuse of illustrations, broadcasting, reproduction by photocopying machine or similar means, and storage in data banks. Under § 54 of the German Copyright Law, where copies are made for other than private use, a fee is payable to "Verwertungsgesellschaft Wort", Munich.

© by Springer-Verlag Berlin Heidelberg 1983
Softcover reprint of the hardcover 1st edition 1983

The use of registered names, trademarks, etc. in this publication does not imply, even in the absence of a specific statement, that such names are exempt from the relevant protective laws and regulations and therefore free for general use.

2153/3130-543210

To Helma and Ginette

Foreword

In introductory solid-state physics texts we are introduced to the concept of a perfect crystalline solid with every atom in its proper place. This is a convenient first step in developing the concept of electronic band structure, and from it deducing the general electronic and optical properties of crystalline solids. However, for the student who does not proceed further, such an idealization can be grossly misleading. A perfect crystal does not exist. There are always defects.

It was recognized very early in the study of solids that these defects often have a profound effect on the real physical properties of a solid. As a result, a major part of scientific research in solid-state physics has, from the early studies of "color centers" in alkali halides to the present vigorous investigations of deep levels in semiconductors, been devoted to the study of defects. We now know that in actual fact, most of the interesting and important properties of solids – electrical, optical, mechanical – are determined not so much by the properties of the perfect crystal as by its imperfections.

There are no better examples of the vital role played by defects than those found in semiconductors: the very existence of the semiconductor industry is based upon the ability to control electrical conductivity and its type by trace additions of shallow donor and acceptor chemical impurities. As optoelectronic devices, semiconductors rely upon trace impurities which serve as luminescence centers. At the same time, defects can serve as unwanted recombination centers limiting the minority carrier lifetimes and degrading device performance.

The study of defects in semiconductors has therefore been an active field of scientific research from the very beginning, when semiconductors first emerged as technologically important materials. Recently, however, we have begun to witness greatly expanded activity in this field. There are many reasons for this. Technologically, defects are of increased concern. The movement toward VLSI submicron semiconductor circuitry has greatly increased the vulnerability of devices to unwanted process-induced defects. At the same time, new devices and new process techniques have increased the demand for precise property control by desired defect introduction. Scientifically, the introduction of new experimental tools such as the various junction capacitance techniques, cooled semiconductor detectors which have made high-resolution optical studies in the infrared possible, dye lasers for selective excitation, optical detection of magnetic resonance, etc., have greatly increased the experimental facts at our disposal. Finally, the strides made by large quantum-mechanical computer techniques and new theoretical methods have made the theoretical treatment of the difficult deep level problem begin to look tractable. There is thus a renaissance in both experiment and theory, causing excitement as major new advances are beginning to be made in this field.

This two-volume series therefore comes at a propitious moment. On the one hand, the field of point defects in semiconductors is a mature, well-developed field, and these first textbooks devoted to the theory and experimental techniques fulfill a real academic need. At the same time, because the field is currently very active and expanding, they should also serve as a valuable introduction for the student or specialist who wants to join this exciting, challenging, and still "new" field of scientific inquiry.

George D. Watkins

Preface

This book treats physical properties of point defects in semiconductors, as they are derived from experiment. We consider electronic and atomic properties of semiconductors and how the presence of defects induces variation of these properties. By defects we mean intrinsic as well as extrinsic defects, i.e., we include impurities.

The first theoretical chapter (Chap.2) is devoted to the Jahn-Teller effect because most of the properties we describe are sensitive to electron-phonon coupling. Electron-phonon coupling is fully developed later, in the chapter on optical properties (Chap.4) which treats photon absorption and emission resulting from transitions between bands, excited states, and localized states associated with a defect. From these optical properties, information can be obtained on energy-level positions and optical cross sections (i.e., wave function, phonon interaction, and symmetry). Chapter 3 discusses paramagnetic resonance, to which defects containing unpaired electrons respond. This type of property appears to be very powerful in identifying defects because it gives direct information on the symmetry and nature of the impurity which may be involved. Carrier transport properties are considered in Chap.5, although they are not very useful for defect identification. However, they are still used to monitor defect behavior when spectroscopic techniques are not applicable. The equilibrium distribution of carriers between bands and defect levels is described here as an introduction to Chap.6, which is on non-radiative transitions. This chapter describes the kinetics of charge transfers between defect levels and bands from which the energy-level position, capture cross section, entropy and enthalpy of ionization can be obtained. Chapter 7 discusses the value of other methods of detecting and analyzing defect characteristics. Finally, Chaps.8 and 9 are devoted to defect creation by irradiation and to defect annihilation, respectively. Irradiation is a tool widely used to study defects because, by a suitable choice of the nature and energy of the irradiating particle, it is possible to some extent to vary the nature and distribution of the defects thus created. It is a means to introduce simple intrinsic defects and therefore to study vacancies, interstitials, and complex defects resulting from their interaction with themselves or with impurities. Defect annihilation gives information on thermodynamic properties of defects, such as migration enthalpies and entropies. Moreover, annealing kinetics studies, sensitive to the space distribution of the defects, can be helpful in defect identification.

The following defect properties are not dealt with in this book: i) large defects, i.e., defects involving typically more than ten lattice sites, because we restrict the subject of this book to point defects; ii) defects in large concentrations, typically, for an atomic concentration, larger than 10^{-2}, because then the point defects are no longer isolated but in interaction.

All the properties we discuss necessitate a knowledge of several notions such as wave functions, energy levels, vibrational properties, entropies and enthalpies of various thermodynamic quantities. These notions can be found in *Point Defects in Semiconductors I. Theoretical Aspects*,* which is summarized in Chap.1. In a way, the present volume can be considered as an application of the notions and concepts developed in the first one.

As in the first volume, we provide an example for each property treated. Whenever possible, we chose to use the same defect (the vacancy) in order to show how the different properties discussed are correlated and to demonstrate that only when several properties are analyzed in conjunction is it possible to provide a complete picture of a defect.

The subjects covered by this second volume correspond to a rather wide range of the physics of semiconductors. For this reason, we asked colleagues to review the chapters corresponding to their speciality. We therefore take the opportunity here to thank A.M. Stoneham, Y. Merle d'Aubigné, G.D. Watkins, D. Pons, and J.W. Corbett for their contributions.

Lille, January 1983 *J. Bourgoin · M. Lannoo*

---

* Springer Series in Solid-State Sciences, Vol. 22 (Springer, Berlin, Heidelberg, New York 1981)

Contents

1. Introduction

The first volume [1.1] contains seven chapters:

1) Atomic Configuration of Point Defects,
2) Effective Mass Theory,
3) Simple Theory of Deep Levels in Semiconductors,
4) Many-Electron Effects and Sophisticated Theories of Deep Levels,
5) Vibrational Properties and Entropy,
6) Thermodynamics of Defects,
7) Defect Migration and Diffusion.

In Chap.1, we define the object under study, i.e., the point defect. After a general definition, we describe the geometrical configurations these defects take in the diamond lattice: interstitial, vacancy, divacancy, substitutional and interstitial impurities, and some complexes formed by the association of intrinsic and extrinsic defects. We also introduce the notion of lattice relaxation and distortion, i.e., the deformation the lattice is submitted to around the defect. The knowledge of the symmetry of a defect is very useful for the treatment of its electronic states because it allows the simplification of the matrix elements of the defect Hamiltonian. For this reason, we introduce the various aspects of group theory useful in this respect, which we illustrate by the example of the vacancy (which has the full symmetry of the lattice, eventually lowered by distortions). Defect symmetry can be determined experimentally using the various spectroscopic properties described in the present volume. This is done by studying the splitting of the associated electronic states under external polarized excitations (such as uniaxial stress or polarized light). This is described, and again illustrated for two states (E and T_2) of the T_d group, to which the covalent lattice belongs.

The next three chapters are devoted to the determination of the electronic states of a defect. Two extreme situations corresponding to the so-called shallow and deep levels are considered, because they have to be treated in completely different manners. First, shallow levels, with which, for instance, the doping impurities are associated, are treated in the effective mass approximation (Chap.2). This treatment, which leads to the solution of a Schrödinger equation similar to the Schrödinger equation of the hydrogen atom, is presented in real space for the sake of simplicity and not, as usually done, in reciprocal space. The simple case corresponding to one band with one extremum is given, in addition to the generalizations to several equivalent extrema and to a degenerate extremum. We also treat in this chapter the case of pairs of defects, because even if defects individually give rise to deep levels, their association in pairs can result in the formation of shallow levels. We thus derive the density of states associated with a distribution of pairs. The chapter ends with a short description of the techniques allowing the characterization of shallow levels; this is illustrated by a few results concerning ground and excited states of doping impurities and pairing effects on ionization energies.

One useful way to treat deep levels is to use the tight-binding theory. Chapter 3 deals with the application of simple tight-binding considerations to the case of point defects. First, the principle of the tight-binding approximation is recalled and the tight-binding matrix elements of a covalent crystal are given and their physical meaning discussed. Secondly, we show how the band structure of a covalent crystal is derived in this approximation and we discuss the type of refinements which are necessary to obtain realistic results. Thirdly, we apply this elementary tight-binding theory to several examples: the vacancy (first in a linear chain and then in the covalent crystal), the interstitial and the substitutional impurity in order to illustrate the degree of importance of the various interactions which have to be considered. A second part of the chapter is devoted to the application of Green's function techniques to the tight-binding theory of defects. First, Green's functions are defined and their relation to densities of states and local densities of states is briefly demonstrated. Then, we describe how Green's functions must be used to treat local perturbations, such as the one representing a deep-level center in a perfect crystal. We apply the above considerations to the so-called Koster-Slater model, which calculates the density of states associated with the introduction of a substitutional impurity in a one-band tight-binding system. Finally, we introduce the method of moments, based on the expansion in series of the resolvent operator associated with Green's function, which we illustrate by treating the case of the vacancy in a linear chain.

The second chapter devoted to the treatment of deep levels (Chap.4) develops more sophisticated theories of deep levels and, in particular, treats many electron effects. The reason theories more refined than the simple tight-binding theory are needed is that electron-electron interactions, ignored in the tight-binding approximation, can play an important role. In a first part of the chapter we describe briefly the main one-electron self-consistent calculations: the CNDO method, the Extended Hückel Theory and the self-consistent Green's function techniques. Self-consistent calculations are necessary because the electron populations of the atoms which form a defect have not neutral free-atom values; the atoms bearing net charges, the dependence of the Hamiltonian matrix elements on these charges must be taken into account. In order to show the necessity of self-consistency and the importance of screening effects on localized states, we present a simplified version of the model of Haldane and Anderson. This model, originally derived for the case of metal impurities in covalent crystals, illustrates the strong dependence of the defect energy level with its charge state. We also show that a quite good approximation of a self-consistent perturbative potential, associated with a defect, used for instance in Green's function calculations, can be provided by the Thomas-Fermi approximation. The second part of the chapter treats many electron effects, the so-called configuration interaction, i.e., analyzes the importance of correlation between electrons. This is done first for the hydrogen molecule (H_2), then for the vacancy within the defect molecule model. The hydrogen molecule being the simplest case possible, it is possible to calculate all matrix elements between Slater determinants, to get the characteristics of the eigenstates, to introduce the configuration interaction and to obtain the restricted as well as the unrestricted Hartree-Fock solutions. Then, the different configurations of the four electrons of a vacancy are derived, as well as the wave functions and matrix elements of the Hamiltonian between the positive, neutral and negative states of the vacancy. This leads us, finally, to discuss the validity of the one-electron theories in terms of the ratio of the intra-atomic interaction to the interatomic Coulomb integral.

The vibrational modes introduced by the presence of a defect and the associated entropy are calculated in Chap.5. Here, it is shown that defects modify the phonon density of states by introducing new modes of vibration which remain localized near the site of the defect. In order to obtain these vibrational modes, the notion of dynamical matrix for the perfect crystal is first introduced. It is then applied to derive the phonon dispersion curve of a linear chain, with one or two atoms per unit cell, and of a covalent crystal. Then, we develop different models corresponding to point defects: the mass defect and the vacancy, first in a linear chain and secondly in a three-dimensional system. For the realistic three-dimensional case, we make use of the Green's function method. Once the vibrational modes are known, it is possible to deduce the corresponding entropy terms. We thus recall the general expression of a vibrational entropy and give an approximate expression of it, using an expansion technique which involves the moments of the density of states. Again, we apply the technique to the case of the vacancy and consider, in particular, the change in entropy for self-diffusion, sum of the migration and formation entropies, which accompanies a relaxation or a distortion around it. We thus find that the entropy change associated with self-diffusion is quite large, in agreement with experimental observations. The chapter ends with a short description of the techniques used to measure entropies.

At thermal equilibrium, the nature and concentration of a given defect is a well defined function of the entropy and enthalpy for the formation of this defect. Chapter 6 begins with a definition of the enthalpy for defect formation and an evaluation of the formation enthalpy of a vacancy. Once this quantity is known, the defect concentrations at equilibrium are derived, for a group IV or for a compound semiconductor, in cases where there is no interaction and where there is interaction between defects; the cases of long-range and short-range interactions are discussed separately. This derivation leads us to deduce thermodynamically the law of equilibrium between different charge states of a defect, law which is usually obtained from Fermi-Dirac statistics. It also leads us to discuss the nature of the defects which should be present at thermal equilibrium in a given semiconductor. The chapter ends with a short description of the techniques used to obtain experimentally enthalpies of defect formation, and with a derivation of the statistical distribution of donor-acceptor pairs, whose distribution has been used in Chap.2 to get the associated electronic density of states.

The final chapter deals with defect migration and diffusion (Chap.7). It starts by describing the various mechanisms by which a defect (or an impurity) migrates in a lattice. These mechanisms are characterized by a jump probability and by a migration energy. We discuss, first, the jump probability in terms of rate and dynamical theories and, second, the way migration energies are evaluated. Because charge-state effects appear to be an important parameter for defect migration, we then describe a model from which the charge state dependence of defect migration can be explained. This model introduces the so-called ionization enhanced diffusion mechanisms. They are the mechanisms by which defect charge-state changes lead to defect migration. In particular, we describe how alternative charge-state changes lead to athermal migration, i.e., without the help of phonons, the energy required to induce the migration being supplied by carrier recombination. The subject of the second part of the chapter is diffusion. We briefly recall Fick's law and the way a diffusion coefficient is measured. We then discuss in turn self-diffusion, substitutional impurity diffusion and interstitial impurity diffusion. From experimental data, we make an attempt to extract the parameter which governs these diffusions and we conclude that charge-state effects are the dominant parameter.

2. Lattice Distortion and the Jahn-Teller Effect

Many point defects are subject to lattice distortion, i.e., the atoms in their neighborhood are displaced with respect to their perfect crystal positions. As a consequence the point-group symmetry is often lowered and this can be observed directly in different experiments such as Electron Paramagnetic Resonance (EPR) (Chap.3) and optical absorption (Chap.4).

Recently, a new important consequence of such distortions has been theoretically predicted and experimentally observed: this is the existence of negative U centers formed by different successive charge states of a given defect. In this chapter we treat in some detail the case of the vacancy in silicon which is the only example for which theoretical calculations have predicted the existence of such a negative U center. To give a correct account of such a theory we shall have to be more precise about the notion of energy levels and, in particular, we shall define the "occupancy levels" which must be used in Fermi-Dirac statistics.

The driving force for distortion around defects is often due to the Jahn-Teller effect [2.1]. Such distortion occurs when the electronic state is degenerate, in which case the nuclear configuration is unstable. Atomic displacements always exist which, by lowering the symmetry, split the degenerate level. As the centroid remains fixed, there is one sublevel which is lowered and the system is thus stabilized by the spontaneous distortion. There are usually several equivalent stable new atomic configurations corresponding to distortions which are equivalent by symmetry.

When a defect is frozen in one of its stable positions, the Jahn-Teller effect is said to be static. When the reorientation of the defect between its different equilibrium configurations begins to be important, the Jahn-Teller effect becomes dynamic. Both limits have been observed experimentally.

We begin this chapter by general considerations on the electron-phonon interaction (Sect.2.1). We recall the basic principles of the Born-Oppenheimer approximation (Sect.2.1.1), define the electron-lattice coupling operator (Sect.2.1.2) and discuss the occupancy levels which give the relative concentrations of different charge states of a given defect (Sect.2.1.3). In Sect.2.2.1 we apply symmetry considerations to the reduction of the Jahn-Teller matrices in T_d symmetry. We then determine the stable distortions (Sect.2.2.2), giving a brief discussion of cases where there is near degeneracy (Sect.2.2.3). Section 2.3 is devoted to the vibronic states (vibrational and electronic). It introduces the static and dynamic Jahn-Teller limits on simple cases. This enables the origin of the Ham effect (consisting of the quenching of electronic operators) to be shown. This section ends with a discussion of the transition from the static to the dynamic limit. Section 2.4 details the case of the silicon vacancy. We recall the static distortions which occur near this defect (Sect.2.4.1) and give a discussion of the relative importance of many electron effects as compared to the Jahn-Teller effect (Sect.2.4.2). We then define the effective force constants to be used for calculating the lattice response to the Jahn-Teller forces (Sect.2.4.3). Finally we discuss the "negative U" nature of this center which has been established recently (Sect.2.4.4).

Textbooks discussing the Jahn-Teller effect for defects in solids are given in [2.2-4] and [Ref.1.1, Chap.5], which should be consulted in order to get an idea of the extensive literature devoted to this question. In this chapter we have not discussed specific examples of this effect, putting the main emphasis on the negative U-center nature of some defects. However, one good illustration of the importance of the Jahn-Teller effect will be found in Sect.4.5 for the GR absorption band in diamond.

2.1 The Electron-Phonon Interaction

In this section we present the basic approximations which allow the separation of electronic and nuclear motion to identify the term which will be at the origin of the Jahn-Teller effect.

2.1.1 The Born-Oppenheimer and Related Adiabatic Approximations

The complete Hamiltonian of the system under study is a function of the whole set of electron coordinates denoted $\underline{r}$ and nuclear coordinates $\underline{R}$. Both types of motion cannot be separated rigorously. However the separation can be achieved approximately using the fact that the nuclear masses M are much larger than the electron mass m. We can expand the solutions in terms of the ratio m/M. The zeroth-order approximation to the solution (i.e., m/M = 0) corresponds to $M \to \infty$, a situation which is physically equivalent to fixed nuclei. To that order the Schrödinger equation can be written

$$H_e(\underline{r},\underline{R})\psi_e(\underline{r},\underline{R}) = E_e(\underline{R})\psi_e(\underline{r},\underline{R}) \quad . \tag{2.1}$$

In this equation $H_e(\underline{r},\underline{R})$ is the total Hamiltonian in the limit $M \to \infty$, i.e., when the kinetic energy of the nuclei has been neglected. Equation (2.1) thus describes the electron motion for nuclei fixed at given positions $\underline{R}$. The total wave function $\Phi(\underline{r},\underline{R})$ for a nondegenerate system is then taken to be of the form

$$\Phi(\underline{r},\underline{R}) = \psi_e(\underline{r},\underline{R})\chi(\underline{R}) \quad . \tag{2.2}$$

The best function $\chi(\underline{R})$ in this trial wave function can be obtained through a variational procedure, equivalent to

$$\int dv\, dV\, \psi_e^*(\underline{r},\underline{R})\delta\chi^*(\underline{R})[H_e(\underline{r},\underline{R}) + T_N - E]\psi_e(\underline{r},\underline{R})\chi(\underline{R}) = 0 \quad , \tag{2.3}$$

where E is the total energy and T_N the nuclear kinetic energy operator; the integration is performed over both electronic and nuclear coordinates. Equation (2.3) simply states that the average value of the total Hamiltonian $H_e + T_N$ is extremum under any first-order change $\delta\chi$ of the trial function. As (2.3) must be verified for any first-order $\delta\chi$, it can be rewritten as

$$\int dv\, \psi_e^*(\underline{r},\underline{R})[H_e(\underline{r},\underline{R}) + T_N]\psi_e(\underline{r},\underline{R})\chi(\underline{R}) = E\chi(\underline{R}) \tag{2.4}$$

where the integration is now performed over the electronic coordinates only. In the adiabatic approximation the contributions resulting from the action of T_N on $\psi_e(\underline{r},\underline{R})$ are dropped and the following simple form is obtained for the equation of nuclear motion

$$[T_N + E_e(\underline{R})]\chi(\underline{R}) = E\chi(\underline{R}) \tag{2.5}$$

which, in practice, is often simplified further by the use of the harmonic approximation [2.5,6].

The case of degeneracy can be treated in the same spirit, except that the total wave function has to be written

$$\Phi(\underline{r},\underline{R}) = \sum_\alpha \chi_\alpha(\underline{R})\psi_{e\alpha}(\underline{r},\underline{R}) \tag{2.6}$$

where the summation is made over the α degenerate (or near degenerate) set. We must again use a variational procedure to get the following set of equations:

$$\sum_{\alpha} \int \psi_{e\beta}^{*}(\underline{r},\underline{R})[H_e(\underline{r},\underline{R}) + T_N]\psi_{e\alpha}(\underline{r},\underline{R})\chi_{\alpha}(\underline{R})dv = E\chi_{\beta}(\underline{R}) \quad . \tag{2.7}$$

This set can be simplified by noticing that the $\psi_{e\alpha}$ are solutions of (2.1) with energies $E_{e\alpha}$. This leads to

$$[T_N + E_{e,\beta}]\chi_{\beta}(\underline{R}) + \sum_{\alpha} H_{\beta\alpha}^{NA}\chi_{\alpha}(\underline{R}) = E\chi_{\beta}(\underline{R}) \quad , \tag{2.8}$$

where the $H_{\beta\alpha}^{NA}$ is the matrix element of the nonadiabaticity operator which is defined by

$$\int dv\ \psi_{e\beta}^{*}(\underline{r},\underline{R})T_N\psi_{e\alpha}(\underline{r},\underline{R})\chi_{\alpha}(\underline{R}) = T_N\chi_{\alpha}(\underline{R})\delta_{\alpha\beta} + H_{\beta\alpha}^{NA}\chi_{\alpha}(\underline{R}) \quad . \tag{2.9}$$

This operator will play an important role, for instance in multiphonon emission capture (see Sect.6.3.3).

2.1.2 Electron-Lattice Coupling

We now discuss the origin of the Jahn-Teller coupling term. For this we start from the perfect crystal equilibrium atomic configuration $\underline{R}_0$. If we introduce a defect in this system we have first to solve (2.1), whose solutions for $\underline{R} = \underline{R}_0$ can be written as

$$H_e^0\psi_{\alpha}^0 = E_e^0\psi_{\alpha}^0 \tag{2.10}$$

where H_e^0 stands for $H_e(\underline{r},\underline{R}_0)$, ψ_{α}^0 for $\psi_{e\alpha}(\underline{r},\underline{R}_0)$ and E_e^0 for $E_e(\underline{R}_0)$. For the perfect crystal configuration $\underline{R}_0$, the point defect usually has a high symmetry so that, in many cases, the energies E_e^0 are degenerate. A point which will be important in the following is that we are mainly concerned with solutions of (2.10) or (2.1) which lie in the forbidden energy gap, i.e., are localized near the defect site (this will be discussed in more detail in Sect.2.1.3). Let us then consider such a localized state, degenerate by symmetry for $\underline{R} = \underline{R}_0$, and discuss its possible splitting for $\underline{R} \neq \underline{R}_0$. This splitting, if it occurs, must be given by the solutions of (2.1) which tend towards the desired energy E_e^0 and wave functions ψ_{α}^0. For a nuclear configuration $\underline{R}$ close to $\underline{R}_0$ we can expand the electronic Hamiltonian $H_e(\underline{r},\underline{R})$ to second order and write

$$H_e(\underline{r},\underline{R}) = H_e^0 + \sum_{s}\left(\frac{\partial H_e}{\partial Q_s}\right)_0 Q_s + \frac{1}{2}\sum_{s,s'}\left(\frac{\partial^2 H_e}{\partial Q_s \partial Q_{s'}}\right)_0 Q_s Q_{s'} + \dots \quad . \tag{2.11}$$

In this expression the Q_s are the normal displacements, i.e., suitable linear combinations of the cartesian coordinates of the atomic displacement vectors from the perfect crystal position (we shall see later how these displacements Q can be defined). The second and third terms in the right-hand side of (2.11) define a perturbation V. We now express the solutions of (2.1) in terms of the ψ_α^0 by conventional first-order perturbation theory on a degenerate state, the unperturbed Hamiltonian being H^0. The splitting of the degenerate level is then given to first order in V, by the diagonalization of the perturbation matrix, whose general element $V_{\alpha\beta}$ is

$$V_{\alpha\beta} = \sum_s Q_s \left\langle \psi_\alpha^0 \middle| \left(\frac{\partial H_e}{\partial Q_s}\right)_0 \middle| \psi_\beta^0 \right\rangle + \frac{1}{2} \sum_{ss'} Q_s Q_{s'} \left\langle \psi_\alpha^0 \middle| \left(\frac{\partial^2 H_e}{\partial Q_s \partial Q_{s'}}\right)_0 \middle| \psi_\beta^0 \right\rangle \quad , \tag{2.12}$$

ψ_α^0, ψ_β^0 belonging to the subspace of the degenerate level. Usually, simplifying assumptions are made concerning the second-order term: only its diagonal contribution, often independent of α in simple cases (if not, it is assumed to be so), is retained. The Q_s can thus be viewed as the normal vibration modes of the system containing the defect.

With such an assumption for the second-order term in (2.12), the splitting is given by the diagonalization of the first-order part since $V_{\alpha\beta}$ can generally be written

$$V_{\alpha\beta} = \left(\frac{1}{2} \sum_s k_s Q_s^2\right) I_{\alpha\beta} + \sum_s \left\langle \psi_\alpha^0 \middle| \left(\frac{\partial H_e}{\partial Q_s}\right)_0 \middle| \psi_\beta^0 \right\rangle Q_s \quad , \tag{2.13}$$

where the k_s are the force constants and I is the unit matrix. Let us call $E_i(Q_s)$ the eigenvalues of the first-order matrix. Each of them has to be minimized with respect to Q_s in order to give the new stable positions of the system. Finally, the adiabatic approximation can be used in the vicinity of each absolute minimum when the different energy branches are not too close together. This will be discussed with specific examples in the following.

The procedure we have just followed describes the physical origin of the Jahn-Teller effect, i.e., the splitting of the degenerate states under low-symmetry distortions. In their original work, JAHN and TELLER [2.1] have used group theory and considered all possible examples for which it can be shown, by symmetry, that such a splitting indeed exists. Some of these cases will be worked out in Sect.2.2.

2.1.3 Occupancy Levels and One-Electron Eigenvalues

Elementary electronic excitations in semiconductors are often discussed in terms of the independent-electron picture. A point defect can create localized gap states, for which the wave function exponentially decreases away from the defect site [Ref.1.1, Chaps.2,3]. In such cases the defect can have different charge states (depending on the occupancy of the localized states) for which the energy-level position varies within the gap. The corresponding shift is not only due to the Coulomb electrostatic interaction but can also result from symmetric relaxation around the defect. Experimentally, the kinetics of charge transfer represent an important factor, i.e., it is important to know whether the charge transfer can occur in the time of the experiment or not.

Such a picture is usually too naive and it is necessary to be more precise concerning the definition of the energy levels. Let us begin by defining the notion of charge state of a defect. For this we consider an isolated point defect in a semiconductor (or an insulator).

The defect introduces a modification $\Delta\rho(\underline{r})$ of the charge density with respect to that in the perfect crystal. If we integrate $\Delta\rho(\underline{r})$ over a volume V centered on the defect we can define the charge

$$q = \int_{V_\infty} \Delta\rho(\underline{r})dV \tag{2.14}$$

which is equal to

$$q = \frac{ne}{\varepsilon} \tag{2.15}$$

where ε is the macroscopic dielectric constant and V_∞ means that we take a volume large enough so that q has reached its limiting value. In semiconductors or insulators $q\varepsilon$ is equal to an integer n times the absolute value e of the electron charge. In this case we say that the defect is in its nth charge state (in a metal the situation is similar but perfect screening occurs corresponding to infinite ε).

With this definition the total electronic energy of a point defect can be labelled as $E_e(n,\underline{R})$ since it depends upon the charge state n for a fixed nuclear configuration $\underline{R}$. For fixed $n,\underline{R}$ E_e can take different values corresponding to the ground and excited states. For each of these states we have (according to the discussion of Sects.2.1.1,2) to minimise $E_e(n,\underline{R})$ to find the corresponding stable lattice configuration, and then expand E_e to second order in the atomic displacements to get the phonon Hamiltonian. Once this is done, it is possible to define the excitation energies as the differences between two such minima. However, before discussing these excitation energies it is convenient to define ionization energies which will correspond, as we shall see next, to the occupancy levels [2.7,8].

We thus define E(n, n + 1) as the following quantity

$$E(n,\ n+1) = E_e(n\ ,\ \underline{R}_n) - E_e(n+1,\ \underline{R}_{n+1}) \quad , \tag{2.16}$$

where $\underline{R}_n$ is the stable atomic configuration for charge state n (it can be different in the ground and excited states). The quantity -E(n, n + 1) represents the ionization energy of the defect. Its value depends on the choice of the origin of energies which can be taken as that of the electron at rest in the vacuum, for instance. From [Ref.1.1, Chap.6] we know that the corresponding equilibrium concentrations of defects c(n) and c(n + 1) in the two charge states are related through:

$$\frac{c(n)}{c(n+1)} = \frac{Z(n)}{Z(n+1)} \exp\left[-\frac{G(n) - G(n+1) - E_F}{kT}\right] \tag{2.17}$$

where Z is a degeneracy factor and G a free energy. Separating the entropy (S) and enthalpy (E) parts, we obtain

$$\frac{c(n)}{c(n+1)} = \frac{Z(n)}{Z(n+1)} \exp\left[\frac{S(n) - S(n+1)}{k}\right] \exp\left[-\frac{E(n,n+1) - E_F}{kT}\right] \quad . \tag{2.18}$$

When we ignore complications due to entropy and degeneracy, we can say that, when the Fermi level raises and crosses the level E(n, n + 1), the concentration c(n + 1) dominates over c(n). Thus E(n, n + 1) can be termed as an occupancy level. However, some care must be taken concerning the degeneracy and entropy factors which may play an important role. In practice E(n, n + 1) is usually calculated as a difference in internal energies rather than enthalpies.

This notion can be generalized to the ionization of two electrons, for instance, defining E(n, n + 2) as

$$E(n,n+2) = \frac{1}{2}\,[E_e(n,\underline{R}_n) - E_e(n+2,\ \underline{R}_{n+2})] \tag{2.19}$$

so that we can write the ratio c(n)/c(n + 2) under the form

$$\frac{c(n)}{c(n+2)} = \frac{Z(n)}{Z(n+2)} \exp\left[\frac{S(n) - S(n+2)}{k}\right] \exp\left\{-\frac{2[E(n,n+2) - E_F]}{kT}\right\} \quad , \tag{2.20}$$

showing again that E(n,n + 2) has the meaning of an occupancy level. This case will be important in the following for the description of the negative U center formed by the vacancy (Sect.2.4.4).

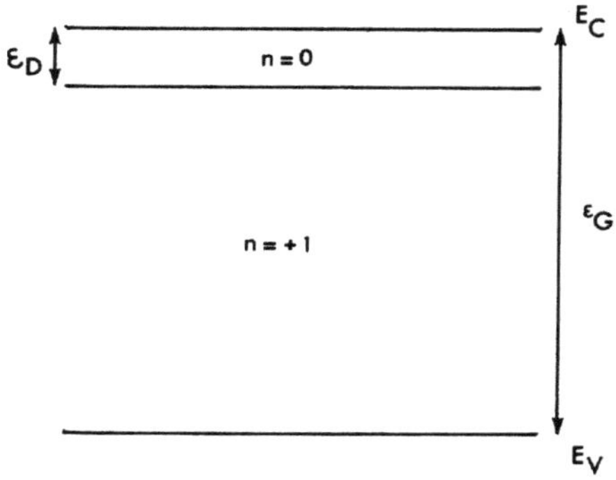

Fig.2.1. Level scheme for hydrogenic donor impurities

One application of the notion of occupancy levels corresponds to noninteracting hydrogenic donors [Ref.1.1, Chap.2]. As shown in this chapter, in [1.1], if there is no significant change in lattice configuration, the ionization energy to the conduction band is ε_D. The corresponding occupancy level is thus $E_C - \varepsilon_D$ (Fig.2.1) which leads to the well-known relation

$$\frac{C(n = 0)}{C(n = 1)} = 2 \exp\left(- \frac{E_C - \varepsilon_D - E_F}{kT}\right) \quad (2.21)$$

where the factor 2 accounts for the spin degeneracy in the case $n = 0$.

To conclude this section, it is important to relate the occupancy levels i) to the excitation energies (for optical transitions) and ii) to the one-electron energies which are determined in actual numerical calculations. The first point is obvious. We can calculate quantities of the type $E(n,n + 1)$ between the ground states in both charge states n and $n + 1$, which we note more precisely $E(n,g,n + 1,g)$. We can also define $E(n,e,n + 1,g)$ as the difference in total energies between an excited state in charge state n and the ground state in charge state $n + 1$ (calculated at the stable atomic positions). The difference $E(n,e,n + 1,g) - E(n,g,n + 1,g)$ is thus clearly equal to the zero-phonon energy for the transition $g \to e$ in charge state n.

The second point is less obvious. In the general case, there is no way of expressing directly an occupancy level $E(n,n + 1)$ as an eigenvalue of a one-electron Schrödinger equation. Instead, as in all the current methods (Hartree or Hartree-Fock, local density [2.9,10], etc.), we must calculate each total energy $E_e(n,\underline{R})$ by summing the one-electron eigenvalues, then substract the electron-electron repulsions which have been counted twice, and finally add the nuclear-nuclear interactions. Clearly, after minimization, to obtain $E_e(n,\underline{R}_n)$, the quantities $E(n,n + 1)$ defined by (2.16) have no simple relationship with any of the one-electron eigenvalues. It is only when we calculate $E_e(n,\underline{R}) - E_e(n + 1,\underline{R})$ in a density functional formalism that we can use Slater's transition argument, i.e., expand each $E_e(n,\underline{R})$ with respect to $E_e(n + \frac{1}{2},\underline{R})$ (the intermediate state) and show that this difference is, to second order, equal to $[-(\partial E_e/\partial m)(m,\underline{R})]_{m=n+\frac{1}{2}}$ [2.11,12]. The change in charge state corresponds to a change in occupation number of the localized orbitals and it has been proved [2.12] that this quantity is just the opposite of the corresponding one-particle eigenvalue, i.e., is the solution of the equation

$$\left[-\frac{\hbar^2}{2m}\Delta + v(\underline{r})\right]\varphi(\underline{r}) = -\left[\frac{\partial E_e}{\partial m}(m,\underline{R})\right]_{m=n+\frac{1}{2}}\varphi(\underline{r}) \tag{2.22}$$

where $v(\underline{r})$ is the one-electron potential energy in the local density approximation. It is only in such cases (for instance, when there is no lattice relaxation or distortion) that (2.22) and $E(n,n + 1)$ can be related to a one-electron energy, i.e., to the eigenvalue of the one-particle equation.

2.2 Symmetry Considerations: The Stable Atomic Configurations

In this section we discuss in some detail the application of the general arguments developed in Sect.2.1.2 to the case of point defects having a T_d (or O_h) symmetry. Symmetry allows the reduction in the number of independent parameters characterizing the distortions. We choose here, as illustration, the case of T_d symmetry but the results are also valid in O_h symmetry. After reducing the Jahn-Teller matrices we minimize the electronic energy to get the stable distortions. Finally, we consider briefly the case of near degeneracy where similar considerations apply.

2.2.1 General Reduction of the Jahn-Teller Matrices in T_d Symmetry

Let us consider the case of a point defect for which the point symmetry group is T_d. It corresponds to many important defects (the vacancy, tetrahedral interstitials, substitutional impurities, etc.). We apply the formalism introduced in [Ref.1.1, Chap.1], and consider a point defect whose ground state is orbitally degenerate by symmetry. From [Ref.1.1, Sect.1.4] we know that its eigenstates ψ_α^0 must be basis functions of an irreducible representation. There are five irreducible representations: A_1, A_2, E, T_1, T_2 [Ref.1.1, Sect.1.4.3] and we consider them in turn. Since A_1 and A_2 are nondegenerate, there is no Jahn-Teller effect (we shall verify later that

Table 2.1. Basis functions for the irreducible representations of T_d

Irreducible representation	Basis function Electronic	Lattice mode	Simple basis function
A_1 (A_2)	$\psi^0_{A_1}$	Q_{A_1}	$\frac{x^2+y^2+z^2}{\sqrt{3}}$
E	$\psi^0_{E,1}$	$Q_{E,1}$	$\frac{2z^2-y^2-x^2}{\sqrt{6}}$
	$\psi^0_{E,2}$	$Q_{E,2}$	$\frac{x^2-y^2}{\sqrt{2}}$
$T_2(T_1)$	$\psi^0_{T_2,1}$	$Q_{T_2,1}$	x
	$\psi^0_{T_2,2}$	$Q_{T_2,2}$	y
	$\psi^0_{T_2,3}$	$Q_{T_2,3}$	z

there can be only a relaxation in that case. Both T_1 and T_2 are triply degenerate and the situation is formally identical in both cases, so that we consider only T_2. Finally, E is twofold degenerate.

Table 2.1 gives the labeling of the basis wave functions for the A_1 (or A_2), E and T_2 (or T_1) cases, together with the simple basis wave functions having the same transformation properties [Ref.1.1, Sect.1.4].

We have shown in Sect.2.1.2 that the splitting of a degenerate level by symmetry-lowering distortions can be obtained, to first order, by diagonalization of the matrix $V_{\alpha\beta}$:

$$V_{\alpha\beta} = \sum_s Q_s \left\langle \psi^0_\alpha \middle| \left(\frac{\partial H_e}{\partial Q_s}\right)_0 \middle| \psi^0_\beta \right\rangle \quad . \tag{2.23}$$

The Q_s are combinations of the atomic displacements, the former being taken to be basis functions of the irreducible representations of T_d. We thus consider displacements of each symmetry, i.e., Q_{A_1} (A_1), $Q_{E,1}, Q_{E,2}$(E) and $Q_{T_2,1}$, $Q_{T_2,2}$, $Q_{T_2,3}$ (T_2), ignoring A_2 and T_1 for the reason given above.

The matrix elements $V_{\alpha\beta}$ can now be simplified on the following grounds:

i) H_e is invariant. Consequently $(\partial H_e/\partial Q_s)_0$ has the same symmetry properties as Q_s itself. It is the same basis function of the same irreducible representation. In fact the matrix which is the representation of the group formed by $(\partial H_e/\partial Q_s)_0$ is just the inverse of the one formed by the Q_s. However the latter being orthogonal, the two must be equal.

ii) $(\partial H_e/\partial Q_s)_0$ is the derivative of the potential energy in H_e with respect to the displacements. It can thus be commuted with ψ_α^0 in the matrix element (2.23). Using the notation of [Ref.1.1, Chap.1], we can therefore write

$$V_{\alpha\beta} = \sum_s Q_s \left[\psi_\alpha^{0*} \psi_\beta^0 \quad , \left(\frac{\partial H_e}{\partial Q_s}\right)_0\right] \quad . \tag{2.24}$$

This situation is strictly identical to the splitting of a degenerate level under a uniaxial stress [Ref.1.1, Sect.1.5], expression (2.24) being equivalent to (1.37) from [1.1] by replacing σ by Q and V by $\partial H_e/\partial Q_s$. The results must then be exactly the same, as well as the argumentation. We reproduce them for the E and T_2 states, with only slight changes in notation.

Twofold Degenerate E State. The Jahn-Teller matrix is identical to [Ref.1.1, Table 1.6],

$$\begin{bmatrix} I_{E,A_1} Q_{A_1} - I_{E,E} Q_{E,1} & I_{E,E} Q_{E,2} \\ I_{E,E} Q_{E,2} & I_{E,A_1} Q_{A_1} + I_{E,E} Q_{E,1} \end{bmatrix} \tag{2.25}$$

where I_{E,A_1} and $I_{E,E}$ are the independent Jahn-Teller coefficients describing the coupling of an E state with A_1 and E modes, respectively. They correspond to the coefficients $A\sqrt{3}$ and $B\sqrt{6}$ of [Ref.1.1, Table 1.6] and are defined by

$$I_{E,A_1} = \frac{1}{2}\left[\left\langle\psi_{E,1}^0\middle|\left(\frac{\partial H_e}{\partial Q_{A_1}}\right)_0\middle|\psi_{E,1}^0\right\rangle + \left\langle\psi_{E,2}^0\middle|\left(\frac{\partial H_e}{\partial Q_{A_1}}\right)_0\middle|\psi_{E,2}^0\right\rangle\right] \tag{2.26}$$

and

$$I_{E,E} = \left\langle\psi_{E,1}^0\middle|\left(\frac{\partial H_e}{\partial Q_{E,2}}\right)_0\middle|\psi_{E,2}^0\right\rangle \quad . \tag{2.27}$$

Triply Degenerate T_2 (or T_1) State. The situation is now slightly more complicated since couplings with the A_1, E and T_2 modes occur. Again, the results can be derived immediately from [Ref.1.1, Table 1.7], giving the following Jahn-Teller matrix

$$I_{T_2,A_1} Q_{A_1} \begin{bmatrix} 1 & 0 & 0 \\ 0 & 1 & 0 \\ 0 & 0 & 1 \end{bmatrix} + I_{T_2,E} \begin{bmatrix} \frac{-Q_{E,1} + \sqrt{3}Q_{E,2}}{2} & 0 & 0 \\ 0 & \frac{-Q_{E,1} - \sqrt{3}Q_{E,2}}{2} & 0 \\ 0 & 0 & Q_{E,1} \end{bmatrix}$$

$$+ I_{T_2,T_2} \begin{bmatrix} 0 & Q_{T_2,3} & Q_{T_2,2} \\ Q_{T_2,3} & 0 & Q_{T_2,1} \\ Q_{T_2,2} & Q_{T_2,1} & 0 \end{bmatrix} \tag{2.28}$$

the coupling coefficients being

$$I_{T_2,A_1} = \frac{1}{3} \operatorname{Tr}\left\{\left(\frac{\partial H_e}{\partial Q_{A_1}}\right)_0\right\} \tag{2.29}$$

$$I_{T_2,E} = \left\langle \psi^0_{T_2,3} \left| \left(\frac{\partial H_e}{\partial Q_{E,1}}\right)_0 \right| \psi^0_{T_2,3} \right\rangle \tag{2.30}$$

$$I_{T_2,T_2} = \left\langle \psi^0_{T_2,1} \left| \left(\frac{\partial H_e}{\partial Q_{T_2,2}}\right)_0 \right| \psi^0_{T_2,3} \right\rangle \quad . \tag{2.31}$$

For both E and T_2 states we have explicitly considered the Q_{A_1} mode. This mode is totally symmetric and gives no splitting. It corresponds to a symmetrical relaxation of the atoms, as opposed to the distortion due to Q_E or Q_{T_2} modes which lowers the symmetry. For the nondegenerate A_1 level the linear term would only be $I_{A_1,A_1} Q_{A_1}$, leading to no distortion.

2.2.2 The Stable Distortions

Let us call $\lambda(Q_s)$ the eigenvalues of the linear Jahn-Teller matrix. As discussed in Sect.2.1.2 we retain the diagonal part of the second order matrix which can always be written under the form

$$\sum_s \frac{1}{2} k_s Q_s^2 I \quad , \tag{2.32}$$

because a symmetrical quadratic form can always be diagonalized. An important point is that the Q_s automatically have well defined symmetry properties, i.e., the Q_s in (2.32) cannot be combinations of two different basis functions. It is enough to consider separately one mode Q_{A_1}, two modes $Q_{E,1}$, $Q_{E,2}$ and three modes $Q_{T_2,1}$, $Q_{T_2,2}$, $Q_{T_2,3}$. If other modes having the same symmetry are important, we must simply add the corresponding contributions which are formally expressed in the same manner. We thus write (2.32) explicitly in the form

$$\left[\frac{1}{2} k_{A_1} Q_{A_1}^2 + \frac{1}{2} k_E (Q_{E,1}^2 + Q_{E,2}^2) + \frac{1}{2} k_{T_2} (Q_{T_2,1}^2 + Q_{T_2,2}^2 + Q_{T_2,3}^2)\right] I \quad . \tag{2.33}$$

The stable distortions are obtained by adding (2.33) to each of the eigenvalues $\lambda(Q_s)$ of the Jahn-Teller matrix and minimizing them separately with respect to each of the Q_s. We shall now do this for the A_1, E and T_2 levels.

a) *The Nondegenerate A_1 (or A_2) Level*

As we have shown, the linear term is $I_{A,A_1} Q_{A_1}$ (A standing for A_1 or A_2). The stable configuration corresponds to all Q_s equal to zero (no distortion), except for Q_{A_1} itself which is obtained by minimizing the quantity

$$E = I_{A,A_1} Q_{A_1} + \frac{1}{2} k_{A_1} Q_{A_1}^2 \quad . \tag{2.34}$$

For the minimum configuration $Q_{A_1 m}$ and for the gain in energy E_{Re} (the relaxation energy), this leads to:

$$Q_{A_1 m} = -\frac{I_{A,A_1}}{k_{A_1}} \quad , \quad E_{Re} = \frac{1}{2} \frac{I_{A,A_1}^2}{k_{A_1}} \quad . \tag{2.35}$$

b) *The Twofold Degenerate Level E*

In this situation, we use the 2×2 Jahn-Teller matrix given by (2.25). The Q_{A_1} contribution, being diagonal, separates and gives a stabilization similar to (2.35) when I_{A,A_1} is replaced by I_{E,A_1}. The T_2 modes do not contribute and we have to diagonalize the 2×2 matrix

$$I_{E,E} \begin{bmatrix} -Q_{E,1} & Q_{E,2} \\ Q_{E,2} & Q_{E,1} \end{bmatrix} \tag{2.36}$$

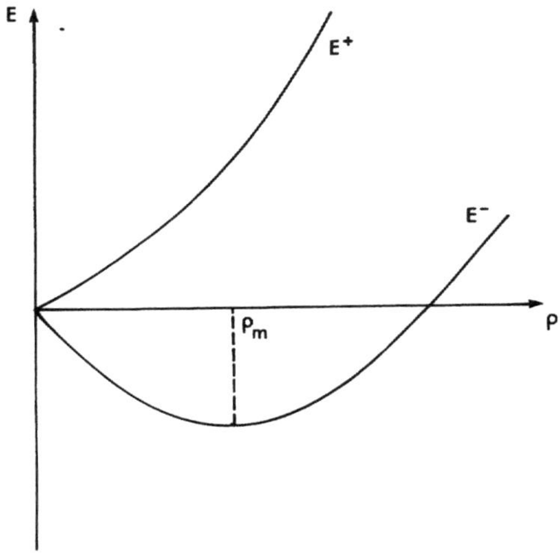

Fig.2.2. The two energy branches for an E state coupled to E modes

whose eigenvalues are

$$\lambda^{\pm} = \pm I_{E,E}\sqrt{Q_{E,1}^2 + Q_{E,2}^2} \quad . \tag{2.37}$$

Addition of the quadratic terms of (2.33) leads to the two energy branches

$$E^{\pm} = \pm I_{E,E}\rho + \frac{1}{2} k_E \rho^2 \tag{2.38a}$$

where

$$\rho = \sqrt{Q_{E,1}^2 + Q_{E,2}^2} \quad . \tag{2.38b}$$

Only one of the $E^{\pm}$ energy curves leads to a stable atomic configuration. When $I_{E,E} > 0$, as shown in Fig.2.2, it is E^- which leads to a minimum configuration ρ_m and to a gain in energy E_{JT} (Jahn-Teller energy)

$$\rho_m = \frac{I_{E,E}}{k_E} \qquad E_{JT} = \frac{I_{E,E}^2}{2k_E} \quad . \tag{2.39}$$

The energy does not depend at all upon the angle θ defined by

$$\theta = \tan^{-1}\left(\frac{Q_{E,2}}{Q_{E,1}}\right) \quad . \tag{2.40}$$

There are an infinite number of stable positions corresponding to all possible values of θ between 0 and 2π. This is a case of complete orientational degeneracy which will be discussed further (Sect.2.3.1) when we consider the dynamic effect.

This degeneracy can be removed by including higher order terms in the Hamiltonian, other than the diagonal ones given by (2.33). Symmetry readily enables it to be shown that the corresponding 2×2 matrix is

$$K\begin{bmatrix}(Q^2_{E,1}-Q^2_{E,2}) & 2Q_{E,1}Q_{E,2}\\ 2Q_{E,1}Q_{E,2} & -(Q^2_{E,1}-Q^2_{E,2})\end{bmatrix} \tag{2.41}$$

which, when added to the matrix (2.36), gives, in terms of ρ and θ, for the modified eigenvalues

$$\lambda^{\pm} = [I^2_{E,E}\rho^2 + K^2\rho^4 - 2I_{E,E}K\rho^3\cos(3\theta)]^{\frac{1}{2}} \quad . \tag{2.42}$$

It is now possible to minimize (2.42) with respect to θ. The condition for obtaining an extremum corresponds to $\sin(3\theta) = 0$. The values of θ giving the minima of the lower branch depend upon the sign of the product $I_{EE}K$. For $I_{E,E}K > 0$ we obtain the three following values [corresponding to $\cos(3\theta)=-1$]

$$\theta = \frac{\pi}{3} \quad , \quad \pi \quad , \quad \frac{5\pi}{3} \quad , \tag{2.43}$$

while for $I_{E,E}K < 0$ we have

$$\theta = 0 \quad , \quad \frac{2\pi}{3} \quad , \quad \frac{4\pi}{3} \quad . \tag{2.44}$$

In all cases there are thus three equivalent minima separated by potential barriers, whose height depends upon the parameter K. There is still an orientational degeneracy but there are now three stable equivalent configurations and not an infinite number as in the previous case.

c) *The Triply Degenerate State T Coupled to E Modes*

We treat this case because it leads to a much simplified situation for the static effect, as well as for the dynamic effect. Again, we drop Q_{A_1} and consider that only $I_{T_2,E}$ is important. Then, the second contribution to the Jahn-Teller matrix in (2.28) has to be considered alone. Using the same notations as for the E state (i.e., using ρ and θ), we find three distinct energy branches E_p given by

$$E_p = I_{T_2,E}\ \rho\cos(\theta + p\,\frac{2\pi}{3}) + \frac{1}{2}\,k_E\rho^2 \tag{2.45}$$

with $p = 0,1,2$. We now have to minimize this expression with respect to ρ and θ. Minimization with respect to ρ gives

$$\rho = -\frac{I_{T_2,E}}{k_E}\cos(\theta + p\,\frac{2\pi}{3})$$

$$E_p = -\frac{(I_{T_2,E})^2}{2k_E}\cos^2(\theta + p\,\frac{2\pi}{3}) \quad . \tag{2.46}$$

Minimization with respect to θ corresponds for $I_{T_2,E} > 0$ to $\theta = \pi/3$, π and $5\pi/3$, and for each value of θ leads to the same minimum values ρ_m and E_{JT}:

$$\rho_m = \frac{I_{T_2,E}}{k_E} \quad , \quad E_{JT} = \frac{I^2_{T_2,E}}{2k_E} \quad . \tag{2.47}$$

For $I_{T_2,E} < 0$, θ takes the values 0, $2\pi/3$ and $4\pi/3$, leading to the same expression for ρ_m and E_{JT}.

The conclusion is that, in the case of a T_2 level coupled to E modes, there are also three equivalent minima in the ρ, θ space, separated by potential barriers. This case is therefore analogous to the case obtained for an E state coupled to E modes.

d) *The Triply Degenerate State Coupled to E and T Modes*

We now proceed to find the absolute minima in the general case, where both $I_{T_2,E}$ and I_{T_2,T_2} are nonvanishing. Instead of calculating the eigenvalues of the 3×3 matrix (2.28) and minimizing them with respect to each Q_s, it is much more interesting to use a method derived by OPIK and PRYCE [2.13]. For this, we must diagonalize the first order matrix $V_{\alpha\beta}$ expressed in the basis of the unperturbed electron states ψ^0_α. Any normalized eigenfunction of this matrix ψ can be expressed as

$$\psi = \sum_\alpha a_\alpha \psi^0_\alpha \tag{2.48}$$

with

$$\sum_\alpha |a_\alpha|^2 = 1 \quad . \tag{2.49}$$

The eigenvalues λ of $V_{\alpha\beta}$ are defined by

$$\lambda a_\alpha = \sum_\beta V_{\alpha\beta} a_\beta \tag{2.50}$$

or equivalently by

$$\lambda = \sum_{\alpha,\beta} a^*_\alpha V_{\alpha\beta} a_\beta \quad , \tag{2.51}$$

and the total energy E_e is the sum of λ and of the quadratic form (2.33), i.e.,

$$E_e = \lambda(Q_s) + \frac{1}{2}\sum_s k_s Q_s^2 \quad . \tag{2.52}$$

The condition that $\partial E_e/\partial Q_s$ is zero can thus be rewritten, using the fact that λ is already extremum with respect to any variation of a_β, as

$$\frac{\partial E_e}{\partial Q_s} = 0 = \sum_{\alpha,\beta} a_\alpha^* \frac{\partial V_{\alpha\beta}}{\partial Q_s} a_\beta + k_s Q_s \quad , \tag{2.53}$$

$V_{\alpha\beta}$ being linear with Q_s, $\partial V_{\alpha\beta}/\partial Q_s$ is independent of Q_s and (2.53) gives directly Q_s in terms of $a_\alpha^* a_\beta$:

$$Q_{E,1} = -\frac{I_{T_2,E}}{k_E}\left(a_3^2 - \frac{a_1^2 + a_2^2}{2}\right)$$

$$Q_{E,2} = -\frac{I_{T_2,E}}{k_E}\frac{\sqrt{3}}{2}(a_1^2 - a_2^2)$$

$$Q_{T_2,1} = -\frac{I_{T_2,T_2}}{k_{T_2}} 2a_2a_3$$

$$Q_{T_2,2} = -\frac{I_{T_2,T_2}}{k_{T_2}} 2a_1a_3$$

$$Q_{T_2,3} = -\frac{I_{T_2,T_2}}{k_{T_2}} 2a_1a_2 \quad . \tag{2.54}$$

We can now use these expressions of Q_s and inject them into (2.50), ending with a system of three equations for a_i. For instance, (2.50) gives

$$a_1\left[\lambda + \frac{I_{T_2,E}^2}{2k_E}(3a_1^2 - 1) + \frac{2I_{T_2,T_2}^2}{k_{T_2}}(1 - a_1^2)\right] = 0 \quad . \tag{2.55}$$

The two other equations are obtained by cyclic permutations. The solutions of these equations can be classified as follows:

i) *$a_i \neq 0$, $a_j = 0$ for $j \neq i$:* Then $a_i^2 = 1$ and $\lambda = I_{T_2,E}^2/k_E$. It is easy to show that the quadratic terms $\sum_s Q_{s/2}^2$ give half the opposite contribution so that we obtain for the Jahn-Teller energy $E_{JT}^{(1)}$ at this extremum

$$E_{JT}^{(1)} = \frac{I_{T_2,E}^2}{2k_E} \quad . \tag{2.56}$$

This is the same solution obtained for pure coupling with the E states and there are also three equivalent solutions for the displacements, since we can take the three possibilities: a_1, a_2 or a_3 equal to zero.

ii) $a_i = 0$, $a_j \neq 0$ *for* $j \neq i$: Thus the two a_j^2 are equal to 1/2. There are six distinct possibilities of this type, whose Jahn-Teller energy $E_{JT}^{(2)}$ is clearly given by

$$E_{JT}^{(2)} = \frac{I_{T_2,E}^2}{8k_E} + \frac{I_{T_2,T_2}^2}{2k_{T_2}} \quad . \tag{2.57}$$

iii) $a_i \neq 0 \ \forall \ i$: All a_i^2 must be equal to 1/3. There are only four distinct situations of this type, leading to the Jahn-Teller energy

$$E_{JT}^{(3)} = \frac{2}{3}\,\frac{I_{T_2,T_2}^2}{k_{T_2}} \quad . \tag{2.58}$$

The four equivalent distortions correspond to (2.54) where the a_i are made equal to $\pm 1/\sqrt{3}$. The modes in this case are pure T_2 mdoes.

We can easily check that solution ii) can never give the absolute minimum in energy.

When $E_{JT}^{(3)} < E_{JT}^{(1)}$, this minimum is given by solution i) which corresponds to pure coupling with the E modes. In the opposite case, the stable solution is given by solution ii) corresponding to pure coupling with the T_2 modes. The detailed results are summarized in Table 2.2.

2.2.3 The Case of Near Degeneracy

Up to now, we have investigated only the case of an electron level whose orbital degeneracy was due to symmetry. However, in practice, there can be other important cases where accidental degeneracy, or even near degeneracy, can occur. By near degeneracy, we mean cases where the zero distortion splitting of the levels is of the same order of magnitudes as the Jahn-Teller splitting itself. There are a variety of different possible cases. We do not review them in detail [2.14] but rather discuss the problem with a simple example: the $A_1 + E$ coupling.

We thus consider a system which, for the perfect crystal configuration $\underline{R}_0$, presents two nearly degenerate levels A_1 and E. We take the energy of the A_1 nondegenerate state as reference and the energy of the twofold gegenerate E state equal to Δ. We want to investigate the influence of possible

Table 2.2. Characteristics of the possible equivalent minima for coupling of a T_2 electronic state with E and T_2 modes. q_1 and q_3 are defined by

$$q_1 = -\frac{I_{T_2,E}}{k_E} \quad \text{and} \quad q_3 = -\frac{2}{3}\frac{I_{T_2,T_2}}{k_{T_2}}$$

Wave function	Pure E minima			Pure T_2 minima			
a_1	0	0	1	$\frac{1}{\sqrt{3}}$	$\frac{1}{\sqrt{3}}$	$-\frac{1}{\sqrt{3}}$	$-\frac{1}{\sqrt{3}}$
a_2	0	1	0	$\frac{1}{\sqrt{3}}$	$-\frac{1}{\sqrt{3}}$	$-\frac{1}{\sqrt{3}}$	$\frac{1}{\sqrt{3}}$
a_3	1	0	0	$\frac{1}{\sqrt{3}}$	$-\frac{1}{\sqrt{3}}$	$\frac{1}{\sqrt{3}}$	$-\frac{1}{\sqrt{3}}$
$Q_{E,1}$	q_1	$-\frac{q_1}{2}$	$-\frac{q_1}{2}$	0	0	0	0
$Q_{E,2}$	0	$-\frac{\sqrt{3}}{2}q_1$	$\frac{\sqrt{3}}{2}q_1$	0	0	0	0
$Q_{T_2,1}$	0	0	0	q_3	q_3	$-q_3$	$-q_3$
$Q_{T_2,2}$	0	0	0	q_3	$-q_3$	$-q_3$	q_3
$Q_{T_2,3}$	0	0	0	q_3	$-q_3$	q_3	$-q_3$
Jahn-Teller energy	$E_{JT}^{(1)} = \frac{I_{T_2,E}^2}{2k_E}$			$E_{JT}^{(3)} = \frac{2}{3}\frac{I_{T_2,T_2}^2}{k_{T_2}}$			

atomic displacements, writing expansion (2.11) for the electronic Hamiltonian. We write the electron wave function ψ as a combination of the unperturbed states which, in this case, are $\psi_{A_1}^0$, $\psi_{E,1}^0$, $\psi_{E,2}^0$. Then, the problem reduces to the diagonalization of the Hamiltonian matrix expressed in this basis. Taking expansion (2.11), H_0 gives a diagonal contribution equal to 0 or to Δ, while the other terms give the matrix $V_{\alpha\beta}$ given by (2.13) where, again, we keep only the diagonal contribution of the second-order terms. The first-order Jahn-Teller matrix can be simplified by symmetry (this is obvious in this simple case), using the methods developed in [Ref.1.1, Sect. 1.4] and we readily obtain

$$[H] = \begin{bmatrix} I_{A,A}Q_{A_1} & GQ_{E,1} & GQ_{E,2} \\ \hline GQ_{E,1} & \Delta + I_{E,A}Q_{A_1} - I_{E,E}Q_{E,1} & I_{E,E}Q_{E,2} \\ GQ_{E,2} & I_{E,E}Q_{E,2} & \Delta + I_{E,A}Q_{A_1} + I_{E,E}Q_{E,1} \end{bmatrix} + \left\{\frac{1}{2} k_{A_1}Q_{A_1}^2 + \frac{1}{2} k_E(Q_{E,1}^2 + Q_{E,2}^2)\right\} I \quad . \tag{2.59}$$

In this expression, I is the unit matrix; the 2×2 E submatrix corresponds to the intralevel Jahn-Teller coupling treated above. The new situation is introduced by the interlevel coupling whose strength depends upon the parameter G. To investigate this effect we assume that G is the dominant term. We then easily obtain the solutions

$$E = \Delta + \frac{1}{2} k_E\rho^2$$

$$E = \frac{\Delta}{2} \pm \sqrt{\left(\frac{\Delta}{2}\right)^2 + G^2\rho^2} + \frac{1}{2} k_E\rho^2 \tag{2.60}$$

where ρ is given by (2.38b); the term in Q_{A_1} has been dropped. The derivative of the lower branch is given by

$$\frac{dE}{d\rho} = \rho\left(k_E - \frac{G^2}{\sqrt{\left(\frac{\Delta}{2}\right)^2 + G^2\rho^2}}\right) \quad \text{for} \quad \Delta > 0 \quad . \tag{2.61}$$

It gives rise to a minimum only if the following condition is fulfilled:

$$\Delta < \frac{2G^2}{k_E} \tag{2.62}$$

i.e., when the level separation Δ at zero distortion is smaller than four times the Jahn-Teller energy at zero Δ. This is the condition for the existence of stable distortions in a case where the electronic ground state A_1 is not degenerate and where there is no Jahn-Teller effect.

More complex cases can be analyzed in the same way [2.14]. In this reference it was also shown that such interlevel coupling may be the reason why the negative vacancy V^- in Si can have a Jahn-Teller effect even if its lowest many-electron state is nondegenerate ($^4A_2$, see [Ref.1.1, Sect.4.2.2]).

2.3 Coupled Electronic and Nuclear Motion: Vibronic States-Static and Dynamic Jahn-Teller Limits

Up to now, we have solved only the first part of the Born-Oppenheimer problem, i.e., we have determined the different branches of the electronic energy $E_e(\underline{R})$ as a function of the nuclear displacements. The characteristic of these branches is that they are degenerate, or near degenerate, for zero distortion. In most cases several stable equivalent situations exist. However, we must still solve the second part of the Born-Oppenheimer problem which concerns the nuclear motion. In practice we have to calculate vibronic states, corresponding to coupled electronic and nuclear motion. We first do this for an E state coupled to E modes with complete angular degeneracy. We then study, in the same case, the possibility of static and dynamic Jahn-Teller effects. After introducing the Ham effect on the T_2 state coupled to E modes, we extend this notion to more complex cases. Finally, we discuss transitions from static to dynamic situations.

2.3.1 The E State Coupled to E Modes (Case of Cylindrical Symmetry)

This case is discussed in detail in [2.15] and in several textbooks [2.2-4] but it is useful to present it here because it is a simple illustration of different possible limiting cases. As we have shown in Sect.2.2.2, the electronic energy $E_e(\underline{R})$ presents two branches $E^{\pm}$. They are pictured in Fig.2.2 in the case where the term containing $\cos(3\theta)$ is negligible.

To begin, we follow the general Born-Oppenheimer treatment. The two branches $E\pm(\rho)$ have to be considered together (they can be near degenerate). This means that we must write the total *vibronic* (for vibrational plus electronic) wave function Φ as a combination of Born-Oppenheimer products of the general form given by (2.6). In our case this leads to

$$\Phi(\underline{r},\rho) = \psi^{+}(\underline{r},\rho)\chi^{+}(\rho) + \psi^{-}(\underline{r},\rho)\chi^{-}(\rho) \tag{2.63}$$

where $\psi^{\pm}$ are the electronic wave functions corresponding to $E^{\pm}(\rho)$. The $\chi^{\pm}(\rho)$ are the vibrational wave functions. The $\psi^{\pm}$ are the eigenfunctions of the 2×2 Jahn-Teller matrix of (2.36) which can be easily expressed in terms of the unperturbed basis functions $\psi^0_{E,1}$ and $\psi^0_{E,2}$ of the E state

$$\psi^{+} = \psi^0_{E,1}\sin\frac{\theta}{2} + \psi^0_{E,2}\cos\frac{\theta}{2}$$

$$\psi^{-} = \psi^0_{E,1}\cos\frac{\theta}{2} - \psi^0_{E,2}\sin\frac{\theta}{2}\ , \tag{2.64}$$

where

$$Q_{E,1} = \rho\cos\theta \quad \text{and} \quad Q_{E,2} = \rho\sin\theta \quad .$$

We now have to solve the general set of equations (2.7) which, in this case, reduce to two equations for χ^+ and χ^-. Since H_e is diagonal in the basis ψ^+, ψ^-, we obtain

$$E^+\chi^+ + \int dv(\psi^+)^* T_N \psi^+ \chi^+ + \int dv(\psi^+)^* T_N \psi^- \chi^- = E\chi^+$$

$$E^-\chi^- + \int dv(\psi^-)^* T_N \psi^- \chi^- + \int dv(\psi^-)^* T_N \psi^+ \chi^+ = E\chi^- \quad . \tag{2.65}$$

The total problem has cylindrical symmetry. Then, the variables ρ and θ can be separated and we can write

$$\chi^+ = f^+(\rho)\exp(ij\theta)$$

$$\chi^- = if^-(\rho)\exp(ij\theta) \quad . \tag{2.66}$$

The total wave function Φ must be invariant when θ changes to $\theta + 2\pi$. However, due to the dependence of ψ^+ and ψ^- on $\theta/2$, they are changed to $-\psi^+$ and $-\psi^-$. This means that j has to be a half-odd integer. With the definitions of $f^\pm$ given in (2.66) and taking into account the fact that

$$T_N = -\frac{\hbar^2}{2M}\left(\frac{\partial^2}{\partial\rho^2} + \frac{1}{\rho}\frac{\partial}{\partial\rho} + \frac{1}{\rho^2}\frac{\partial^2}{\partial\theta^2}\right) \quad , \tag{2.67}$$

we can transform (2.65) into

$$\left[E^+(\rho) - \frac{\hbar^2}{2M}\left(\frac{\partial^2}{\partial\rho^2} + \frac{1}{\rho}\frac{\partial}{\partial\rho} - \frac{j^2+\frac{1}{4}}{\rho^2}\right)\right]f^+(\rho) - \frac{\hbar^2}{2M}\frac{j}{\rho^2}f^-(\rho) = Ef^+(\rho)$$

$$\left[E^-(\rho) - \frac{\hbar^2}{2M}\left(\frac{\partial^2}{\partial\rho^2} + \frac{1}{\rho}\frac{\partial}{\partial\rho} - \frac{j^2+\frac{1}{4}}{\rho^2}\right)\right]f^-(\rho) - \frac{\hbar^2}{2M}\frac{j}{\rho^2}f^+(\rho) = Ef^-(\rho) \quad . \tag{2.68}$$

This set of coupled differential equations can be simplified further by writing $f^\pm(\rho)$ in the form

$$f^\pm(\rho) = \rho^{-\frac{1}{2}}y^\pm(\rho) \tag{2.69}$$

which leads to

$$\left[-\frac{\hbar^2}{2M}\frac{d^2}{d\rho^2} + E^\pm(\rho) + \frac{\hbar^2}{2M}\frac{j^2}{\rho^2}\right]y^\pm(\rho) - \frac{\hbar^2}{2M}\frac{j}{\rho^2}y^\mp(\rho) = Ey^\pm(\rho) \quad . \tag{2.70}$$

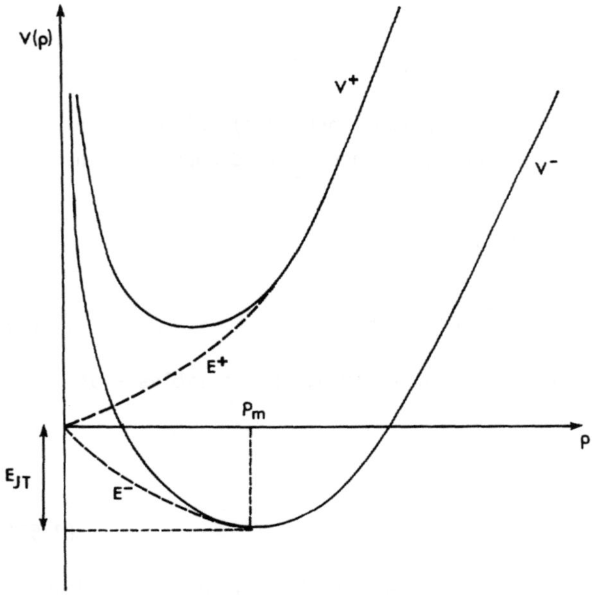

Fig.2.3. (----)$E^{\pm}(\rho)$; (——) potential energy $V^{\pm}(\rho)$

The second term on the left couples the two energy branches $E^{+}(\rho)$ and $E^{-}(\rho)$ given by (2.38). For strong coupling, the two branches are well separated and the coupling term is negligible, at least for low values of j. The two equations (2.70) are then decoupled, corresponding to the limit where the vibronic wave function Φ is a simple Born-Oppenheimer product. The effective potential $V^{\pm}(\rho)$ for nuclear motion in the two branches is equal to $E^{\pm}(\rho) + (\hbar^2/2M)(j^2/\rho^2)$ and is plotted in Fig.2.3.

The low-lying states of this system can be easily obtained in the strong coupling limit. In the vicinity of the absolute minimum ρ_m the term $(\hbar^2/2M)(j^2/\rho^2)$ is small (for small j). The equation for y^- becomes

$$\left[-\frac{\hbar^2}{2M}\frac{d^2}{d\rho^2} + \frac{1}{2}k(\rho-\rho_m)^2\right]y^- = (E + E_{JT})y^- \quad , \tag{2.71}$$

whose solutions are harmonic oscillator wave functions centered on ρ_m, of energies

$$E = -E_{JT} + (n + \tfrac{1}{2})\hbar\omega \tag{2.72}$$

with

$$\omega = \sqrt{\frac{k}{m}} \quad . \tag{2.73}$$

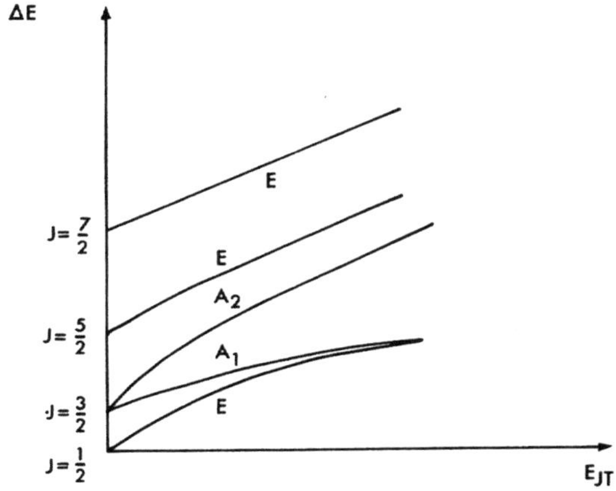

Fig.2.4. Vibronic levels vs the strength of the Jahn-Teller coupling for E states coupled to E modes (arbitrary units)

The addition of the term $(\hbar^2/2M)(j^2/\rho^2)$, treated by first-order perturbation theory, roughly gives a shift of the levels of order $(\hbar^2/2M)(j^2/\rho_m^2)$. (These sublevels are twofold degenerate since j can take positive and negative half-odd integer values; their ratios are in the following sequence: 1/4, 9/4, 25/4, 49/4, ...). This model is valid only in the strong coupling limit, i.e., when $\hbar\omega/E_{JT} \ll 1$ and also when $\hbar\omega \gg (\hbar^2/2M)j^2/\rho_m^2)$. However, since $k\rho_m^2/2$ is equal to E_{JT}, this second condition, equivalent to $(\hbar\omega/4E_{JT}) j^2 \ll 1$, is automatically verified if the first one is, at least for low values of j. The results in this limit are given on Fig.2.4 and correspond to $A \to 0$.

2.3.2 Static and Dynamic Jahn-Teller Effects

We introduce the static and dynamic limits on the case of an E state coupled to E modes, when higher order terms are included. The degeneracy in θ is lifted and there are now three equivalent minima in $Q_{E,1}$, $Q_{E,2}$ space. Expansion of (2.42) leads to a corrective term of the form $A(\rho)\cos(3\theta)$, which would also be the case when anharmonic terms are included. The equation for nuclear motion on the lower energy branch becomes

$$\left[-I\rho + \frac{1}{2}k\rho^2 + A(\rho)\cos(3\theta) - \frac{\hbar^2}{2M}\left(\frac{\partial^2}{\partial\rho^2} + \frac{1}{\rho}\frac{\partial}{\partial\rho} + \frac{1}{\rho^2}\frac{\partial^2}{\partial\theta^2}\right)\right]\chi^- = E\chi^- \quad . \tag{2.74}$$

We consider again the case of strong coupling, i.e., $E_{JT} = I^2/2k$ large compared to $\hbar\omega = \hbar\sqrt{k/M}$. For low-lying states, the motion will be confined in the neighborhood of $\rho \simeq \rho_m$ and we can approximate $\rho^{-2}(\partial^2/\partial\theta^2)$ and $A(\rho)\cos(3\theta)$

by $\rho_m^{-2}(\partial^2/\partial\theta^2)$ and $A(\rho_m)\cos(3\theta)$. Then the variables ρ and θ can be separated in (2.74). Substituting for χ^- the form $\sqrt{\rho}y^-(\rho)Y(\theta)$ we obtain

$$\left(-\frac{\hbar^2}{2M}\frac{d^2}{d\rho^2}-\frac{\hbar^2}{8M\rho^2}-I\rho+\frac{1}{2}k\rho^2\right)y^-=E_\rho y^-$$

$$\left[-\frac{\hbar^2}{2M\rho_m^2}\frac{d^2}{d\theta^2}+A(\rho_m)\cos(3\theta)\right]Y(\theta)=E_\theta Y(\theta)\quad . \tag{2.75}$$

The total energy is the sum of E_ρ and E_θ. The solutions of the equation in ρ are

$$E_\rho \simeq -E_{JT}+(n+\tfrac{1}{2})\hbar\omega-\frac{\hbar^2}{8M\rho_m^2}\quad . \tag{2.76}$$

The new situation arises for the equation in θ for which two limiting cases can be considered: the static case, when $|A(\rho_m)|$ is strong enough i.e., larger than $\hbar\omega$, and the dynamic case, when $|A(\rho_m)|$ is very small. Let us consider these two limits in some detail.

a) *The Static Limit*

We assume $A(\rho_m)$ to be positive. Then, there are three minima at $\theta=\pi/3$, π, $5\pi/3$ with an angular Jahn-Teller energy E_{JT}^θ given by

$$E_{JT}^\theta = A(\rho_m)\quad . \tag{2.77}$$

If $A(\rho_m)$ is large, we can expand the potential energy $A(\rho_m)\cos(3\theta)$ to second order, near each of these minima. In each case we obtain an harmonic oscillator wave function (expressed in terms of the variable $\rho_m\theta$), centered on the corresponding minima, with the vibrational frequency

$$\omega_\theta=\left(\frac{9A}{M\rho_m^2}\right)=\omega\,\frac{3}{2}\sqrt{\frac{2E_{JT}^\theta}{E_{JT}}} \tag{2.78}$$

where $\omega=\sqrt{k/M}$ is the frequency for motion in ρ. The energy levels of each oscillator are thus

$$E_\theta=-E_{JT}^\theta+(n+\tfrac{1}{2})\hbar\omega_\theta\quad . \tag{2.79}$$

The total vibronic wave function for the ith minimum is of the form

$$\Phi_i=\psi_i^-\sqrt{\rho}y^-(\rho)Y_{i,n}(\rho_m\theta)\quad , \tag{2.80}$$

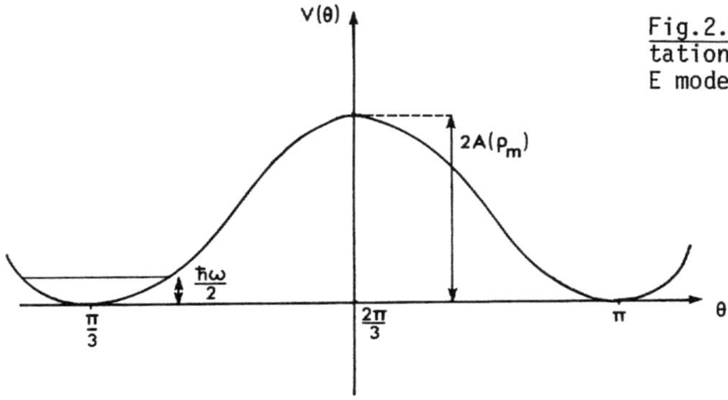

Fig.2.5. Barrier for rotation (Estate coupled to E modes)

where $Y_{i,n}(\rho_m\theta)$ is the nth harmonic oscillator centered on the ith minimum, corresponding to $\theta = \pi/3$, π or $5\pi/3$.

This separation between the three potential wells is valid only in the static case, when $E^{\theta}_{JT} \gg \hbar\omega_{\theta}$, which condition can also be rewritten $\sqrt{E^{\theta}_{JT}\, E_{JT}} \gg (3/\sqrt{2})\hbar\omega$ using (2.78), requiring E^{θ}_{JT} to be at least of order $\hbar\omega$. In this case the three potential wells give the same energy levels, expressed by (2.79). Therefore, in the static case, each of the lowest vibronic levels will be triply degenerate.

It is interesting to investigate the slow rotation limit where the dynamic situation can be viewed as tunneling from one potential well to the other one. Here, again, $\hbar\omega_{\theta}$ is assumed to be small compared to $A(\rho_m) = E^{\theta}_{JT}$, but we consider that there is a finite probability that the system makes a transition from one potential well to another one. In other words, there is a splitting of the triply degenerate level. This splitting can be obtained to first order by writting Φ as a combination of the degenerate Φ_i. This leads to diagonalization of a 3×3 matrix $H_{ij} - E\,S_{ij}$ (H_{ij} being the matrix element of the total Hamiltonian between the two states Φ_i and Φ_j and S_{ij} the corresponding overlap $\langle\Phi_i|\Phi_j\rangle$). All nondiagonal terms are equal (we denote by γ their.common value) and we consider the limit where they are small. Then, the eigenvalues are close to the diagonal terms H_{ii} (whose values are also independent of i). We can rewrite γ as $(H_{ij} - H_{ii}\,S_{ij}) - (E - H_{ii})S_{ij}$. Since H_{ij} and S_{ij} are small, the splitting $E - H_{ii}$ is a first-order term in these quantities. To that order, we can neglect $(E - H_{ii})S_{ij}$ and write

$$\gamma = H_{ij} - H_{ii}\,S_{ij} \quad . \tag{2.81}$$

In this expression, H is the exact Hamiltonian, equal to $T_N + H_e$. Furthermore, Φ_i is taken to be a Born-Oppenheimer product $\bar{\psi}_i\chi_i$ corresponding to the ith potential well (i.e., ψ_i^- is the electronic wave function at the ith minimum). Then γ reduces to

$$\gamma = \langle\psi_i^-|\psi_j^-\rangle\gamma_0 \tag{2.82}$$

with

$$\gamma_0 = \langle\chi_i|T_N + H_e|\chi_j\rangle - \langle\chi_i|T_N + H_e|\chi_i\rangle\langle\chi_i|\chi_j\rangle \quad . \tag{2.83}$$

The first term $\langle\psi_i^-|\psi_j^-\rangle$ is the overlap between the electronic wave functions ψ_i^- and ψ_j^- relative to the minima i and j. From (2.64) and taking $\theta = \pi/3$ and π for instance, we easily find that this overlap is equal to -1/2. The term γ_0, on the other hand, corresponds to tunneling between the vibrational states χ_i and χ_j, i.e., between the two potential wells of Fig.2.5 (the motion in ρ is the same, only the motion in θ is important).

The calculation of γ_0 thus reduces to the determination of the splitting of the potential wells of Fig.2.5. This is a standard problem to treat in the WKB approximation [2.17]. It can be shown [2.3] that when $\hbar\omega_\theta \ll A(\rho_m)$, then

$$\gamma_0 = -\frac{\hbar}{2\pi}\,\omega_\theta\,\exp\left[-\frac{1}{\hbar}\int_{\pi/3}^{\pi}\sqrt{2MA(\rho_m)(\cos 3\theta - 1)}\right]d\theta \quad . \tag{2.84}$$

Integration of (2.84) yields

$$\gamma = \frac{\hbar}{4\pi}\,\omega_\theta\,\exp\left(-\frac{4E_{JT}^{\theta}}{\hbar\omega_\theta}\right) \quad . \tag{2.85}$$

This is an approximate expression but it shows the essential behavior of the system. First, γ is positive contrary to what happens in conventional situations. The diagonalization of the 3×3 matrix gives two levels: one nondegenerate at 2γ, the other twofold degenerate at $-\gamma$ (taking the diagonal energy as the origin of energies). It can be shown [2.16] that they are respectively of A_1 and E symmetry. Thus, in this case, the symmetrical A_1 state is higher in energy, although usually it is lower (this is due to the negative overlap between the electronic parts).

A final important point is that the total vibronic ground state is of E symmetry. Thus, the final degeneracy is the same as the initial electronic degeneracy over the whole range of parameters as shown in Fig.2.4. This conclusion will also hold in other cases.

b) *The Dynamic Limit*

Let us start from the situation discussed in Sect.2.3.1 where E^{θ}_{JT} is equal to zero. All levels are then twofold degenerate and labeled by the quantum number j. We can now investigate the limit where E^{θ}_{JT} is small by first-order theory. This leads to a 2×2 matrix for each given j, corresponding to the matrix elements of $A(\rho_m)\cos(3\theta)$ between the angular wave functions $\exp(i|j|\theta)$ and $\exp(-i|j|\theta)$. There can be first-order splitting only if $2|j| = 3$. This effect is clearly seen on Fig.2.4. We have drawn on this figure the continuous transition from the dynamic to the static limit, taken from [2.16].

The qualitative conclusions we have obtained can be generalized. A useful illustration of the theory developed in this section is the GR1 center in diamond which will be discussed in detail in Chap.4.

2.3.3 The Ham Effect

We now introduce an important consequence of the Jahn-Teller coupling, the so-called Ham effect [2.18]. It corresponds to the fact that the matrix elements of an electronic operator in the total vibronic ground state are reduced (or quenched) with respect to those in the original degenerate electronic state.

The simplest illustration of this situation corresponds to the T_2 level coupled to an E mode. The electron energies are obtained from the Jahn-Teller-matrix (2.28) which reduces to

$$I_{T_2,E}\begin{bmatrix} q_1 & 0 & 0 \\ 0 & q_2 & 0 \\ 0 & 0 & q_3 \end{bmatrix} \tag{2.86}$$

where q_1, q_2, q_3 are particular combinations of $Q_{E,1}$, $Q_{E,2}$ given, respectively, by

$$\left.\begin{aligned} q_1 &= \frac{1}{2}(-Q_{E,1} + \sqrt{3}\,Q_{E,2}) \\ q_2 &= -\frac{1}{2}(Q_{E,1} + \sqrt{3}\,Q_{E,2}) \\ \text{and} \quad & \\ q_3 &= Q_{E,1} \end{aligned}\right\} \tag{2.87}$$

We can also build other combinations of $Q_{E,1}$ and $Q_{E,2}$, which are respectively orthogonal to the q_i and which we label q_i', defined by

$$\left.\begin{aligned} q_1' &= \frac{1}{2}(\sqrt{3}\,Q_{E,1} + Q_{E,2}) \\ q_2' &= \frac{1}{2}(\sqrt{3}\,Q_{E,1} - Q_{E,2}) \\ q_3' &= Q_{E,2} \end{aligned}\right\} . \qquad (2.88)$$

With these notations, the electronic energy E_{ei} of the ith branch becomes

$$E_{ei} = E_{e0} + I_{T_2,E}q_i + \frac{1}{2}k_E(q_i^2 + q_i'^2) \quad . \qquad (2.89)$$

It corresponds to the electronic wave function $\psi^0_{T_2,i}$. This electronic energy can also be rewritten in the form

$$E_{ei} = E_{e0} - E_{JT} + \frac{1}{2}k_E[(q_i - \rho_m)^2 + q_i'^2] \qquad (2.90)$$

where ρ_m and E_{JT} are given by (2.47). The vibronic wave function in this case is a simple Born-Oppenheimer product

$$\Phi_i = \psi^0_{T_2,i}\chi_i \qquad (2.91)$$

where the wave function χ_i for nuclear motion is the solution of (2.5)

$$\left\{-\frac{\hbar^2}{2M}\left(\frac{\partial^2}{\partial q_i^2} + \frac{\partial^2}{\partial q_i'^2}\right) + \frac{1}{2}k_E[(q_i - \rho_m)^2 + q_i'^2]\right\}\chi_i = (E - E_{e0} + E_{JT})\chi_i \quad . \qquad (2.92)$$

Thus, χ_i is the product of two single oscillator wave functions corresponding to q_i and q_i', centered respectively on ρ_m and 0. It differs from the case of the coupling of an E state with E modes in that, here, both oscillators have the same force constant k_E and the same vibrational frequency $\omega = \sqrt{k_E/M}$.

The three branches give three equivalent minima. Each total vibronic energy is then triply degenerate. Thus, contrary to out initial statement, the Jahn-Teller coupling *does not lift* the degeneracy of the ground state. Instead, starting from a triply degenerate electronic state we end with a triply degenerate "vibronic" state (which is also a T_2 state). Figure 2.4 shows that this was also true for the E electronic state (except in the static limit), the vibronic ground state also belonging to the E representation. This is also verified, as we shall see, for T_2 states coupled with T_2 modes.

Let us now consider an electronic operator O having nondiagonal matrix elements O_{ij} between the basis functions $\psi^0_{T_2,i}$ of the orbitally degenerate electronic state. Its matrix elements, within the triply degenerate vibronic ground state, are

$$\bar{O}_{ij} = \langle\Phi_i|O|\Phi_j\rangle \tag{2.93}$$

or, equivalently, using (2.91)

$$\bar{O}_{ij} = O_{ij}\langle\chi_i|\chi_j\rangle \quad . \tag{2.94}$$

We have thus simply to evaluate the overlap between χ_i and χ_j. This overlap can be written, for instance, in the form

$$\langle\chi_1|\chi_2\rangle = \frac{\int dQ_{E,1}dQ_{E,2}\,\exp\left\{-\frac{M\omega}{2\hbar}\,[(q_1-\rho_m)^2+(q_2-\rho_m)^2+q_1'^2+q_2'^2]\right\}}{|\int dQ\,\exp\left(-\frac{M\omega}{\hbar}\,Q^2\right)|^2} \quad . \tag{2.95}$$

The exponential term in the numerator can be expressed, using (2.87) and (2.88) as

$$\exp - \frac{M\omega}{\hbar}\left[q_3'^2 + \left(q_3 + \frac{\rho_m}{2}\right)^2 + \frac{3}{4}\,\rho_m^2\right] \quad . \tag{2.96}$$

Such overlaps also occur in the theory of optical lineshape (Chap.4), nonradiative transitions (Chap.6), and also polaron problems. Expressing the integral with respect to q_3' and $q_3 + \rho_m/2$, we immediately obtain

$$\langle\chi_1|\chi_2\rangle = \exp\left(-\frac{3}{4}\,\frac{M\omega}{\hbar}\,\rho_m^2\right) \quad . \tag{2.97}$$

Remembering that $M\omega^2\rho_m^2/2$ is equal to E_{JT} we can conclude that the final matrix elements $\bar{O}_{ij}$ are given by

$$\bar{O}_{ij} = O_{ij}\,\exp\left(-\frac{3}{2}\,\frac{E_{JT}}{\hbar\omega}\right) \quad . \tag{2.98}$$

This relation expresses the Ham effect; $\exp[-3/2(E_{JT}/\hbar\omega)]$ is the Ham reduction factor. A trivial application of this relation is that, in the limit $E_{JT}/\hbar\omega \to \infty$, any matrix element of the angular momentum $\underline{L}$ vanishes. This is normal since the system is frozen in one of the potential wells, so that $\underline{L}$ is completely quenched (Sect.3.3.1).

2.3.4 Extension to More Complex Cases

First, we investigate the case of the coupling of a T_2 level and T_2 modes in the static limit, then the E level with E coupling.

a) T_2 Level with T_2 Modes

We strictly follow the same analysis as in Sect.2.3.3. We start from the Jahn-Teller matrix given by (2.28), where we retain only the terms in I_{T_2,T_2}. The analysis in Sect.2.2.2 shows that there are four equivalent minima for values of $Q_{T_2,i}$ as shown in Table 2.2. We thus define new combinations q_i of the normal modes:

$$\begin{bmatrix} q_1 \\ q_2 \\ q_3 \\ q_4 \end{bmatrix} = \begin{bmatrix} \frac{1}{\sqrt{3}} & \frac{1}{\sqrt{3}} & \frac{1}{\sqrt{3}} \\ \frac{1}{\sqrt{3}} & -\frac{1}{\sqrt{3}} & -\frac{1}{\sqrt{3}} \\ -\frac{1}{\sqrt{3}} & -\frac{1}{\sqrt{3}} & \frac{1}{\sqrt{3}} \\ -\frac{1}{\sqrt{3}} & \frac{1}{\sqrt{3}} & -\frac{1}{\sqrt{3}} \end{bmatrix} \begin{bmatrix} Q_{T_2,1} \\ Q_{T_2,2} \\ Q_{T_2,3} \end{bmatrix} \quad . \tag{2.99}$$

We also build other combinations q_i', q_i'', such that any set q_i, q_i' and q_i'' is orthonormal. Let us lable E_{ei} the energy of the lower branch in the vicinity of the ith minimum. The expansion of E_{ei} to second order in the modes can be expressed as [2.3]

$$E_{ei} \simeq E_{e0} - E_{JT} + \frac{1}{2} k_T \left[q_i^2 + \frac{2}{3} (q_i'^2 + q_i''^2) \right] \quad . \tag{2.100}$$

Again, for strong coupling, we can associate with each potential well a vibronic state Φ_i which is a Born-Oppenheimer product $\psi_i \chi_i$. Here ψ_i is the electronic wave function at the minimum, whose coefficients are given in Table 2.2. χ_i is the product of three harmonic oscillator wave functions. There are four such states giving rise to a fourfold degenerate level in the static limit. In this respect, the situation is analogous to what we obtained for the E level coupled to E modes. To find the first-order splitting of this level we have to diagonalize a matrix of general terms $\langle\Phi_i|H|\Phi_j\rangle - E\langle\Phi_i|\Phi_j\rangle$, as discussed for the E state. Arguments similar to those used in that case allow the nondiagonal terms γ to be reduced to a quantity which, again, is found to be positive. Diagonalization of this

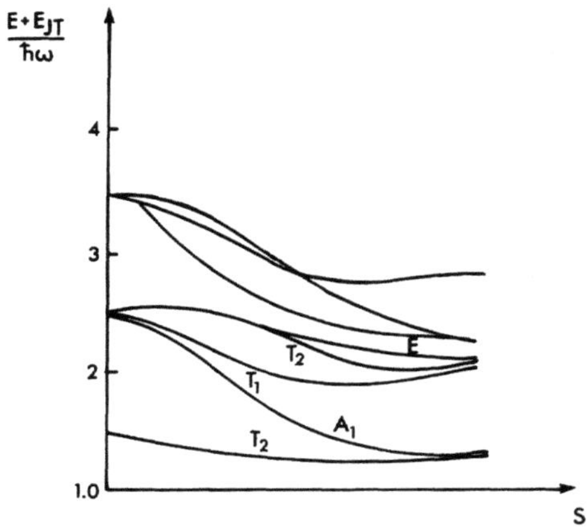

Fig.2.6. Transition from the dynamic to the static limit in the case of a T_2 level coupled to T_2 modes. (From [2.19]). S is the strength of the Jahn-Teller coupling

completely symmetrical matrix leads to a triply degenerate ground state, of symmetry T_2 (at $-\gamma$), and to a nondegenerate excited state, of symmetry A_1 (at $+3\gamma$). Again, the ground state has the same symmetry as the initial electronic state. In this strong coupling limit it is found that [2.19]

$$4\gamma \simeq 1.32\ E_{JT}\ \exp\left(-\ 1.24\ \frac{E_{JT}}{\hbar\omega}\right) \tag{2.101}$$

which is similar to (2.85) for the E state. The part in $\exp(-1.24\ E_{JT}/\hbar\omega)$ results from the overlap between Gaussians centered on different positions (the transition from this static limit to the complete dynamic limit is given on Fig.2.6).

The Ham effect should also occur in that case, but it is not as simple as for coupling of the T_2 state with E modes. First, we have to determine the matrix elements of an electronic operator O between the vibronic basis states Φ_i. The diagonal terms $\bar{O}_{ii}$ can be written as

$$\bar{O}_{ii} = \langle\psi_i|O|\psi_i\rangle \tag{2.102}$$

which can be expressed easily knowing the matrix elements of O in the basis $\psi^0_{T_2,i}$ (the four ψ_i are given in Table 2.2). As regards the nondiagonal terms $\bar{O}_{ij}$, they are of the form

$$\bar{O}_{ij} = \langle\psi_i|O|\psi_j\rangle\langle\chi_i|\chi_j\rangle \tag{2.103}$$

which now contains a reduction factor in $\exp(-1.24\ E_{JT}/\hbar\omega)$. The, the knowledge of these matrix elements enables us to write the complete matrix $\bar{O}$ and also its submatrix confined to the T_2 ground state. We do not detail these points [2.2] but simply notice that some operators are quenched and others are submitted only to partial quenching.

b) *E Level with E Modes*

Again the Ham effect will occur with complications analogous to those we have just described. Details can be found in [2.2].

2.3.5 Transitions from Static to Dynamic Situations

There are many experimental consequences of the existence of a Jahn-Teller coupling. The most obvious one is the lowering of symmetry corresponding to the static situation. It manifests itself directly in EPR (Chap.3) and optical spectra (Chap.4) which give in many cases a direct picture of the local symmetry around the defect. Another consequence comes from the quenching of electronic operators such as $\underline{L}$ (the Ham effect) which leads to observed parameters lower than the predicted ones. A final effect corresponds, for a given defect, to a possible continuous transition from the static to the dynamic limit.

In most cases, the strong coupling limit leads to equivalent potential wells separated by potential barriers. At low temperature the state of a given defect is frozen in one of the valleys. Even in that limit, zero-point motion leads to fluctuations in the electronic part $\psi_i(\underline{r},\underline{R})$ of the total wave function Φ_i corresponding to the ith minimum. This can have some effect on the superhyperfine structure observed in spin resonance [2.3]. There is also some probability for transitions from one potential well to another equivalent one, leading to a reorientation of the defect. The rate of reorientation $1/\tau$ is related to the parameter γ, defined in (2.81), by

$$\frac{1}{\tau} = \frac{\gamma}{\hbar} \quad . \tag{2.104}$$

When the temperature is raised, higher excited vibronic states become populated and there can be thermal activation over the barrier (i.e., behavior in $\exp(-\Delta/kT)$ where Δ is the barrier height). There can even be situations where the excited states cannot be described from the static limit but rather are dynamic states. The following situation can, in principle, be observed: at low temperature a Jahn-Teller effect with a reorientation time τ_0 is static; when the temperature increases $1/\tau$, thermally activated, increases; finally for higher temperature $1/\tau$ becomes infinite and the defect is again isotropic.

Obviously, this fact has important experimental consequences. A static Jahn-Teller situation is found in experiments whose characteristic time is shorter than the reorientation time τ; a dynamic Jahn-Teller situation is found in the opposite case. The same defect at the same temperature can be apparently distorted in some experiments and undistorted in other experiments. There is also a transition from a static to a dynamic situation when the temperature is raised. One consequence of this is, for instance, the motional narrowing of EPR lines discussed in [2.3]. To illustrate the effect, consider two possible distortions of a given defect with g factors (Sect. 3.2) equal to g_1 and g_2. Calling $\Delta\omega$ the quantity $(g_1 - g_2)\,\beta H/\hbar$, there are two possible situations: i) when $\tau\Delta\omega \gg 1$, the defect stays long enough in each state so that two lines are observed (each of width $2/\tau$ and separated by $\Delta\omega$) ii) when $\tau\Delta\omega \ll 1$, then only one line is observed (corresponding to $g_1 + g_2$, of width $\tau\Delta\omega^2$).

2.4 The Vacancy in Silicon

Vacancies in covalent semiconductors are subject to Jahn-Teller distortions. This is the case in diamond (Chap.4 and references therein) as well as in silicon (Sect.3.7). Here we concentrate on the second example because it illustrates that a point defect can be a negative U center. We shall first recall briefly the conclusions of the one-electron picture for the vacancy and show that it must experience Jahn-Teller distortions in its different charge states. We then discuss the relative importance of Jahn-Teller and many-electron splitting. We also examine the question of the effective force constants describing the lattice response to Jahn-Teller forces. Finally we present the evidence, given in [2.7,8], that the vacancy in silicon is one of the first identified negative U centers.

2.4.1 Static Distortions Near the Vacancy

To investigate the possibility of a Jahn-Teller effect at the vacancy let us first recall its electronic structure discussed in [Ref.1.1, Sects.3.1.5 and 4.2.2]. We begin with a one-electron calculation. Figure 2.7 recalls the local situation in the molecular model of the vacancy, with its four dangling bonds. The corresponding level structure is given in Fig.2.8a, with a lower nondegenerate state A_1 and an upper triply degenerate level T_2. The neutral charge state V^0 contains four electrons, two in a A_1 state,

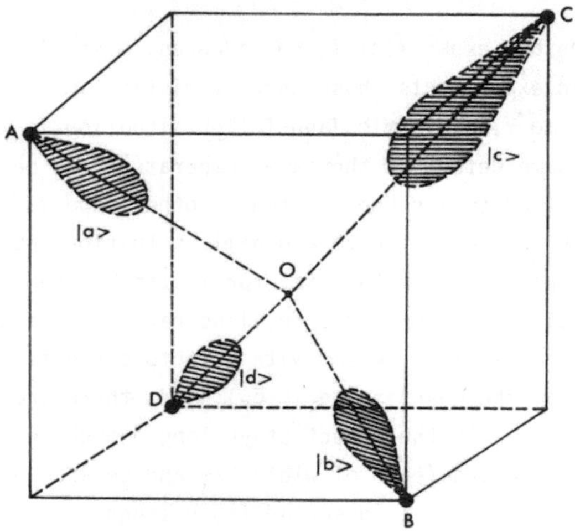

Fig.2.7. The four dangling bond states at the vacancy

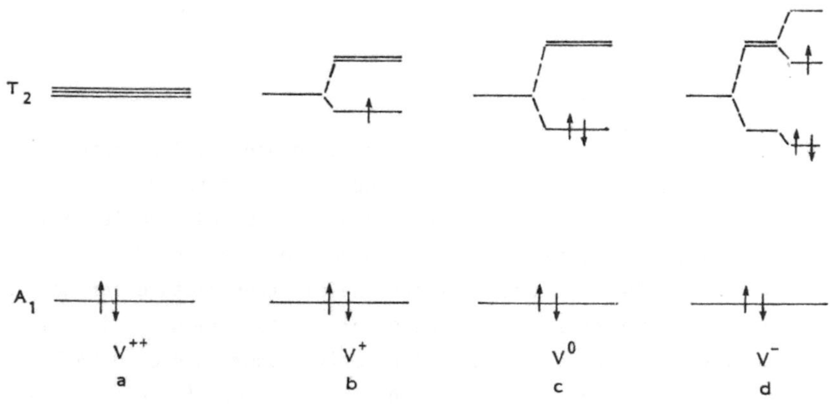

Fig.2.8a-d. Jahn-Teller splitting for the different charge states of the vacancy

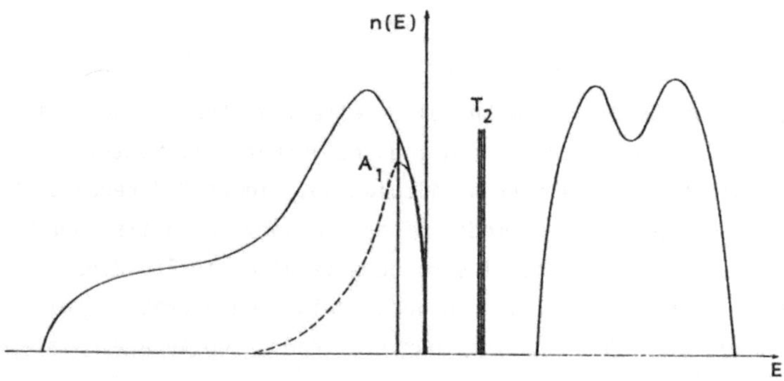

Fig.2.9. A_1 and T_2 electronic states of the vacancy

two in T_2 states. The population of the levels for other charge states is trivial to obtain and will be recalled later.

The relation between this simple defect-molecule model and the real situation has been discussed in [Ref.1.1, Sect.3.2.7]. The results of this are recalled in Fig.2.9 which gives the density of states n(E). The T_2 state appears as a true bound state in the fundamental energy gap, while the A_1 state becomes a resonance near the top of the valence band. The T_2 eigenfunctions become delocalized, the weight of the dangling bonds being reduced to 50%. A_1 being in the valence band is always populated with two electrons. There are thus 2-n electrons in T_2 states for the nth charge state of the vacancy.

To analyze the possibility of Jahn-Teller distortions we have to solve first the many-electron problem, starting from the occupied one-electron orbitals. Let us for the moment forget this problem and apply the Jahn-Teller theorem directly to the one-electron picture, as was done originally by WATKINS [2.20,21]. (This procedure is valid when many-electron effects are negligible as we shall see in Sect.2.4.2.) Consider first the V^{++} state (Fig.2.8a): there are two electrons in the A_1 level and the ground state is nondegenerate: there is no distortion. To obtain V^{+} we add one electron on the T_2 level (Fig.2.8b) which, because it is triply degenerate, leads to Jahn-Teller distortion as discussed in Sect.2.2. The stable distortions are either produced by E or T modes depending upon the strength of their coupling.

Before discussing in detail the local symmetry of the defect after distortion we must be more precise about Q_s. It is reasonable to think that the creation of the vacancy will induce perturbations mainly on its nearest neighbors. Thus the linear Jahn-Teller terms will mainly exist for atomic displacements of these neighbors. We build symmetrical combinations of these displacements which behave as basis functions for the irreducible representations of T_d. Among these we retain the A_1, E and T_2 modes, eliminating modes which correspond to translations and rotations of the defect molecule [2.22,23] the elimination of rotation modes is clearly valid for free molecules but, strictly speaking, is not justified here). The remaining modes are pictured on Fig.2.10 and their detailed expression is given in [2.23]. If we now look at the possible stable distortions for the T_2 state we get two possibilities, corresponding to E or T modes.

i) *E modes:* There are three equivalent distortions (Sect.2.2.2) leading to the Jahn-Teller energy of (2.47). With the $Q_{E,1}$, $Q_{E,2}$ of Fig.2.10, this leads to three tetragonal distortions along the x, y or z directions. One of them

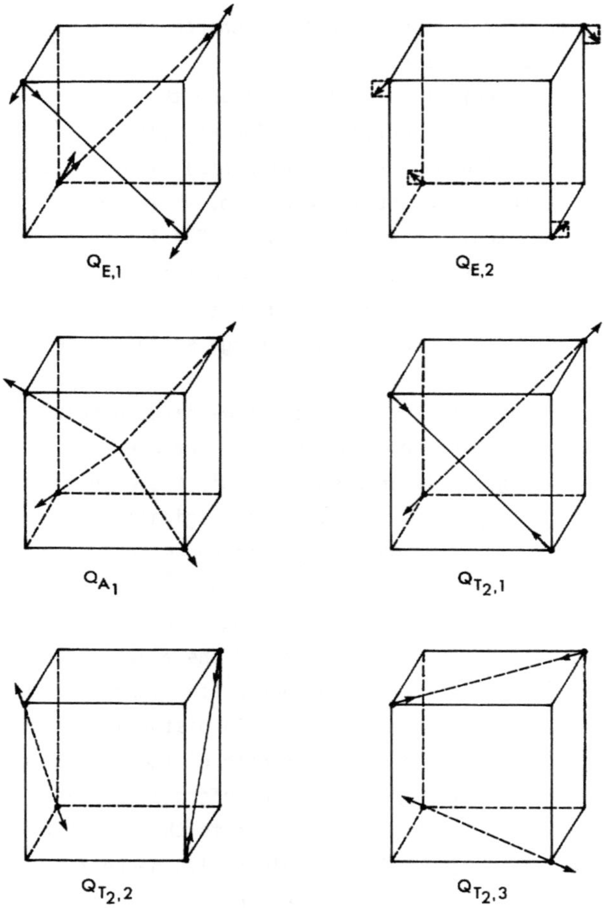

Fig.2.10. Lattice modes around the vacancy

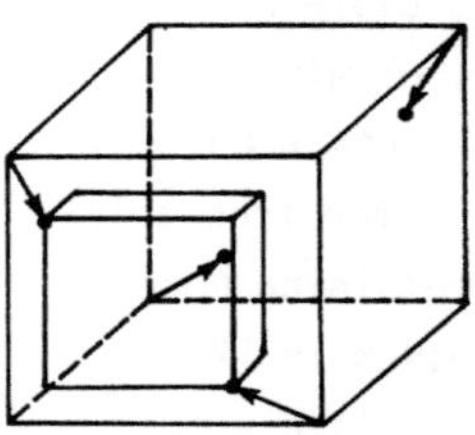

◄ Fig.2.11. Tetragonal distortion

Fig.2.12. Trigonal distortion ►

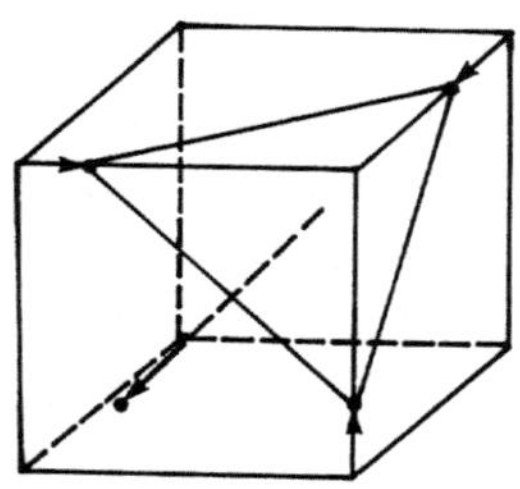

is given in Fig.2.11 and the symmetry is lowered to D_{2d}. As for the case of splitting under a <100> uniaxial stress [Ref.1.1, Sect.1.5] the electronic state T_2 is split into a lower nondegenerate component and an upper twofold degenerate one (Fig.2.8).

ii) *T_2 modes:* If they dominate, we obtain four equivalent distortions defined in Table 2.2 with a Jahn-Teller energy equal to $E_{JT}^{(3)}$. With the $Q_{T_2,1}$, $Q_{T_2,2}$ and $Q_{T_2,3}$ modes of Fig.2.10, this leads to four distortions along <111> axes. One of these, along the <111> direction, is pictured in Fig.2.12, and the symmetry is lowered to C_{3v}. Again the level is split into a nondegenerate and a twofold degenerate level, but with basis states different from the tetragonal case (Table 2.2).

We can now discuss the picture proposed by WATKINS [2.20,21]. Coming back to the V^+ case, the Jahn-Teller theorem predicts "tetragonal" or "trigonal" distortions. Experimentally EPR data (Sect.3.7) show that tetragonal distortions dominate [2.20,21]. The level splitting is thus that given in Fig.2.8b with three equivalent configurations of symmetry D_{2d}. We can then add one electron to obtain the neutral vacancy V^0. This state corresponds to Fig. 2.8c; there is no change in the resulting symmetry except that the distortion must be increased in magnitude since there are now two electrons in the lower localized state. Going from V^0 to V^- drastically changes the situation because the added electron must occupy the higher twofold degenerate state. This leads to an extra Jahn-Teller distortion which must now be trigonal in order to split the level. The result is a "mixed" (tetragonal plus trigonal) distortion which is experimentally observed. As already pointed out, if many electron effects are not negligible, this mixed distortion has to be interpreted in terms of near degeneracy (Sect.2.23 and [2.14]).

2.4.2 The Relative Importance of the Many-Electron Effects and the Jahn-Teller Effect

The simple one-electron picture we have just described works surprisingly well (Chap.4 and [2.20,21]). This has led WATKINS [2.24,25] to conclude that many-electron effects are negligible in contradiction with theoretical estimations which concluded that these effects should be more important than the Jahn-Teller effect itself [2.26-28]. However, there has been recent progress [2.29] towards a better understanding of that problem, which we now describe.

We have to build the many-electron states derived from a given configuration and work out their relative energies. The principles of this technique have been discussed in detail in [Ref.1.1, Sect.4.2]. Let us consider the case of V^0 and recall the predicted lower many-electron states in that case. They are given in Fig.2.13 together with the corresponding eigenstates

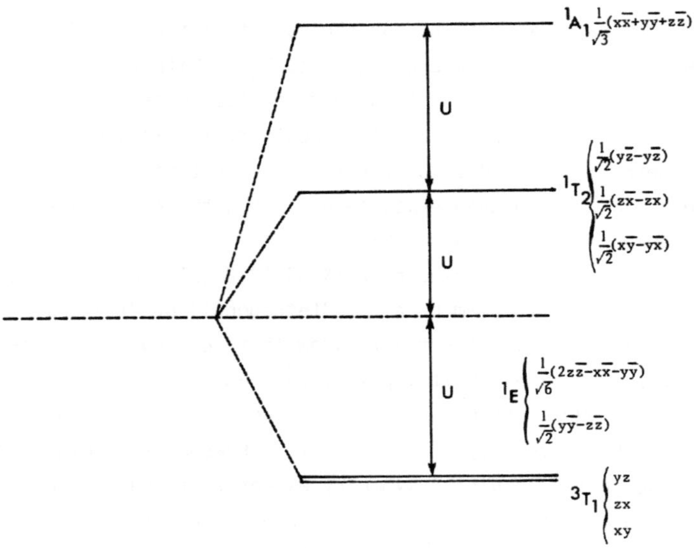

Fig.2.13. Many-electron states for the neutral vacancy

($x\bar{x}$ for instance denotes a Slater determinant with one electron of spin up x and one of spin down $\bar{x}$, where x is a one-electron orbital of the T_2 state [Ref.1.1, Sect.4.2.2]). The corresponding splitting can be described in terms of one parameter U [2.30], which corresponds to the singlet-triplet splitting.

To perform a combined many-electron, plus Jahn-Teller, treatment we have to choose a common basis set in order to express the Hamiltonian matrices H_{me} and H_{JT} describing these two phenomena. We choose for this the many electron states in which H_{me} is diagonal, i.e., those given in Fig.2.13. With such a basis H_{JT} is nondiagonal; to evaluate its influence we impose a distortion having tetragonal symmetry (this is the experimentally observed one) and take it to be along the Z direction. To simplify we also assume that the one-electron contribution to the Jahn-Teller terms is dominant, i.e., we ignore complications due to many-electron effects when we evaluate the distortions.

We thus call δ the change in energy of the one-electron z state under a given tetragonal distortion in the OZ direction. The same distortion changes the energies of the x and y states by $-\delta/2$. In this model H_{JT} has only diagonal elements in a Slater determinant basis. A determinant $z\bar{z}$ will change its energy by 2δ, zy or $z\bar{y}$ by $\delta/2$, and yx by $-\delta$. This defines H_{JT} as a diagonal matrix in the single Slater determinant basis. It is then a simple matter to express it in the basis of Fig.2.13. This gives three independent matrices (using the basis states of Fig.2.13)

$$\left.\begin{aligned}
{}^1A_1 + {}^1E &\Rightarrow \begin{bmatrix} 2U & \sqrt{2}\delta & 0 \\ \sqrt{2}\delta & \delta-U & 0 \\ 0 & 0 & -\delta-U \end{bmatrix} \\
{}^1T_2 &\Rightarrow \begin{bmatrix} -\delta+U & 0 & 0 \\ 0 & \frac{\delta}{2}+U & 0 \\ 0 & 0 & \frac{\delta}{2}+U \end{bmatrix} \\
\text{and} \quad {}^3T_1 &\Rightarrow \begin{bmatrix} -\delta+U & 0 & 0 \\ 0 & \frac{\delta}{2}+U & 0 \\ 0 & 0 & \frac{\delta}{2}+U \end{bmatrix}
\end{aligned}\right\} \quad . \tag{2.105}$$

These matrices are easily diagonalized for any value of the Jahn-Teller parameter δ and of the multiplet splitting parameter U. The corresponding energy E, in units of U, is plotted in Fig.2.14 versus the (dimensionless) parameter $x = |\delta|/U$ (we consider here only negative values of δ). For high values of x the Jahn-Teller effect dominates and all curves tend asymptotically towards straight lines. For $x \to 0$, we recover the multiplet splitting at zero distortion. In this limit we give in Fig.2.14 the straight lines which are tangent at $x = 0$ to the exact curve. The ^1E level becomes the most stable over the whole range of values and Fig.2.14 shows that the limit between the two different regimes where either multiplet or Jahn-Teller effects dominate, corresponds to $|\delta| = U$ (i.e., to the intersection of the dashed lines which represent the limiting behavior).

Let us now investigate the actual situation for V^0 in silicon. The tetragonal Jahn-Teller distortion and energy have been estimated recently both theoretically and from experimental data (see [2.7,8] and next subsection) in the limit of vanishing multiplet effects. It is found that for the stable configuration of V^0, the term δ is equal to -0.68 eV. On the other hand, recent calculations of U [2.29] lead to a value of 0.33 eV, in good agreement with our predictions in [Ref.1.1, Sect.4.2.2]. We are thus in a case where x is equal to two, i.e., in a region where the Jahn-Teller effect dominates. In fact, for x = 2 the relative error introduced by the neglect

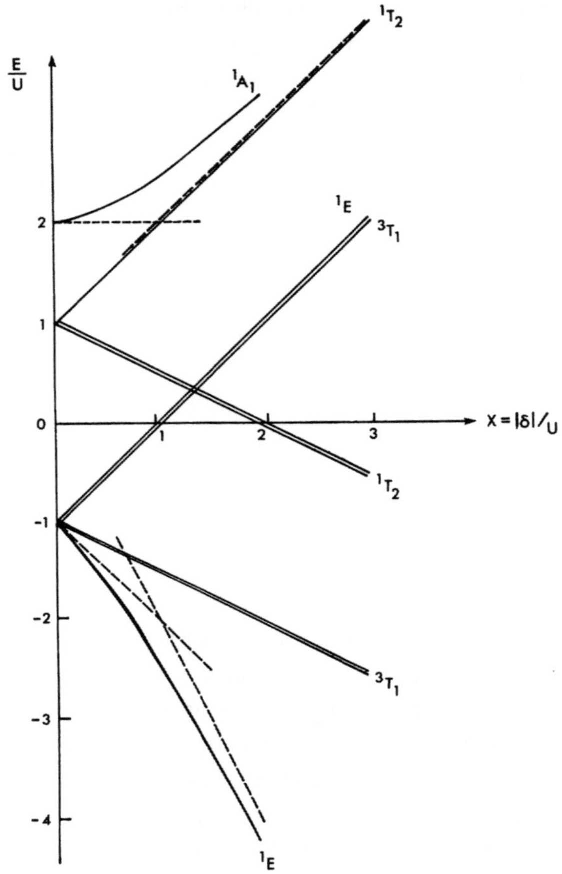

Fig.2.14. Splitting of the many-electron states of V^0 by the Jahn-Teller effect. (——) exact curves, (----) straingt line asymptotes

of U is only 6% compared to the Jahn-Teller stabilization of the level, which is 2δ, (-1.36 eV). Addition of the many-electron effects introduces only a small stabilization, amounting to 0.25 U, (about 0.08 eV).

Up to now, we have discussed only the case of V^0. The case of V^+ and V^{++} are much more trivial since, in both cases, there can be no multiplet splitting in the ground-state configuration. Thus, for V^+, there will be only Jahn-Teller splitting as predicted in the simple one-electron picture. The case of V^- is less obvious and we do not discuss it here. Returning to V^0, we must notice that, although the many-electron effects are not important for the ground state, this is not so for the structure of the excited levels (for a more detailed discussion of this point, see [2.29]).

2.4.3 Effective Force Constants Near the Vacancy

Now that we have seen that many-electron effects are not important (at least for V^{++}, V^{+}, V^{0}), we can apply Watkin's picture directly. The Jahn-Teller effect on a T_2 level will be characterized by a linear coupling coefficient I_{T_2E} (which will be estimated in the next section) and also by a force constant k_E corresponding to the harmonic response of the system. We have seen why the linear terms are likely to be important only for the nearest neighbors, leading to $Q_{E,1}$ and $Q_{E,2}$ pictured in Fig.2.10. This is not true for the harmonic terms whose general form is given by

$$\frac{1}{2}\sum_{\alpha} M\omega_{\alpha}^{2}q_{\alpha}^{2} \tag{2.106}$$

where the ω_α and q_α are the eigenvalues and eigenmodes of the phonon Hamiltonian of the crystal, plus the vacancy. Let us then consider a tetragonal distortion Q_Z along the z direction. The total change in energy of the lattice to second order in the displacements is

$$\delta E = I_{T_2,E}Q_Z + \frac{1}{2}\sum_{\alpha} M\omega_{\alpha}^{2}q_{\alpha}^{2} \quad . \tag{2.107}$$

This is not the simple form $I_{T_2,E}Q_Z + k_E Q_Z^2/2$ which was considered in the theory of Sect.2.2. However, we shall show that we can reduce the problem to this simple form at least for energetic considerations.

We want to find the minimum of δE with respect to the atomic displacements. The q_α being a complete orthonormal set, we can write

$$Q_Z = \sum_{\alpha} c_{\alpha}q_{\alpha} \quad , \tag{2.108}$$

and then minimize δE with respect to each q_α. We get in this way the minimum values

$$q_{\alpha m} = -\frac{I_{T_2,E}}{M\omega_{\alpha}^{2}}\,c_{\alpha}$$

$$\delta E_m = -\frac{1}{2}I_{T_2,E}^{2}\sum_{\alpha}\frac{|c_{\alpha}|^{2}}{M\omega_{\alpha}^{2}} \quad . \tag{2.109}$$

We can obtain the same result by writing

$$\delta E = I_{T_2,E}Q_Z + \frac{1}{2}k_E Q_Z^{2} \tag{2.110}$$

and then minimize with respect to Q_z alone, defining the effective force constant k_E for mode E by

$$k_E^{-1} = \sum_\alpha \frac{|c_\alpha|^2}{M\omega_\alpha^2} \quad . \tag{2.111}$$

This effective force constant k_E, as well as k_{A_1}, have been determined either using cluster models [2.31,8] or a Green's function technique [2.32]. The results are given in Table 2.3, where it can be seen that they are substantially lower than 1/2 $M\omega_M^2$ (where ω_M is the optical mode frequency).

Table 2.3. Effective force constants in eV/Å^2

	Ref.31	Ref.8	Ref.32
k_{A_1}	3.72	7.47	4.69
k_E	5.08	14.81	9.34
$M\omega_M^2$		27.935	

2.4.4 The Negative U Center Formed by V^{++}, V^+, V^0, in Silicon

We now present a very important conclusion of recent theoretical [2.7,8] and experimental works [2.8,33], that successive charge states of the vacancy can form a so-called negative U center. The general concept of such centers was introduced some time ago [2.34] but mainly for disordered systems and not specifically for point defects in crystalline semiconductors.

We want to calculate the relative concentrations of the three charge states $V^{++}(n=2)$, $V^+(n=1)$, $V^0(n=0)$ at any temperature T and for any Fermi level position E_F. For this we use (2.18) and (2.19) in which, for simplicity, we drop the degeneracy factors and entropy parts. This gives

$$\frac{C(V^{++})}{C(V^+)} = \exp\left(\frac{E(1,2) - E_F}{kT}\right) \tag{2.112}$$

and also

$$\frac{C(V^+)}{C(V^0)} = \exp\left(\frac{E(0,1) - E_F}{kT}\right) \quad . \tag{2.113}$$

We see that V^{++} dominates over V^+ for $E_F < E(1,2)$, the reverse situation being obtained for $E_F > E(1,2)$. The same is true for the pair V^+, V^0 the occupancy level being now E(0,1).

It is interesting to express each occupancy level $E(n,n+1)$ as the sum of a contribution $E_0\,(n,n+1)$ (calculated for the atoms in perfect crystal configuration $\underline{R}_0$) and a term $\Delta E\,(n,n+1)$ coming from the atomic displacements (relaxation or distortion)

$$E(n,n+1) = E_0(n,n+1) + \Delta E(n,n+1) \quad . \tag{2.114}$$

The evolution of $E_0(n,n+1)$ with n can be understood very easily. This corresponds to the change in the ionization energy of the defect with the charge state. When n is decreased by one unit (i.e., when one more electron is bound to the defect), each electron experiences one more electronic repulsion. Then the ionization energy decreases when n decreases and $E_0(n,\, n+1)$ is raised within the gap. The quantity $E_0(n-1,n) - E_0(n,n+1)$ corresponds to the Coulomb repulsion between two electrons at the defect and is usually labeled U. This is a repulsive energy which was calculated for the vacancy [2.7,8] to have a value of about 0.25 eV, irrespective of the charge state (see [Ref.1.1, Sect.4.1.2]). The ordering of the levels $E_0(1,2)$ and $E_0(0,1)$ is given in Fig.2.15. We find the conventional situation where, when the Fermi level is raised in the gap, we find first V^{++}, then V^{+} and finally V^{0} as shown in Fig.2.15.

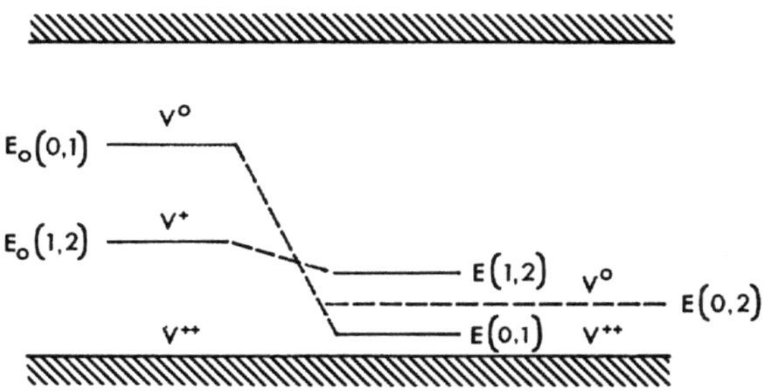

Fig.2.15. Occupancy levels without (on the left) and with (on the right) lattice distortion

The new feature, in the vacancy case, is brought by the term $\Delta E(n,n+1)$ whose origin is the Jahn-Teller distortion. To discuss this term we must remember that $E(n,n+1)$ is the difference between the two total energies $E_e(n,\underline{R}_n)$ and $E_e(n+1,\, \underline{R}_{n+1})$. For the vacancy in silicon the only displacement is the tetragonal distortion, labeled Q for simplicity. The total energies for V^{++}, V^{+} and V^{0} can be expanded to second order in Q, giving

$$E_e(2,Q) = E_e(2,Q=0) + \frac{1}{2}kQ^2$$
$$E_e(1,Q) = E_e(1,Q=0) - IQ + \frac{1}{2}kQ^2$$
$$E_e(0,Q) = E_e(0,Q=0) - 2IQ + \frac{1}{2}kQ^2 \quad . \tag{2.115}$$

These expressions (2.115) are the basis for understanding the negative U center [2.7,8]. The terms linear in Q can be understood in the following way: for V^{++} there is no electron in the T_2 gap state, thus no degeneracy and no Jahn-Teller term; for V^+ there is one electron in T_2 and thus, a linear Jahn-Teller term $-IQ$; for V^0 there are two electrons in T_2 and the linear term is $-2IQ$ to a good approximation. As regards the quadratic restoring term $1/2\ kQ^2$ it is assumed to be independent of the charge state, k being the effective force constant defined in Sect.2.4.3. These approximations seem to be correct for the tetragonal distortion of V^+ and V^0 [Ref.2.2, p.856] but not for the relative influence of the trigonal distortion in the sequence V^+, V^0, V^-.

We now minimize the three expressions (2.115) with respect to Q (to get $E_e(n,\underline{R}_n)$ in the general notation). This gives

$$E_e(2,Q_2) = E_e(2,Q=0)$$
$$E_e(1,Q_1) = E_e(1,Q=0) - \frac{I^2}{2k}$$
$$E_e(0,Q_0) = E_e(0,Q=0) - \frac{4I^2}{2k} \quad . \tag{2.116}$$

The occupancy levels are given by the differences in these total energies and we get

$$E(1,2) = E_0(1,2) - E_{JT} \qquad E(0,1) = E_0(0,1) - 3E_{JT} \tag{2.117}$$

where E_{JT} denotes the quantity $I^2/2k$. They are naturally expressed as in (2.114), the distortion energies being $-E_{JT}$ and $-3E_{JT}$. As discussed above, $E_0(0,1)$ lies above $E_0(1,2)$ by about 0.25 eV. However, this order can be reversed for E(0,1) and E(1,2) if E_{JT} is large enough. This is actually the case for the vacancy in silicon which corresponds to the situation of Fig. 2.15, with an inverted order for E(1,2) and E(0,1). For such an inverted situation to occur, the following inequality must be satisfied

$$U - 2E_{JT} < 0 \quad , \tag{2.118}$$

where the Coulomb parameter U is the difference $E_0(0,1) - E_0(1,2)$. The quantity $U-2E_{JT}$ is the effective Coulomb parameter which can be positive or negative depending upon the magnitude of E_{JT}.

Let us examine the consequences of the situation presented in Fig.2.15. Equations (2.112) and (2.113) show that i) when $E_F < E(0,1)$, then V^{++} dominates ii) when $E_F > E(1,2)$ then V^0 dominates. The difficulty lies in the range $E(0,1) < E_F < E(1,2)$ for which we know only that it is either V^0 or V^{++} which dominate, but never V^+. However, if we take the ratio of (2.112) and (2.113) we can discuss the relative concentration of V^{++} and V^0. We have to define, as in (2.19), the occupancy level $E(0,2)$:

$$E(0,2) = \frac{1}{2}\,[E(0,1) + E(1,2)] \quad . \tag{2.119}$$

When $E_F < E(0,2)$, V^{++} dominates and, when $E_F > E(0,2)$, V^0 dominates. The general conclusion is that V^+ is never the dominant charge state at equilibrium and that the occupancy level for the pair V^0, V^{++} is $E(0,2)$ defined by (2.119). This is summarized by Fig.2.15.

We now give the numbers calculated by BARAFF et al. [2.8]. Taking the origin of energies at E_V, they find $E_0(1,2)$ equal to 0.32 eV. As we have seen before, U is equal to 0.25 eV, so that $E_0(0,1)$ lies at 0.57 eV. They have calculated the distortion energy and find E_{JT} of the order of 0.19 eV. This is enough to change the sign of the effective Coulomb parameter which is equal to -0.13 eV. The final value of $E(1,2)$ is 0.13 eV and $E(0,1)$ is at 0 eV.

All these conclusions are in agreement with experiment (EPR and DLTS). These observations are discussed in Sects.3.8.5 and 6.4.3.

3. Electron Paramagnetic Resonance

Electron paramagnetic resonance is the resonant absorption of electromagnetic radiation by systems composed of unpaired electrons placed in a magnetic field. The ground states of partially filled electron orbitals are spin degenerate. In a magnetic field, because there are several possible orientations for the magnetic moment associated with the total spin, the degeneracy is lifted. Energy levels associated with each orientation arise and absorption occurs when transitions are induced between them.

In semiconductors electrons of the valence shells have spin and angular momenta that cancel each other due to the strong electrical forces associated with the bonds and semiconductors are diamagnetic. But paramagnetism can arise from point defects. Indeed a local magnetic moment can be associated with a defect which traps an electron (or a hole), and which consequently possesses an unpaired electron, or with an impurity which presents a partially filled shell (this is the case of transition elements for instance).

When electron paramagnetic resonance (EPR) works in a semiconductor, it proves to be the most powerful technique to identify defects [3.1-3]. The reason is that an EPR spectrum contains detailed microscopic information on the electronic structure of these defects. Thanks to EPR a reasonable understanding has been obtained in many semiconductors for a large amount of impurities, impurity related and intrinsic defects.

A good knowledge of practically all the simple defects created by irradiation has been reached for silicon and could also probably be obtained in 2-6 compounds. Unfortunately EPR does not work well in all semiconductors.

For materials in which the spin-lattice relaxation time is too short, the EPR lines are broad and difficult to detect (this may be the case of germanium). In other materials (gallium arsenide, for instance) all atoms have nuclear spins which give rise to hyperfine interactions; the EPR signal is broad and these interactions cannot be resolved.

In this chapter we describe the kind of information EPR can provide on the electronic structure of a point defect [3.4]. For this we consider in turn the main interaction terms in the spin Hamiltonian. Each type of interaction will be illustrated using an example taken from the experimental work performed in electron irradiated silicon by WATKINS. We finally treat in detail the case of the vacancy in silicon for which the EPR observations can be directly related to the simple LCAO considerations developed in [Ref. 1.1, Chap.3] and [3.5].

3.1 The Hamiltonian

The electron Hamiltonian we have considered so far contains two terms: the kinetic energy term and a potential (Coulomb) energy term, i.e.,

$$H_o = \sum_i \left[-\frac{\hbar^2}{2m} \Delta_i + V(\underline{r}_i) \right] + \sum_{i>j} \frac{e^2}{r_{ij}} \quad . \tag{3.1}$$

The true Hamiltonian H contains additional terms due to the interaction between the electron orbital moment, the electron spin and the nuclear magnetic moments of the neighbors of the defect. The interaction of the spin $\underline{S}$ of an electron with its orbital momentum $\underline{L}$ is the spin-orbit interaction H_{LS}; the interaction between spins themselves is the spin-spin interaction H_{SS}; the interaction of $\underline{S}$ and $\underline{L}$ with the nuclear spins $\underline{I}$ present within the orbit of the electron give the hyperfine interactions H_{SI} and H_{LI}. Moreover, when an external magnetic field $\underline{H}$ is applied, as in this chapter, the coupling of $\underline{L}$, $\underline{S}$ and $\underline{I}$ with $\underline{H}$ gives rise, respectively, to the electronic H_{Ze} and nuclear H_{Zn} Zeeman interactions. Hence the true Hamiltonian H must be written

$$H = H_o = H_{LS} + H_{SS} + H_{SI} + H_{LI} + H_{Ze} + H_{Zn} \quad . \tag{3.2}$$

In this chapter we consider each of these interactions separately. We derive their expressions and detail the information they provide on a defect. Because they are small compared to H_o ($\simeq$ 1 eV): $H_{LS} \simeq 10^{-2}$ eV, $H_{SS} \simeq 10^{-4}$ eV, $H_{SI} \simeq H_{LI} \simeq 10^{-6}$ eV, $H_{Ze} \simeq 10^{-4}$ eV, $H_{Zn} \simeq 10^{-7}$ eV, they can be treated as perturbations on H_o (whose eigenvalues will be noted E_o). For spherically symmetric systems the eigenfunctions ψ_o of H_o, defined by the principal quantum number n and the orbital quantum numbers ℓ and m

$$|\psi_0\rangle = |n, \ell, m\rangle \quad (3.3)$$

are spin and orbitally degenerate. This degeneracy can be lifted by the additional interactions considered and a state $|\psi\rangle$ of H must be specified either by the use of two new quantum numbers j and m_j or m_ℓ and m_S which are, respectively, the quantum numbers associated with the total angular momentum

$$\underline{J} = \underline{L} + \underline{S} \quad (3.4)$$

or with the components of $\underline{L}$ and $\underline{S}$ parallel to the magnetic field $\underline{H}$. Finally, when the system contains a nuclear spin $\underline{I}$, its associated quantum number m_I must also be specified. EPR transitions are allowed for certain changes in the values of these quantum numbers [3.1,6].

3.2 Electronic Zeeman Interaction

3.2.1 Zeeman Interaction

An electron bonded to a center possesses an angular momentum

$$\underline{L} = \underline{r} \times \underline{p} \quad (3.5)$$

associated with its motion. With such a momentum is associated the dipolar magnetic moment

$$\underline{\mu}_\ell = -\beta\underline{L} \quad (3.6)$$

where β is the Bohr magneton. The expression of β can be derived simply from a classical argument using the fact that the electron, moving on a circular orbit of radius r with a velocity v, is equivalent to a current (in cgs units)

$$j = -\frac{e}{c\tau} \quad (3.7)$$

where -e is the electron charge and c the velocity of light. The period of the movement, τ, is such that

$$v\tau = 2\pi r \quad . \quad (3.8)$$

Classical electrodynamics tells us that a magnetic dipole moment amplitude

$$|\underline{\mu}_\ell| = j\pi r^2 \quad (3.9)$$

is associated with this circular current. It is obvious, using the above expressions, that $|\underline{\mu}_\ell|$ is proportional to the angular momentum $|\underline{L}|$ given by

$$|\underline{L}|\hbar = mvr \quad . \quad (3.10)$$

The coefficient of proportionality is

$$\beta = \frac{e\hbar}{2mc} \quad . \tag{3.11}$$

The electron also possesses an intrinsic angular momentum, the spin $\underline{S}$, with which is also associated the magnetic moment

$$\underline{\mu}_S = -2\beta\underline{S} \quad . \tag{3.12}$$

The total magnetic moment of the electron is therefore

$$\underline{\mu}_T = -\beta(\underline{L} + 2\underline{S}) \quad . \tag{3.13}$$

In paramagnetic systems, for which the electronic levels are not completely filled, the quantity $\underline{L} + \underline{S}$ is nonzero and gives rise to the following interaction with a magnetic field $\underline{H}$

$$H_{Ze} = -\underline{\mu}_T \cdot \underline{H} \quad , \tag{3.14}$$

that is

$$H_{Ze} = \beta\underline{H} \cdot (\underline{L} + 2\underline{S}) \quad . \tag{3.15}$$

For a many-electron atom, the Zeeman Hamiltonian H_{Ze} is still given by (3.15), but where $\underline{L}$ is the total orbital momentum and $\underline{S}$ the total spin. The eigenstates of the free atom Hamiltonian H_0 (3.1) are characterized by the values of L and S and have (2L + 1)(2S + 1) degeneracy. In the approximation of the Russell-Saunders coupling, the spacing in energy between these states is assumed to be large compared to the spin-orbit splitting term. The effect of the latter is thus to lift the (2L + 1)(2S + 1) degeneracy. However $\underline{J} = \underline{L} + \underline{S}$ still commutes with the Hamiltonian so that the new eigenstates can be written $|L,S,J,M_J\rangle$. The magnetic field will thus split the resulting (2J + 1) degeneracy. It is shown in all quantum mechanical text books that the resulting Zeeman Hamiltonian within each L,S,J degenerate level can be written

$$H_{Ze} = g\beta\underline{H} \cdot \underline{J} \quad , \tag{3.16}$$

where g is the so-called Landé factor, given by [3.6]

$$g = 1 + \frac{1}{2}\,\frac{J(J + 1) + S(S + 1) - L(L + 1)}{J(J + 1)} \quad . \tag{3.17}$$

For a paramagnetic defect in a crystal, the nuclear quadrupolar electric and magnetic dipole moments of the atoms in the neighborhood of the defect

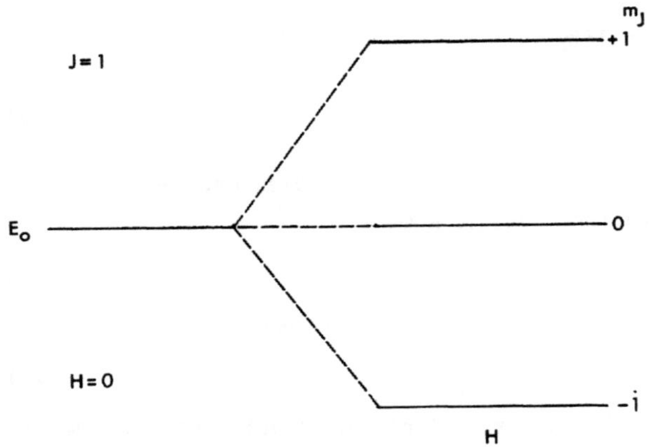

Fig.3.1. Zeeman splitting of a level with J = 1

add to the effects of the uniform applied field (Sect.3.4) and give rise to hyperfine structures which are very sensitive to the details of the defect wave function.

The eigenvalues ΔE_{Ze} are readily obtained using the eigenvalues m_J of the component of $\underline{J}$ along $\underline{H}$

$$\Delta E_{Ze} = g\beta|\underline{H}|m_J \quad . \tag{3.18}$$

Because m_J varies between -J and J, the degenerate level is split by the action of the applied magnetic field into 2J + 1 sublevels (Fig.3.1) separated by the energy

$$\delta E_{Ze} = \Delta E_{Ze}(m_J) - \Delta E_{Ze}(m_J - 1) \quad , \tag{3.19}$$

that is

$$\delta E_{Ze} = g\beta|\underline{H}| \quad . \tag{3.20}$$

3.2.2 Spin Resonance

Transitions can be induced between two adjacent levels for which $\Delta m_J = \pm 1$ by the photon energy

$$h\nu = g\beta|\underline{H}| \quad . \tag{3.21}$$

A resonance is observed at this frequency. The rate of absorption, i.e., the amount of energy absorbed per unit time, is the product

$$P = h\nu W\delta n \tag{3.22}$$

where W is the quantum probability per unit time that a transition is induced and $\delta n = n_i - n_j$ the difference in the electron populations n_i and n_j on the two adjacent levels j and j - 1 = i between which the resonance occurs. The ratio n_j/n_i at thermal equilibrium is given by the Boltzmann factor

$$\frac{n_j}{n_i} = \exp\left(-\frac{\delta E_{Ze}}{kT}\right) \quad , \tag{3.23}$$

which can be written

$$\frac{n_j}{n_i} = 1 - \frac{\delta E_{Ze}}{kT} \tag{3.24}$$

since, usually $\delta E_{Ze} \ll kT$. From this relation and considering a two-level system of total population

$$n = n_j + n_i \quad , \tag{3.25}$$

we obtain for δn, to first order

$$\delta n = \frac{n}{2}\frac{\delta E_{Ze}}{kT} \quad . \tag{3.26}$$

Actually, the absorption of the photons $h\nu$ tends to equalize the populations n_j and n_i while the thermal process tends to bring them back to their equilibrium values. Relation (3.26) is therefore valid only if the photon absorption does not disturb significantly the thermal equilibrium populations. This condition can be stated as follows: let W_+ and W_- be the transition probabilities of an electron from states i to j and from j to i, respectively. For photon induced transitions

$$W_+ = W_- = W \tag{3.27}$$

the rate equation describing the change in population can be written

$$\frac{d\,\delta n}{dt} = -2W\,\delta n \quad . \tag{3.28}$$

For the thermal process the transition probabilities W_+^t and W_-^t are different and the corresponding rate equation is

$$\frac{d\,\delta n}{dt} = n(W_-^t - W_+^t) - \delta n(W_+^t + W_-^t) \quad , \tag{3.29}$$

which can be written

$$\frac{d\,\delta n}{dt} = \frac{\delta n_o - \delta n_1}{T_1} \tag{3.30}$$

where

$$T_1 = \frac{1}{W_+^t + W_-^t} \tag{3.31}$$

is a characteristic time, called the spin-lattice relaxation time, associated with the approach to thermal equilibrium. The quantity

$$\delta n_o = n\,\frac{W_-^t - W_+^t}{W_-^t + W_+^t} \tag{3.32}$$

is the difference in the populations of the two levels due to the thermal process; it is therefore biven by (3.26) derived above. Hence, the rate equation describing the evolution of the populations under both photon absorption and thermalization is

$$\frac{d\,\delta n}{dt} = -2W\,\delta n + \frac{\delta n_o - \delta n}{T_1} \quad . \tag{3.33}$$

The steady-state solution of this equation

$$\delta n = \frac{\delta n_o}{1 + 2WT_1} \tag{3.34}$$

reduces to $\delta n = \delta n_o$ when $2WT_1 \ll 1$.

The probability W that a transition is induced is calculated from elementary quantum mechanics. Under the action of a time dependent perturbation $F\exp(i\omega t)$ produced by an electromagnetic wave, W is time independent [3.6] and given by

$$W = \frac{2\pi}{\hbar}\,|\langle\psi_i|F|\psi_j\rangle|^2\;\delta(E_j - E_i - h\nu) \quad . \tag{3.35}$$

In this case the perturbation F is the interaction of magnetic moments with the sinusoidal magnetic field of an electromagnetic wave

$$\underline{h} = \frac{1}{2}\,\underline{h}_o(e^{i\omega t} + e^{-i\omega t}) \tag{3.36}$$

and

$$W = \frac{2\pi}{\hbar}\,\frac{h_o^2}{4}\,|\langle\psi_i|\mu|\psi_j\rangle|^2\delta(E_j - E_i - h\nu) \tag{3.37}$$

where μ is the component of $\underline{\mu}$ along $\underline{h}$. Using (3.22), we finally obtain the following expression for the power dissipated

$$P = \pi^2 \nu h_o^2 |<\psi_i|\mu|\psi_j>|^2 \delta(E_j - E_i - h\nu)\delta n \quad . \tag{3.38}$$

In practice the power dissipated takes the form

$$P = \frac{\pi}{2\hbar} h_o^2 \omega |<\psi_i|\mu|\psi_j>|^2 \frac{\delta n_o}{1 + 2WT_1} f(\omega) \tag{3.39}$$

where $f(\omega)$ is the line-shape function (normalized to unity, to take account of the broadening of the δ function). In the limit $2WT_1 \ll 1$ and using the fact that, at resonance, $\delta E_{Ze} = \hbar\omega$, this expression reduces to

$$P = \frac{\pi}{4} \frac{\omega^2}{kT} |<\psi_i|\mu|\psi_j>|^2 \, n f(\omega) h_o^2 \quad . \tag{3.40}$$

3.2.3 Observation of Resonance

In order to observe resonance, the sample is placed in the alternating magnetic field, of frequency ω, of a cavity. The impedance of the cavity containing the sample is measured as a function of ω. Let L_o and R_o be, respectively, the inductance and the resistance of the coil equivalent to the cavity. When the sample, of susceptibility χ, is in the coil the inductance becomes $L_o(1 + 4\pi\chi)$ and the impedance is

$$Z = R_o + iL_o\omega(1 + 4\pi\chi) \tag{3.41}$$

(the sample is assumed to fill the total volume of the cavity, i.e., the filling factor is taken to be equal to 1). Because the susceptibility is a complex quantity

$$\chi = \chi' - i\chi'' \quad , \tag{3.42}$$

the introduction of the sample in the coil changes its inductance (through χ') and its resistance (through χ''). The change of resistance ΔR is given by

$$\frac{\Delta R}{R_o} = \frac{L_o\omega}{R_o} 4\pi\chi'' \tag{3.43}$$

usually written as

$$\frac{\Delta R}{R_o} = 4\pi\chi'' Q \quad , \tag{3.44}$$

introducing the so-called quality factor $Q = L_o\omega/R_o$. This factor Q is of the order of $10^3 - 10^4$ in the microwave range, (i.e., for a cavity) at which the measurements have to be performed.

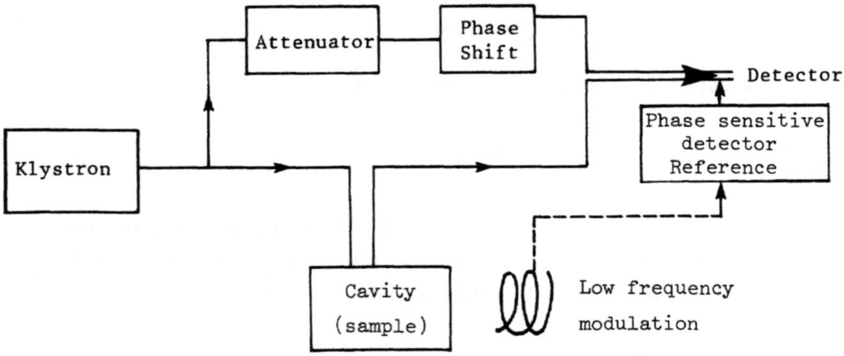

Fig.3.2. Schematic representation of the EPR experiment

This absorption of energy is usually measured at low temperature (usually 77 K or below) because it is necessary that the difference in the populations of the two levels be nonnegligible. Of course, the observation of the resonance will be possible only in a material for which the relaxation time T_1 is not too large.

The way the absorption at the resonance is detected consists in measuring the loss of the quality factor Q of the cavity containing the sample. A cavity is used because, for the magnetic field easily available (1 - 10 KG), the resonance occurs in the GHz range: for $g \simeq 2$ (the value of the g factor of the free electron) 3500 G correspond to a resonance frequency of GHz (the so-called X band) and 8500 G to 24 GHz (the K band). Since it is difficult in practice to vary the frequency of a klystron in a wide enough range, the frequency is kept constant and the magnitude of the magnetic field is made to vary.

For values of the magnetic field away from the resonance the cavity is critically coupled so that no signal is reflected. At resonance, the critical coupling being destroyed, a small signal is reflected which is detected (Fig.3.2). The detection is usually performed with a phase sensitive detector locked on the low-frequency modulation of the magnetic field. A "bucking" arm is included in the system to feed the detector with a signal from the klystron. This bucking signal is needed to resolve the effects of absorption and dispersion. In addition, it provides a steady level at the detector to insure appropriate detector efficiency [3.7].

The resonance is detected as a loss of the quality factor of the cavity due to the absorption of energy, via χ''. Indeed χ'' is given by

$$\chi'' = \pi\delta n|\langle\psi_i|\mu|\psi_j\rangle|^2\delta(E_j - E_i - h\nu) \quad , \tag{3.45}$$

as can be shown by comparing (3.38) with the following relation

$$P = \frac{h_o^2}{2}\,\omega\chi'' \tag{3.46}$$

derived by writing that the power dissipated by the introduction of the sample in the coil

$$P = \frac{1}{2}\,\Delta R I^2 \quad , \tag{3.47}$$

where the current I which flows in the coil is related to the magnetic field it induces by

$$\frac{1}{2}\,L_o I^2 = \frac{h_o^2}{8\pi} \quad . \tag{3.48}$$

3.3 Spin Orbit Coupling

3.3.1 Quenching of Orbital Motion

An electron of a free atom or bonded to a defect may possess an orbital momentum $\underline{L}$. Once the defect is placed in a crystal, the motion of this electron is affected by the electrostatic interactions with the neighboring atoms (called ligands). The interactions can be described by the introduction of an electrostatic potential (the crystal field) in the region occupied by the orbit of the electron. This electrostatic potential reflects the symmetry of the defect and its neighborhood. Depending on the symmetry of this field the electronic levels are split. This splitting as well as the remaining degeneracy can be predicted by group theory [Ref.1.1, Sect.1.4]. Kramer's theorem maintains that, in the absence of an applied magnetic field, for a system containing an unpaired electron, the remaining degeneracy is at least two. The effect of the crystal field is then to lift, at least partially, the degeneracy associated with $\underline{L}$.

VAN VLECK [3.8] has shown that removal of orbital degeneracy by the crystal field is equivalent to quenching the angular momentum. Quenching $\underline{L}$ means that the expectation value of its component along $\underline{H}$ is zero

$$\langle L_z\rangle = \langle\psi|L_z|\psi\rangle = 0 \quad . \tag{3.49}$$

The angular momentum is quenched for nondegenerate states because, being solutions of a real differential equation, their eigenfunctions can be chosen real. Indeed, for any wave function ψ the expectation value of L_z can be written

$$\langle L_z \rangle = \frac{\hbar}{i} \int \psi^* \left(x \frac{\partial}{\partial y} - y \frac{\partial}{\partial x} \right) \psi \, dv \quad . \tag{3.50}$$

For a real wave function, the integral is real and $\langle L_z \rangle$ is imaginary. However, since the diagonal matrix elements of any Hermitian operator such as L_z must be real, the above equality is verified only if

$$\langle L_z \rangle = 0 \quad . \tag{3.51}$$

Hence, whenever the crystalline field leaves an electronic state nondegenerate, the orbital momentum of that state is quenched. This quenching can be illustrated as follows. Consider an electron localized on a free defect, in a p atom-like state. There are three degenerate states associated with a p orbital level (for which $\ell = 1$). The three p functions are products of coordinates x, y and z with a spherically symmetric function f(r):

$$P_x = xf(r) \quad , \quad P_y = yf(r) \quad , \quad P_z = zf(r) \quad . \tag{3.52}$$

When the defect is placed in a cubic crystal field these three states have the same energy in the absence of applied magnetic field since all the directions x,y,z are equivalent. However, when the defect is placed in an environment of lower symmetry these directions are no longer equivalent and the states have different energies. For instance, in such a crystal field, the ground state, say p_x, can be written as a linear combination of $m_\ell = +1$ and $m_\ell = -1$ states:

$$P_x = \frac{1}{\sqrt{2}} \left[\frac{x + iy}{\sqrt{2}} f(r) + \frac{x - iy}{\sqrt{2}} f(r) \right] \quad . \tag{3.53}$$

In (3.53) the first term corresponds to $m_\ell = +1$, the second to $m_\ell = -1$. The action of L_z on such states gives +1 and -1, respectively, and it is easily verified that $\langle p_x | L_z | p_x \rangle$ is equal to zero.

3.3.2 Effective Spin Hamiltonian

Normally the orbital momentum is quenched. Then the Zeeman interaction H_{Ze} reduces, in first order, to its spin part

$$H_{Ze} = g\beta \underline{H} \cdot \underline{S} \quad . \tag{3.54}$$

For a free electron the value of g:$g_o = 2.0023$ is isotropic. However, for an electron bonded to a defect the g value deviates from g_o. Classically this deviation

$$\Delta g = g - g_o \tag{3.55}$$

can be viewed as a measure of the effective moment of the magnetic dipole associated with the spin. Indeed, the magnetic dipole

$$\underline{\mu}_s = -g_o\beta\underline{S} \tag{3.56}$$

induces orbital dipolar moments $\underline{\mu}_i$ on the atoms (comprising the defect) and the effective Zeeman interaction is

$$H_{Ze} = (g_o\beta\underline{S} - \sum_i \underline{\mu}_i)\cdot\underline{H} \quad . \tag{3.57}$$

The amplitude of the induced dipolar moments is, of course, proportional to the amplitude of the moment which induces them, i.e., $\underline{\mu}_i$ can be written

$$\sum_i \underline{\mu}_i = -\Delta g\beta\underline{S} \tag{3.58}$$

introducing the coefficient of proportionality Δg. The value of Δg is a function of the magnetic field orientation since $\underline{\mu}_i$ reflects the neighborhood of the defect. As a result Δg is a tensor and H_{Ze} must be written as

$$H_{Ze} = \beta\underline{H}\cdot(\tilde{g}\underline{S}) \tag{3.59}$$

where

$$\tilde{g} = g_o + \Delta\tilde{g} \quad . \tag{3.60}$$

This is equivalent to the concept of the effective spin Hamiltonian whose virtue is to describe the behavior of electronic levels in a solid using the same terms as for a free atom.

3.3.3 Quantitative Treatment of the g Tensor

In order to account for the magnetic interaction between spin and orbital dipole moments, the Hamiltonian contains an additional term; for an electron in a single atom

$$H = g_o\beta\underline{S}\underline{H} + \beta\underline{L}\underline{H} + \lambda(r)\underline{L}\underline{S} \tag{3.61}$$

r being the distance to the nucleus. The function $\lambda(r)$, which is related to the electric field seen by the electron, takes important values near the nucleus only. We have shown in Sect.3.3.1 that in a crystal field which quenches $\underline{L}$, first-order perturbation treatment results in the cancellation of the second term in H. However, the terms involving $\underline{L}$ contribute in a second-order perturbation treatment when the spin-orbit interaction (third term) is taken into account. Noting ψ_o the orbital part of the ground-state wave function (of energy E_o) and ψ_n the corresponding part of the nth excited state wave function (of energy E_n), second-order perturbation theory gives for the ground-state spin manifold the effective Hamiltonian

$$H_{eff} = g_o\beta\underline{S}\underline{H} - \sum_{n\neq 0} \frac{|\langle\psi_o|\beta\underline{L}\underline{H} + \lambda(r)\underline{L}\underline{S}|\psi_n\rangle|^2}{E_n - E_o} \quad . \tag{3.62}$$

Expanding the summation and retaining only the terms containing both $\underline{S}$ and $\underline{H}$ we get [3.9]

$$H_{eff} \simeq \beta\underline{S}\tilde{g}\underline{H} \tag{3.63}$$

with

$$\tilde{g} = g_o - 2\tilde{\Lambda} \tag{3.64}$$

where $\tilde{\Lambda}$ is a tensor whose components are

$$\Lambda_{ij} = \mathrm{Re} \sum_{n\neq 0} \frac{\langle\psi_o|L_i|\psi_n\rangle\langle\psi_n|\lambda(r)L_j|\psi_o\rangle}{E_n - E_o} \quad . \tag{3.65}$$

Expression (3.64) gives the g shift derived classically in Sect.3.3.2. Physically it expresses the fact that the ground state contains a small amount of angular momentum induced by the spin-orbit interaction.

Since the excited states of a defect and often even the ground state are unknown, it is difficult to derive information from the amplitude of the g shift. However, in cases where the electronic wave function is highly localized, the approximation which consists in considering only molecular states (the ground and first excited states) provides an order of magnitude for Δg [3.10] (this is illustrated in Sect.3.7).

3.3.4 Analysis of the g Tensor

As shown in Sect.3.3.2,3 the g tensor reflects the environment of the defect, i.e., it provides information on its symmetry. In the general case, the g tensor is an ellipsoid whose principal axes are noted g_1, g_2 and g_3 or $g_{\parallel}$ ($=g_1$) and $g_{\perp}$ in the case where $g_2 = g_3$. In a cubic crystal, in order to obtain these values and the position of the ellipsoid compared to the crystallographic directions, the sample is rotated around a <110> direction chosen to be perpendicular to the magnetic field $\underline{H}$. In this way, as illustrated in

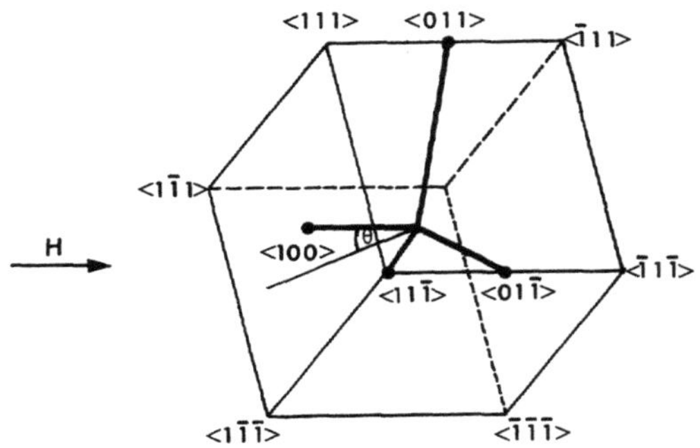

Fig.3.3. The principal axes of the g tensor are obtained by a rotation about a <011> axes perpendicular to the magnetic field H, from <011̄> to <100> through an angle θ

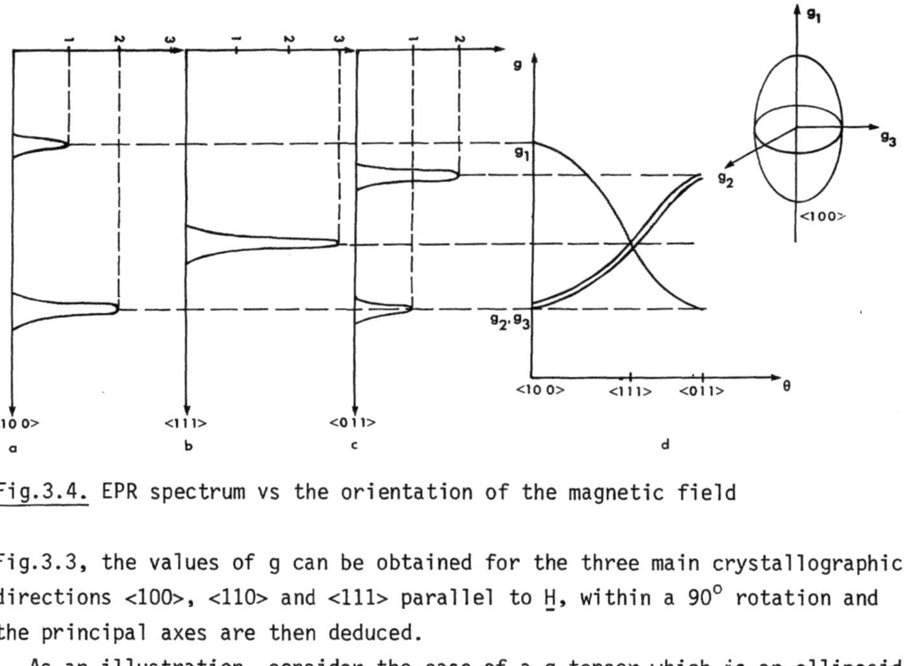

Fig.3.4. EPR spectrum vs the orientation of the magnetic field

Fig.3.3, the values of g can be obtained for the three main crystallographic directions <100>, <110> and <111> parallel to H, within a 90° rotation and the principal axes are then deduced.

As an illustration, consider the case of a g tensor which is an ellipsoid of revolution along a <100> direction. When H is parallel to [100] a resonance occurs for g_1. The two other orientations [010] and [001], not equivalent to [100] since perpendicular to H, give rise to a resonance at another g value: $g_2 = g_3$. The amplitude of the resonance is then twice as large than the amplitude A of the resonance at g_1 since all the orientations are equally populated. Consequently, the spectrum looks like the one depicted in Fig.3.4a. When H becomes parallel to a <111> direction, all the <100> orientations are equivalent and give rise to a single resonance; the amplitude of the resonance is 3A (Fig.3.4b). When H is parallel to the [011] direction, i.e., perpendi-

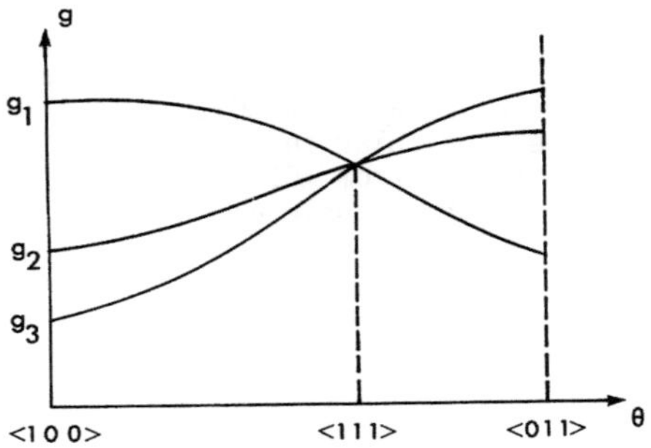

Fig.3.5. g against the orientation of $\underline{H}$ in the general case

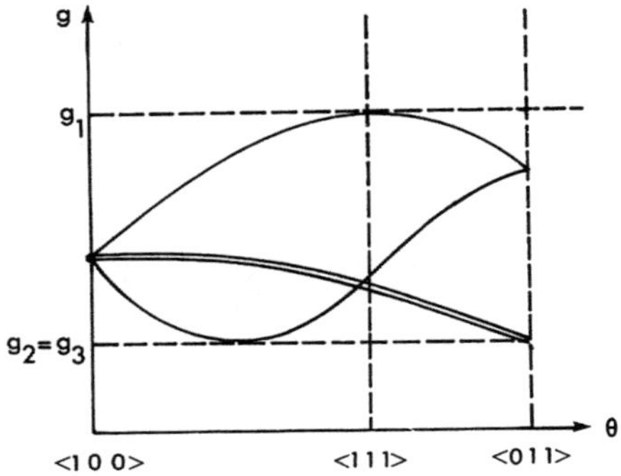

Fig.3.6. g against the orientation of $\underline{H}$ for an ellipsoid of revolution along <111>

cular to the [100] direction, resonance occurs at g_2 and at another value of g for the [101] and [110] equivalent directions, giving rise to the spectrum depicted in Fig.3.4c. The full variation of g with orientation θ is schematized in Fig.3.4d. In the case where the ellipsoid is not an ellipsoid of revolution ($g_2 \neq g_3$), it is obvious that the variation of g with θ will be as shown in Fig.3.5. In the same way, it can be easily seen that a defect for which the g tensor is an ellipsoid of revolution around a <111> direction gives rise to the variation g (θ) depicted in Fig.3.6. Of course, more complicated cases exist when the g tensor is not oriented along one of the main crystallographic directions.

The change in the g value is usually small. It can be measured accurately by comparison with a standard sample for which the g value (g_0) is known. Measurement of the shift $\Delta H = H_0 - H$ of the magnetic field at the resonance between the signal studied (such that $h\nu = g\beta H$) and the standard sample (such that $h\nu = g_0\beta H_0$) gives g from

$$g = \frac{g_0}{1 - \frac{\Delta H}{H_0}} \quad . \tag{3.66}$$

3.4 Hyperfine Interaction

There are magnetic fields other than the applied one with which the spin of the electron interacts. For instance, magnetic dipoles due to the spins $\underline{I}$ of nuclei present in the electron orbit induce magnetic fields. Their inter-

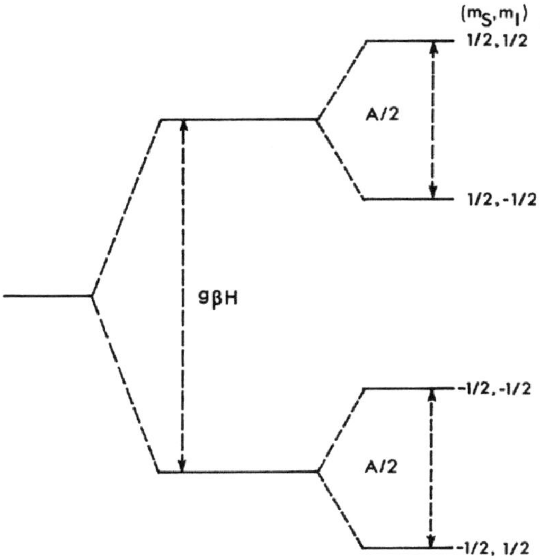

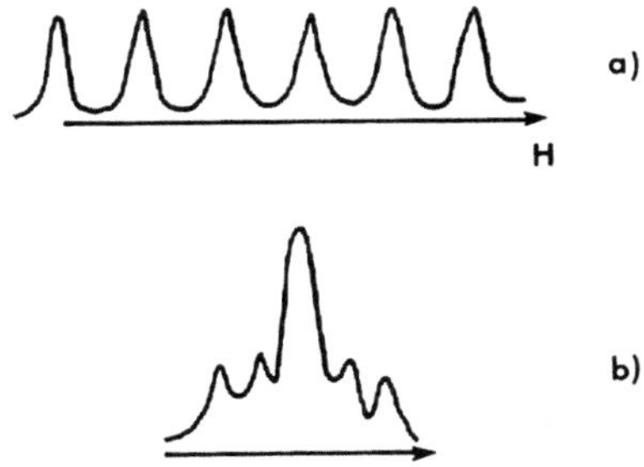

▲ Fig.3.8. Experimental observation of hyperfine structure for Al^{27} in Si

◄ Fig.3.7. An example of hyperfine structure

action with the electronic spin $\underline{S}$, called hyperfine interaction [3.11] is proportional to $\underline{I}$

$$H_{HF} = \underline{S}\tilde{A}\underline{I} \quad . \tag{3.67}$$

The eigenvalues of this Hamiltonian are (when A is a scalar $\ll g\beta H$)

$$\Delta E_{HF} = Am_S m_I \tag{3.68}$$

where m_I is the nuclear azimuthal quantum number which can take (2I + 1) values between -I and I. As we shall see next $\tilde{A}$ is in general a tensor, the hyperfine tensor, which contains the magnitude of the nuclear dipole and all the distance and orientation-dependent parts.

The result of a hyperfine interaction on the energy level scheme is depicted in Fig.3.7 in the case of nuclear and electronic spins of 1/2. It can easily be seen on this figure that, because $\Delta m_I = 0$, the resonance occurs at the energies

$$\Delta E_{Ze} \pm 2\Delta E_{HF} = g\beta|\underline{H}| \pm \frac{A}{2} \quad . \tag{3.69}$$

Two absorption lines must be observed, symmetric to the line $g\beta H$.

The study of the hyperfine interaction is an extremely powerful method to identify the nature of the atom (impurity) involved in a defect and to provide detailed information on the associated wave function. When an electron is bonded to an atom having a nuclear spin I, each level is split into (2I + 1) sublevels which give rise to (2I + 1) identical resonance lines.

Because a nuclear spin is characteristic of a given atom, such a spectrum is the signature of this atom. Consider for instance the spectrum of Fig.3.8a observed in aluminium-doped silicon irradiated at helium temperature with energetic electrons [3.12]. Such a spectrum consisting of $(2I + 1) = 6$ identical lines is associated with an atom having a nuclear spin $I = 5/2$. Since Al^{27} possesses a spin $I = 5/2$ and is 100% abundant, the spectrum is undoubtedly due to a defect containing Al.

The hyperfine interaction also gives the character of the wave function. The reason is that this interaction, being due to the coupling of the magnetic moment of the electron $\underline{\mu}_e$ with a nuclear magnetic moment $\underline{\mu}_n$, is composed of two terms: i) the so-called contact interaction term

$$H_c = \frac{8\pi}{3} g\beta g_n\beta_n |\psi(o)|^2 \underline{I} \cdot \underline{S} \tag{3.70}$$

which is isotropic and ii) the anisotropic dipolar interaction term:

$$H_d = -g\beta g_n\beta_n \left[\frac{\underline{I} \cdot \underline{S}}{r^3} - 3\,\frac{(\underline{I} \cdot \underline{r})(\underline{S} \cdot \underline{r})}{r^5}\right] \quad . \tag{3.71}$$

These terms are derived [3.1] from the classical dipole-dipole interaction between two dipoles $\underline{\mu}_e$ and $\underline{\mu}_n$ separated by a distance $\underline{r}$

$$H_d = \frac{\underline{\mu}_e \cdot \underline{\mu}_n}{r^3} - 3\,\frac{(\underline{\mu}_e \cdot \underline{r})(\underline{\mu}_n \cdot \underline{r})}{r^5} \quad , \tag{3.72}$$

to which is added a "Fermi contact" term

$$H_c = -\frac{8\pi}{3} (\underline{\mu}_e \cdot \underline{\mu}_n)\delta(\underline{r}) \tag{3.73}$$

to account for the fact that for close distance between dipoles the dipole approximation leading to (3.72) is no longer valid.

The contact interaction is nonzero when the electron has a finite probability to be on the nucleus: the electron wave function possesses an as orbital character. For an s function, $\tilde{A}$ is isotropic and equal to:

$$A = \frac{8\pi}{3} g g_n \beta\beta_n |\psi_s(o)|^2 \tag{3.74}$$

and the measurement of A is means, by comparing with tabulated Hartree-Fock wave functions, to know if the function is a 1s, 2s, etc., function. The dipolar interaction is zero for a spherical electronic cloud and, hence, reflects the p (or eventually d) character of the wave function. For a p function pointing along the z axis, A is axially symmetric around z

$$A_{zz} = -2A_{xx} = -2A_{yy} \tag{3.75}$$

and equal to

$$A_{zz} = \frac{4}{5} gg_n\beta\beta_n\left\langle\frac{1}{r^3}\right\rangle \quad . \tag{3.76}$$

Again, from tabulated Hartree-Fock wave functions, values of $<r^{-3}>$ can be estimated and the orbital character deduced. More generally, for a wave function of axial symmetry, it can be shown [3.13] that

$$A_{||} = 0 + 2P \tag{3.77}$$

and

$$A_{\perp} = 0 - P \quad , \tag{3.78}$$

where 0 represents the isotropic s contribution (3.74), and

$$P = g\beta g_n\beta_n \left\langle\frac{z^2 - \frac{1}{2}(x^2 + y^2)}{r^5}\right\rangle \tag{3.79}$$

is the anisotropic p contribution. From the measurement of $A_{||}$ and $A_{\perp}$, the coefficients c_s and c_p of the molecular orbital wave function ψ built on ψ_s and ψ_p orbitals

$$\psi = c_s\psi_s + c_p\psi_p \tag{3.80}$$

can be obtained. The wave function overlapping several sites i is constructed from a sum of atomic orbitals

$$\psi_i = a_i\varphi_{is} + b_i\varphi_{ip} \tag{3.81}$$

on each of these sites. Then

$$\psi = \sum_i \eta_i\psi_i \quad , \tag{3.82}$$

where the coefficient η_i is the fraction of the total wave function localized on the orbital ψ_i. The hyperfine interaction at site i, resulting mainly from ψ_i, gives thus an estimate of η_i^2, i.e., of $|a_i|^2$ from the isotropic part of the interaction and of $|b_i|^2$ from the anisotropic part.

Hyperfine interaction can be observed with several types of neighbors when they possess a nuclear spin. For instance, a silicon crystal contains in addition to Si^{28} atoms with nuclear spin $I = 0$, Si^{29} atoms with nuclear spin $I = 1/2$ (5% abundant) which can give rise to hyperfine interaction. Consider again, as illustration, the spectrum associated with Al described in (Fig.

3.8). Each individual resonance line is accompanied by two pairs of satellites (Fig.3.8b). They are due to the hyperfine interaction with two kinds of Si^{29} neighbors. The A tensor is isotropic [3.10,14] indicating that the Al atom is in an s state. Since the electronic structure of an Al^0 atom is $1s^2$ $2s^2$ $2p^6$ $3s^2$ 3p, this strongly suggests that the Al atom is in a 3s state, i.e., doubly ionized (Al^{++}). This is confirmed by the magnitude of the hyperfine constant A, in reasonable agreement with the value of $|\psi(o)|^2$ for a 3s state. This Al atom is certainly in an interstitial position since a substitutional atom would give rise to a shallow level.

What is the geometrical site of this interstitial? WATKINS' measurements, which give isotropic g and A tensors, suggest from symmetry arguments that interstitial Al^{++} is in a site of tetrahedral symmetry. Indeed, for the Al^{++} to exist in a site of tetrahedral symmetry the g and A tensors must be invariant under the transformations of the point group T_d, i.e., be isotropic. For the Al^{++} to exist in the hexagonal interstitial site, the g and A tensors must be invariant under the transformations of the C_{3v} group, i.e., be axially symmetric about a <111> axis. Finally, interstitial sites of lower symmetry would correspond to tensors of lower symmetry. BROWER [3.14] observed additional lines on sides of the hyperfine lines due to hyperfine interaction with more distant Si^{29} atoms. He made a careful analysis of the angular dependence and intensity of these so called superhyperfine resonances which allowed him to determine the symmetry class, the number of equivalent sites and the numerical values for the respective superhyperfine tensors. The conclusion is that the Al^{++} interstitial is undoubtedly in an environment of T_d symmetry. Hyperfine interactions of course exist with more distant neighbors. However, for these neighbors the amplitude of the wave function is small and the interactions are too weak to be resolved; they contribute only to the width of the lines. The way to resolve such weak interactions is possibly through the ENDOR technique, which is the subject of the next section.

3.5 Nuclear Zeeman Interaction—Double Resonance

Electron nuclear double resonance (ENDOR), introduced by FEHER [3.15], consists in the observation of a nuclear magnetic resonance. The interaction of a nuclear magnetic moment $\underline{I}$ with the magnetic field gives rise to the nuclear Zeeman interaction

$$H_{Zn} = -g_n \beta_n \underline{H} \cdot \underline{I} \qquad (3.83)$$

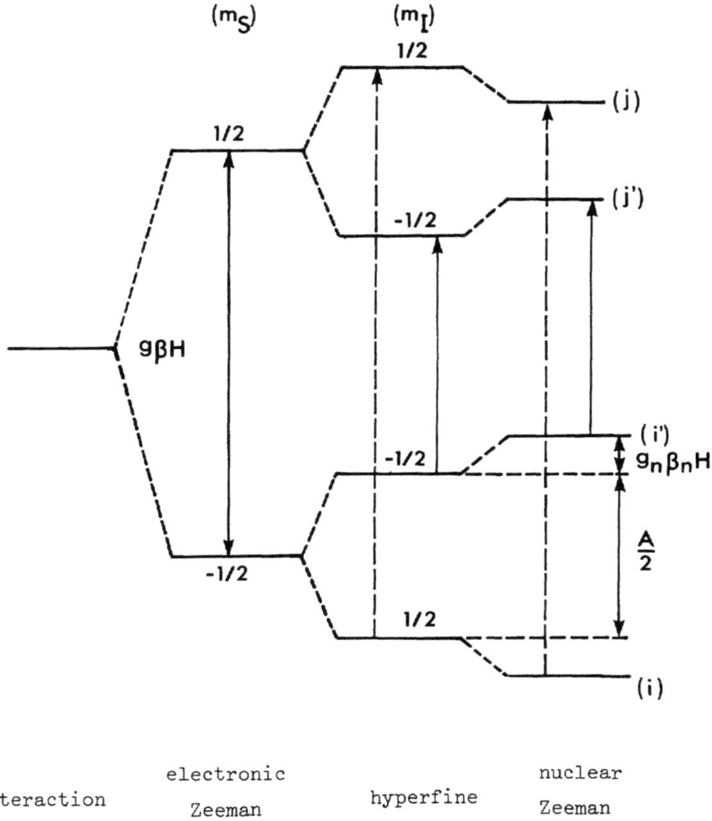

Fig.3.9. Splitting of the energy levels due to the different interactions

where $\underline{H}$ is the total magnetic field experienced by the nuclear moment. In a simple isotropic case, the eigenvalues of this Hamiltonian

$$\Delta E_{Zn} = -g_n\beta_n|\underline{H}|m_I \tag{3.84}$$

induce shifts $\pm\Delta E_{Zn}$ of the energy levels but do not change the energies of the EPR resonances, given by (3.69). This is illustrated for the case $I = 1/2$ in Fig.3.9.

The ENDOR technique consists in applying first a microwave power large enough to equalize the populations of the two levels at the frequency of the resonance (levels i and j for instance, for which $h\nu = g\beta H + A/2$, Fig.3.9). The result is that the intensity of the EPR line is considerably reduced if not canceled. Then the application of a radiofrequency power at the frequency

$$h\nu_n = \frac{A}{2} - g_n\beta_n|\underline{H}| \tag{3.85}$$

induces the transition j → j'. Because this transition modifies the population of the j state the EPR resonance i → j is again observed. Similarly the transition i → i' at energy

$$h\nu_n' = A/2 + g_n\beta_n|\underline{H}| \tag{3.86}$$

can be observed via the EPR signal. When the electron resonance can be saturated, the ENDOR method allows the direct measurement of nuclear magnetic moments [3.15,16]

$$h(\nu_n' - \nu_n) = 2g_n\beta_n|\underline{H}| \tag{3.87}$$

and of the hyperfine constant

$$h(\nu_n + \nu_n') = A \quad . \tag{3.88}$$

3.6 Spin-Spin Interaction. Fine Structure

Up to now we have considered the common case of a defect containing only a single unpaired electron of spin S = 1/2. However there are cases where a defect contains two or more electrons coupled together. This is the case, for instance, of multivacancy defects (Fig.3.10) or of the neutral state of some defects [3.17]. Then $\underline{S}$ is larger than 1/2 and it is necessary to include higher order terms in S in the spin hamiltonian, the important one being

$$H_{FS} = \underline{S}\cdot\underline{\tilde{D}}\cdot\underline{S} \quad . \tag{3.89}$$

For defects such as multivacancies, which contain two unpaired electrons situated on distant bonds, the tensor $\tilde{D}$ originates from the interaction between the magnetic dipoles $\underline{S}_1$ and $\underline{S}_2$ of these two electrons. This interaction can therefore be written, in analogy with (3.72),

$$H_{dd} = (g\beta)^2\left[\frac{\underline{S}_1\cdot\underline{S}_2}{r^3} - \frac{3(\underline{S}_1\cdot\underline{r})(\underline{S}_2\cdot\underline{r})}{r^5}\right] \tag{3.90}$$

which leads [3.18] to

$$D_{ij} = \frac{1}{2}g^2\beta^2\left\langle\frac{r^2\delta_{ij} - 3x_ix_j}{r^5}\right\rangle \quad , \tag{3.91}$$

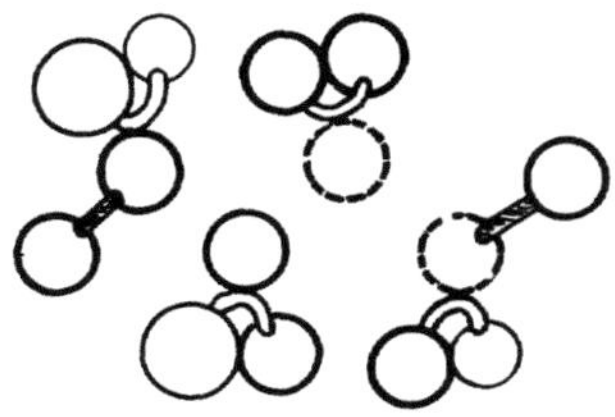

Fig.3.10. A model for the four-vacancy center in silicon (after [3.20]). The symmetry of the EPR spectrum [3.19] suggests an array of vacancies in a [110] plane, the unpaired electron (in black) ending up at the two extremes of the center

the bracket denoting an average taken over the wave functions of the two electrons. In the case of multivacancy centers, the measurement of the magnitude of D is a way to evaluate r, the distance between the unpaired electrons, i.e., the size of the defect and consequently the number of vacancies involved in this defect [3.19].

The fine structure term introduces additional structures in the spectrum. This can easily be seen for instance in the simple Hamiltonian

$$H_* = g\beta \underline{S} \cdot \underline{H} + \underline{S}\tilde{D}\underline{S} \tag{3.92}$$

written as

$$H_* = g\beta \underline{S} \cdot \underline{H} + D_{\parallel} S_z^2 + D_{\perp}[S(S + 1) - S_z^2] \tag{3.93}$$

in case of axial symmetry along the z axis. With the magnetic field along the z axis the eigenvalues of H_* are, for a S = 1 center

$$\Delta E(M = \pm 1) = \pm\, g\beta H + D_{\parallel} + D_{\perp} \tag{3.94}$$

and

$$\Delta E(M = 0) = 2D_{\perp} \quad . \tag{3.95}$$

The introduction of the fine structure term produces splitting of the EPR spectrum. This fine structure can be easily distinguished from a hyperfine splitting because it goes through zero when the sample is rotated in the magnetic field while the hyperfine splitting does not.

3.7 EPR of Impurities and Vacancy–Impurity Pairs in Silicon

We give here a detailed treatment of the g shift and the hyperfine tensor for some impurity related defects which present the common characteristic that the unpaired electron is located on one or two dangling bonds. We essentially follow the arguments given by WATKINS and CORBETT [3.5,21], except for some differences in presentation. We mainly examine the case of one electron in one dangling bond from which all the necessary results can be easily derived.

3.7.1 Evaluation of the g Shift

As discussed in Sect.3.3.3, the shift $\Delta\tilde{g}$ in the $\tilde{g}$ tensor from its free electron value is, to first order, due to the spin-orbit interaction. However the latter is now a sum of different atomic contributions, so that

$$(\Delta\tilde{g}_0)_{ij} = -2\mathrm{Re}\sum_{n\neq 0} \frac{\langle 0|(\underline{L})_i|n\rangle\langle n|\sum_N \lambda(r_N)(\underline{L}_N)_j|0\rangle}{E_n - E_o} \tag{3.96}$$

[$(\underline{L})_i$ and $(\underline{L}_N)_j$ are, respectively, the ith and j^{th} components of $\underline{L}$ and $\underline{L}_N$], where the sum over N is made over all nuclei of the system, $\underline{L}_N$ being the orbital momentum operator on nucleus N, r_N the distance between the electron and this nucleus. It is to be noted that the same choice of gauge must be consistently used for all the terms in the Hamiltonian. Here we choose, as in [3.21], a gauge centered on atom 0. Then $\underline{L}$ is equal to $\underline{L}_0$, the orbital momentum operator on atom 0. We consider first an unpaired electron in an isolated dangling bond. The local situation is depicted in Fig.3.11 in terms of hybridized orbitals. The calculation of $\Delta\tilde{g}_0$ being quite complex, we use the molecular model described in [Ref.1.1, Chap.3], which can be considered as a good starting point for a qualitative description. Then $|0\rangle$ is a pure dangling bond state localized on atom 0. On the backbonds we can form bonding and antibonding combinations B_k and A_k, of energies $-E_B$ and E_A (the origin being the dangling bond energy, E_B and E_A being positive). The state $|0\rangle$ is an atomic state and the major contribution to $\Delta\tilde{g}$ will come from the spin-orbit coupling term corresponding to atom 0. In such a case we can write (3.96) in the simplified form

$$(\Delta\tilde{g}_0)_{ij} = 2\lambda\left(\frac{\sum_k\langle 0|L_i|B_k\rangle\langle B_k|L_j|0\rangle}{E_B} - \frac{\sum_k\langle 0|L_i|A_k\rangle\langle A_k|L_j|0\rangle}{E_A}\right) \tag{3.97}$$

where λ is the corresponding atomic spin-orbit coupling constant and $\underline{L}$ is $\underline{L}_0$. We have thus to determine matrix elements such as $[2(1 \pm S)]^{-\frac{1}{2}}$ $\langle 0|L_i|k \pm k'\rangle$ where S is the overlap $\langle k|k'\rangle$. The calculation of $\langle 0|L_i|k\rangle$ is easy since it reduces to an intraatomic contribution between p orbitals. This is not true of $\langle 0|L_i|k'\rangle$, but we can formally simplify the treatment using the following argument: it is always possible to expand $|k'\rangle$ in terms of the complete set of eigenstates of atom $|0\rangle$; only the p states with axial symmetry along the bond will contribute to $(\Delta\tilde{g}_0)_{ij}$. Thus $\langle 0|L_i|k'\rangle$ is proportional to $\langle 0|L_i|k\rangle$ and we call γ the ratio of these two terms. Noting that

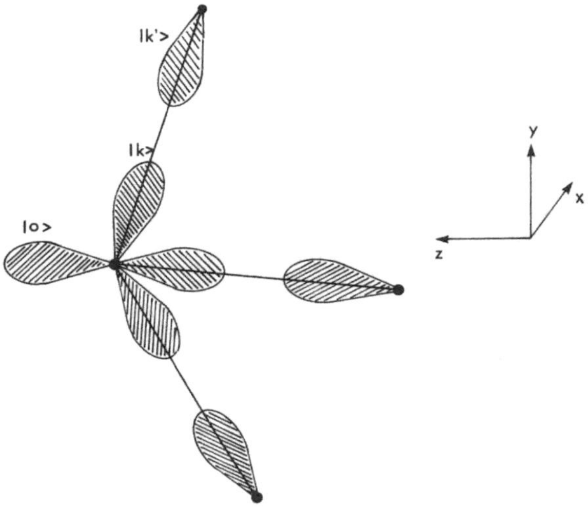

Fig.3.11. An isolated dangling bond |0> and its three backbonds

in the molecular model E_B and E_A are given respectively by $E_0(1 + S)^{-1}$ and $E_0(1 - S)^{-1}$, where E_0 is equal to $\langle k|H|k'\rangle$, we can now write

$$(\Delta\tilde{g}_0)_{ij} = \frac{4\lambda\gamma}{E_0}\sum_k \langle 0|L_i|k\rangle\langle k|L_j|0\rangle \quad . \tag{3.98}$$

The sign and order of magnitude of γ can be roughly estimated by noting that in the expansion of $|k'\rangle$ the most important term is presumably $\langle k|k'\rangle|k\rangle$, in view of the strong overlap S between these two orbitals. Thus γ is probably of the order of S, i.e., about 0.7. It is clear that $(\Delta\tilde{g}_0)_{ij}$ will have axial symmetry along the dangling bond. Also, the expression of $(\Delta\tilde{g}_0)_{ij}$ will involve only the p part of the states $|0\rangle$ and $|k\rangle$. It can be rewritten, $|0\rangle$ and $|k\rangle$being sp^3 hybrids,

$$(\Delta\tilde{g}_0)_{ij} = \frac{9\lambda\gamma}{4E_0}\sum_k \langle p_z|L_i|p_k\rangle\langle p_k|L_j|0\rangle \quad , \tag{3.99}$$

where $|p_z\rangle$ is the p state along the Z axis of Fig.3.11 and p_k the p state along bond k. Using the fact that $\sum_k |p_k\rangle\langle p_k|$ can be expressed as

$$\sum_k |p_k\rangle\langle p_k| = \frac{1}{9}|p_z\rangle\langle p_z| + \frac{4}{3}(|p_x\rangle\langle p_x| + |p_y\rangle\langle p_y|) \quad . \tag{3.100}$$

The matrix elements of $\underline{L}$ in a p atomic basis being

$$\langle i|L_j|k\rangle = i\varepsilon_{ijk} \quad , \tag{3.101}$$

for $i \neq j \neq k$, with ε_{ijk} = +1 for an even permutation of xyz, = -1 for an odd one, we obtain

$$\Delta\tilde{g}_0 = \frac{3\lambda\gamma}{E_0} \begin{bmatrix} 0 & 0 & 0 \\ 0 & 1 & 0 \\ 0 & 0 & 1 \end{bmatrix} \quad . \tag{3.102}$$

A first trivial conclusion is that $\Delta\tilde{g}_0$ will have two components: $\Delta g_{0\parallel}$ equal to zero and $\Delta g_{0\perp}$ which is positive. The order of magnitude of $\Delta g_{0\perp}$ can be estimated using $\lambda \sim 0.02$ eV [3.21] and taking $\gamma \sim 0.7$, as discussed above. The value of E_0 can be deduced from the average separation between bonding and antibonding states [equal to $2E_0(1 - S^2)^{-1}$ in the model] which is of the order 7.5 eV (i.e., 2β of [Ref.1.1, Table 3.1]). With $S \sim 0.7$ this leads to $E_0 \sim 3$ eV. Assuming a reduction factor of 60%, due to delocalization of the dangling bonds, we get

$$\Delta g_{0\perp} \sim 0.0085 \tag{3.103}$$

close to the value derived by WATKINS and CORBETT [3.21].

We can now extend this result to the case of two coupled dangling bonds (Fig.3.12). We assume the unpaired electron to be in a state $(|0\rangle \pm |0'\rangle)/\sqrt{2}$. To derive the result in that case we notice the general fact that if the wave function can be written as a sum of nonoverlapping terms centered on different sites, then one can choose a different vector potential for each site [3.1]. This allows us to express $\Delta\tilde{g}$ as

$$\Delta\tilde{g} = \frac{1}{2}\,(\Delta\tilde{g}_0 + \Delta\tilde{g}_{0'}) \quad , \tag{3.104}$$

the sum of the tensors corresponding to the isolated dangling bonds $|0\rangle$ and $|0'\rangle$, respectively. This will necessarily be diagonal in the new system of axes given in Fig.3.12. The new diagonal components $\Delta\tilde{g}_{i'i'}$ are related to the diagonal components $(\Delta\tilde{g}_0)_{ii}$ in the local axes of Fig.3.11 through

$$\Delta\tilde{g}_{i'i'} = \Delta g_{0\parallel}|\langle i'|z\rangle|^2 + \Delta g_{0\perp}(1 - |\langle i'|z\rangle|^2) \quad , \tag{3.105}$$

where $|i'\rangle$ and $|z\rangle$ are unit vectors along the corresponding direction. As $|\langle z'|z\rangle|^2$, $|\langle y'|z\rangle|^2$ and $|\langle x'|z\rangle|^2$ are, respectively, equal to 1/3, 2/3 and 0 we get, using the fact that $\Delta g_{0\parallel}$ is zero,

$$\begin{aligned} \Delta g_{zz} &= \frac{2}{3}\,(\Delta g_0)_\perp \\ \Delta g_{yy} &= \frac{1}{3}\,(\Delta g_0)_\perp \\ \Delta g_{xx} &= (\Delta g_0)_\perp \end{aligned} \tag{3.106}$$

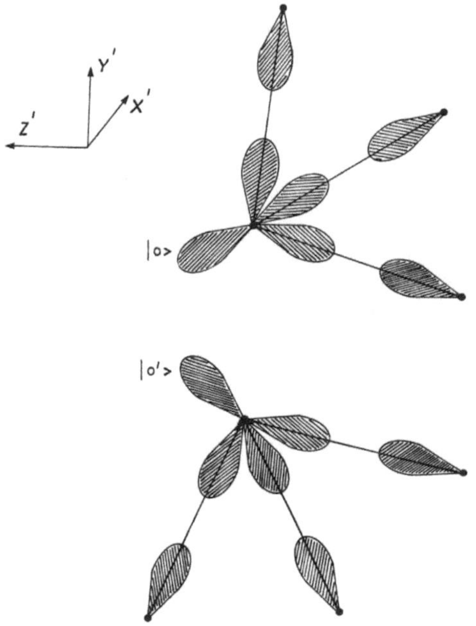

Fig.3.12. The case of two coupled dangling bonds

which is the result found by WATKINS and CORBETT in their calculation on the Si-A center [3.21b].

3.7.2 The Hyperfine Tensor

The dangling-bond atom has a 4.7% probability to be a Si^{29} nucleus. This one has a nuclear spin I equal to 1/2 and, as discussed in Sect.3.4, there will be a hyperfine interaction term given by (3.67). Considering the isolated dangling bond case, the discussion in Sect.3.4 enables us to say that the hyperfine tensor A_o is axially symmetric and has two components given by

$$(A_o)_{\parallel} = a + 2b$$

$$(A_o)_{\perp} = a - b \quad , \tag{3.107}$$

with a given by (3.74)

$$a = \frac{16\pi}{3} g_n \beta\beta_n \, |\langle\psi_{3S}|0\rangle|^2 |\psi_{3S}(0)|^2 \quad , \tag{3.108}$$

and b by $A_{zz/2}$ in (3.76)

$$b = \frac{4}{5} g_n \beta\beta_n |\langle\psi_{3p}|0\rangle|^2 \left\langle\frac{1}{r^3}\right\rangle_{3p} \quad . \tag{3.109}$$

Numerical calculation using free-atom wave functions leads to [3.21a]

$$a \sim 1380|\langle\psi_{3s}|0\rangle|^2 \times 10^{-4}\ \mathrm{cm}^{-1}$$

$$b \sim 33.7|\langle\psi_{3p}|0\rangle|^2 \times 10^{-4}\ \mathrm{cm}^{-1} \quad . \qquad (3.110)$$

Both values must eventually be multiplied by a reduction factor to account for the delocalization of the dangling-bond wave function (about 0.5). For an sp^3 dangling bond $|\langle\psi_{3s}|0\rangle|^2$ is 1/4 and $|\langle\psi_{3p}|0\rangle|^2$, 3/4. Taking account of these values gives the following order of magnitude

$$a \sim 170 \times 10^{-4}\ \mathrm{cm}^{-1}$$

$$b \sim 12.6 \times 10^{-4}\ \mathrm{cm}^{-1} \quad . \qquad (3.111)$$

3.7.3 Experimental Results

Part of the experimental information is given in Table 3.1 reproduced from [3.21a].

Table 3.1. Values of g along the principal axes, components of the hyperfine tensor, number of hyperfine lines and approximate hyperfine axes for three centers observed by EPR in electron irradiated silicon [3.21]

Center	g	Principal axes	Si^{29} hf $[10^{-4}\ \mathrm{cm}^{-1}]$	Number of hyperfine Si^{29} and approximate hf axis
	2.0026	$1\ 1\ \bar{1}$	$A_{\parallel} = 130 \pm 2$	(1) $1\ 1\ \bar{1}$
SiB	2.0085	$\sim 0\ 1\ 1$	$A_{\perp} = 71 \pm 2$	
	2.0107	$2\ \bar{1}\ 1$		
	2.0005	$1\ 1\ \bar{1}$	$A_{\parallel} = 150 \pm 0.5$	(1) $1\ 1\ \bar{1}$
SiE	2.0112	$\sim 0\ 1\ 1$	$A_{\perp} = 98.5 \pm 0.5$	
	2.0096	$2\ \bar{1}\ 1$		
	2.0092	$0\ 1\ 1$	$A_{\parallel} = 153 \pm 0.5$	(1) $1\ 1\ \bar{1}$
SiA	2.0026	$0\ \bar{1}\ 1$	$A_{\perp} = 128.8 \pm 0.5$	(1) $1\ 1\ \bar{1}$
	2.0033	$1\ 0\ 0$		

We do not reproduce the whole discussion of [3.21a]. We first notice that SiB and SiE have $\tilde{g}$ and $\tilde{A}$ tensors compatible with one dangling-bond state. The $\Delta g_{\perp}$ is in good agreement with (3.103); in fact it is slightly smaller indicating a delocalization of the dangling-bond state. The magnitudes of $A_{\parallel}$ and $A_{\perp}$ are of the order of magnitude of (3.111) but indicate a slight reduction in a and increase in b, corresponding to an enhanced p character of the dangling-bond state. The Si-A center is due to substitutional oxygen which, by rebonding, leads to an unpaired electron on two dangling bonds. The g tensor is in rough agreement with (3.106) and the hyperfine tensor is again compatible with what was obtained for dangling bonds.

3.8 The Vacancy in Silicon

3.8.1 EPR Spectrum for V^+

Irradiation with energetic electrons (Chap.8) at low temperature (near liquid helium temperature) of p-type silicon doped with aluminum reveals a spectrum [3.22] composed of three equally intense, sharp (1.5 G at 4 K) lines. Each line contains four pairs of weak satellites symetrically disposed on either side (Fig.3.13).

Orientation dependence studies show that the three central lines are described by a g tensor axially symmetric around a <100> direction (Fig. 3.4) for which

$$g_{\parallel} = 2.0087 \qquad (3.112)$$

and

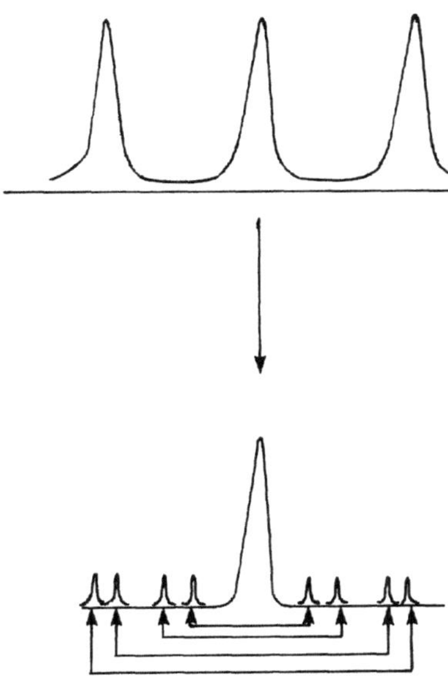

Fig.3.13. EPR spectrum for V^+

$$g_{\perp} = 1.9989 \quad . \tag{3.113}$$

Because the three lines are equally intense, this suggests that they arise from a defect which has three possible <100> orientations, nonequivalent in the magnetic field but equally populated.

In order to verify this suggestion, WATKINS applied a uniaxial stress along a <100> direction. To be effective, this stress has to orient the defect preferentially and therefore to populate or depopulate certain orientations. WATKINS indeed observed a decrease in the amplitude of one line as compared to the others.

The satellites originate from hyperfine interactions with Si^{29} nuclei since their relative amplitude reflects the isotopic abundance of Si^{29}. Four pairs of satellites indicate that there are four types of Si sites in the vicinity of the defect. The hyperfine tensor A is found to be axially symmetric

$$A_{\parallel} = 44 \times 10^{-4}\ \mathrm{cm}^{-1} \tag{3.114}$$

and

$$A_{\perp} = 30 \times 10^{-4}\ \mathrm{cm}^{-1} \tag{3.115}$$

for each of these sites. The axis of this tensor is tilted $\sim 7^{\circ}$ from a <111> direction (toward the <100> axis associated with g), different for each site.

3.8.2 Microscopic Model for V^+

These observations strongly suggest that the defect is the vacancy V^+. Indeed, when an atom is removed from its lattice site to form a vacancy, four bonds are broken. The electrons, the number of which is odd since the defect is paramagnetic, are distributed over these four dangling bonds. The resulting unpaired spin interacts with the four neighboring Si atoms which are in four <111> directions, nonequivalent in the magnetic field. The tilting of the hyperfine axis from a <111> direction indicates a slight distortion of the four neighbors. (This is consistent with the weak anisotropy of the g tensor.) A distortion around the vacancy is not surprising in view of the LCAO model (developed in [Ref.1.1, Sect.4.2.2]): the electronic levels consist of an A_1 level and of a triply degenerated T_2 level (Fig.2.8a). These levels, once filled with three electrons (the most reasonable odd number of electrons in a p-type material), result in a singly occupied T_2 level. Due to the Jahn-Teller effect (Sect.2.4) the degeneracy of the T_2 level is lifted (Fig.2.8b) by a distortion. The g tensor being axially symmetric around a <100> axis, this distortion is D_{2d}, i.e., tetragonal.

We now proceed to the evaluation of the hyperfine tensor which is the easiest to estimate. For a tetragonal distribution along a <100> axis, the unpaired electron wave function is an x-like combination of the four dangling bonds [Ref.1.1, Chap.3]. It has thus an equal weight 1/4 on each dangling bond. The components of the hyperfine tensor on each of the four nearest neighbors of the vacancy can thus be deduced from a and b given in (3.111), multiplied by a corrective factor equal to 1/4. This leads to $A_{\parallel}$ and $A_{\perp}$ given by

$$A_{\parallel} = 49 \times 10^{-4}\ \mathrm{cm}^{-1}$$
$$A_{\perp} = 39 \times 10^{-4}\ \mathrm{cm}^{-1} \qquad (3.116)$$

which are fairly close to the experimental values. In fact, to obtain (3.116), we have assumed sp^3 orbitals and a delocalization of 50%. We could, as was done by WATKINS [3.22], have treated $|\langle\psi_{3s}|0\rangle|^2$, $|\langle\psi_{3p}|0\rangle|^2$ and the localization as adjustable parameters to reproduce the experimental values. Such an analysis tends to indicate that each dangling bond is 15% s, 85% p, with a localization of about 66%. Thus the simple dangling-bond model quite successfully accounts for the hyperfine tensor.

The situation for the g tensor is not so trivial. In principle we can apply the calculation of Sect.3.7.1. However, we now have to evaluate $\Delta\tilde{g}$ for an unpaired electron delocalized on four dangling bonds, which is the average of the individual tensors. This simply takes the average of the three components, so that $\Delta\tilde{g}$ is now isotropic

$$\Delta\tilde{g} = \frac{2}{3}(\Delta g_0)_{\perp} \begin{bmatrix} 1 & 0 & 1 \\ 0 & 1 & 0 \\ 0 & 0 & 1 \end{bmatrix} \quad , \qquad (3.117)$$

giving rise to a positive shift. This is in complete disagreement with the experimental results. This point was not discussed by WATKINS [3.22] who gave only the hyperfine tensor analysis. Here we want to see if the g tensor discrepancy can be explained by slightly refining the model. For this we start from the fact that, for a <100> tetragonal distortion, the unpaired electron is in the state $|x\rangle$, the x-like combination of the dangling bond states, the other components $|y\rangle$ and $|z\rangle$ of the T_2 state being degenerate at a higher energy (Fig.2.8b). If that splitting, which we call Δ, is small, transitions from the $|x\rangle$ to the $|y\rangle$ and $|z\rangle$ states could appreciably contribute to the g tensor, even if the overlap between the dangling bonds is not large. This contribution can simply be added to the previous one and is of the form ($\underline{L}$ being centered at the vacancy site)

$$\Delta\tilde{g}' = -\frac{2}{\Delta}\,\mathrm{Re}\left\{\langle x|(L)_i|y\rangle\langle y|\sum_N \lambda(r_N)(\underline{L}_N)_j|x\rangle + \langle x|(L)_i|z\rangle\langle z|\sum_N \lambda(r_N)(\underline{L}_N)_j|x\rangle\right\} \quad . \tag{3.118}$$

We calculated these matrix elements using Slater functions orthogonalized to the core 2p functions, and found that the contribution (3.118) is too small by about an order of magnitude and has the wrong sign. It is thus necessary to consider a second type of correction. In addition to their diagonal contribution, the dangling bonds of two second nearest neighbors, $|0\rangle$ and $|0'\rangle$ (Fig.3.12), will also give rise to terms of the form

$$\Delta\tilde{g}_{oo'} = 2\,\mathrm{Re}\left\{\sum_n \frac{\langle 0|\sum_N \lambda_N \underline{L}_N|n\rangle\langle n|\underline{L}|0'\rangle}{E_o - E_n}\right\} \quad , \tag{3.119}$$

where $|n\rangle$ are again bonding and antibonding like states. Such contributions can be important, since there are close circuits connecting $|0\rangle$ and $|0'\rangle$ through a common second nearest neighbor. They are not easy to calculate. We parametrize (3.119) along the natural axes of Fig.3.12, with values $g_{x'}$, $g_{y'}$ and $g_{z'}$. For a system of two dangling bonds (SiA center) this gives

$$\Delta g_1 = g_{x'} , \quad \Delta g_2 = -g_{z'} , \quad \Delta g_3 = g_{y'} \quad . \tag{3.120}$$

For V^+ the x like state is a combination of four dangling bonds, and we get

$$\Delta g_{||} = 2(g_{x'} - g_{y'} - g_{z'})\Delta g_{\perp} = 2g_{y'} \quad . \tag{3.121}$$

We now choose a set of values such that

$$g_{x'} = 0 \quad , \quad g_{y'} = -0.0045 \quad , \quad g_{z'} = 0.0025 \tag{3.122}$$

and calculate the resulting corrections given by (3.120,121). In this way we get, for V^+, $\Delta g_{||}$ and $\Delta g_{\perp}$ equal to 2.0118 and 1.999 respectively, and for the SiA center, Δg_1, Δg_2, Δg_3 equal to 2.0108, 2.0025 and 2.0035. This can be extended to other defects involving dangling bonds, giving a coherent picture of the $\underset{\sim}{g}$ tensor for six such centers [3.23].

3.8.3 Charge States of the Vacancy

The spectrum associated with V^+ appears as a weak structure superposed on the spectrum associated with Al^{++} (Sect.3.4). Its magnitude can be increased by illumination with a light of energy $h\nu < 0.35$ eV. Originally WATKINS [3.22] argued that this light produces only free holes which induce the transition $V^o + h \rightarrow V^+$,V^o being the species present at equilibrium. But later (Sect. 3.8.5) WATKINS recognized, from a comparison between EPR and transient capacitance data, that the transition involved was in fact $V^{++} \rightarrow V^+ + h$, i.e., that the species present at equilibrium in p type material is V^{++}. In n type material when a band-gap light, which produces free electrons, is shined on the sample the neutral vacancy traps an electron and becomes V^-. In this new state the vacancy is again paramagnetic: a new spectrum is observed whose characteristies are [3.12]

$$g_1 = 2.0151 \quad , \quad g_2 = 2.0028 \quad , \quad g_3 = 2.0038 \tag{3.123}$$

and

$$A_{||} = 133\times 10^{-4}\ \mathrm{cm}^{-1} \quad , \quad A_{\perp} = 112\times 10^{-4}\ \mathrm{cm}^{-1} \quad . \tag{3.124}$$

The hyperfine structure differs from the hyperfine structure observed for the V^+ spectrum. It consists only of two pairs of satellites, indicating that the unpaired electron is now localized on only two of the four neighbors. Again this is consistent with the notion of bond formation in the LCAO model for the V^- state. Two bonds are formed between the four neighbors, leaving the unpaired electron resident in an antibonding orbital. Since these orbitals extend between only two neighbors, hyperfine interaction is observed only for these two atoms. Population of the degenerate level with an electron (Fig.2.8d) results in an additional distortion which, in this case, is trigonal.

3.8.4 Jahn-Teller Distortion

As discussed in [Ref.1.1, Sect.4.2.2], the validitiy of the one-electron picture to describe the electronic structure can be questioned. The experimental results obtained by Watkins strongly indicate that it is valid. One justification of its validity, proposed by WATKINS [3.24], is that the Jahn-Teller energy is large enough to override the electron-electron interaction. We have discussed this point in some detail in Sect.2.4.2, showing that even if both effects are of comparable magnitude, the ground state energy of V^o is practically not affected by the electron-electron interactions [3.25].

More information concerning the Jahn-Teller energy can be obtained from the analysis of stress induced alignment [3.26]. There are three equivalent orientations for a distorted configuration axially symmetric along a <100> direction. They are equally populated in the absence of external perturbation. In the temperature range 14-20 K, Watkins observed an increasing broadening of the lines because the vacancies move from one configuration, oriented in a given <100> direction, to another. The vacancies are brought out and into resonance as they randomly reorient and the g value changes accordingly. From the variation of the rate of reorientation versus temperature, Watkins measured the activation energy associated with this reorientation, i.e., the energy barrier which separates two different configurations. The value he found, of the order of 10-20 meV, strongly suggests that the reorientation is indeed of electronic origin, i.e., involves bond switching rather than atomic jumps.

The method of analysis for defect alignment under externally applied uniaxial stress has been described by ELKINS and WATKINS [3.27]. As shown in Sect.2.2, the variation of energy versus the amplitude Q of a given distortion can be written as

$$E_d = - VQ + \frac{1}{2} kQ^2 \quad , \tag{3.125}$$

where V is the Jahn-Teller coupling coefficient and k the force constant associated with the mode of distortion considered. The distortion for which the system is stable is obtained by minimizing $E_d(Q)$. It occurs for the Jahn-Teller stabilization energy

$$E_{JT} = - \frac{V^2}{2k} \quad . \tag{3.126}$$

To evaluate this energy, V and k must be measured. Estimation of k can be made from elastic properties or from phonon dispersion curves if it is assumed that the local modes of vibration around a vacancy are the same as in the perfect crystal. Estimation of V is performed through the study of the variation of the vacancy population in a given orientation versus the amplitude of the external stress. Application of a uniaxial stress changes the energy of each orientation by

$$\Delta E_i = -VQ_i \tag{3.127}$$

where Q_i is the amplitude of the induced distortion. This energy is obtained directly from the relative intensities between the EPR lines i and j

$$\frac{a_i}{a_j} = \exp -\left(\frac{\Delta E_i - \Delta E_j}{kT}\right) \quad . \tag{3.128}$$

The amplitude Q_i is estimated from the elastic properties.

The results obtained by Watkins for the three charge states V^+, V^o and V^- of the vacancy are given in Table 3.2.

Table 3.2. Jahn-Teller coupling coefficients (V) and energies (E_{JT}) estimated by Watkins for the vacancy in silicon. In case (i) the force constant k (see text) is taken to be the perfect crystal one (as estimated by SWALIN [3.28] from compressibility data). In case (ii) Watkins used for values of k: $k_{D_{2d}}$ = 1.3 eV Å^{-2} and $k_{C_{3v}}$ = 0.97 eV Å^{-2} deduced from calculation of local modes of vibrations performed around the vacancy by LARKINS and STONEHAM [3.29]

	Case	Mode D_{2d} V [eV A^{-1}]	E_{JT} [eV]	Mode C_{3v} V [eV A^{-1}]	E_{JT} [eV]	Total E_{JT} [eV]
V^+		2.9	0.4	-	-	0.4
V^o	i	6.0	1.5	-	-	1.5
V^-		4.5	0.9	3.8	1.2	2.1
V^+		2.1	1.7	-	-	1.7
V^o	ii	4.3	7.1	-	-	7.1
V^-		3.2	4.0	2.1	2.3	6.3

The values of the Jahn-Teller energies are somewhat larger than those given in Sect.2.4.4. This is not surprising when use is made of the k values from LARKINS and STONEHAM [3.29] since these are smaller than those given in [3.30,31] by a factor of 2-3 (note that normalization of Q is not the same as in Sect.2.4.4, so that the k differ by a factor of 4). Another possible source of discrepancy comes from uncertainty in the experimental determination of V, which depends upon the estimation of Q_i induced by the applied stress.

3.8.5 Energy Levels

Let us now complete this picture of the vacancy by describing the energy levels associated with its different charge states. Watkins attempted to locate them by correlating the type of charge state with the positions of the Fermi level. In p-type material he observed the V^+ state; he also suggested that the V^o state can be present, depending on the Fermi-level posi-

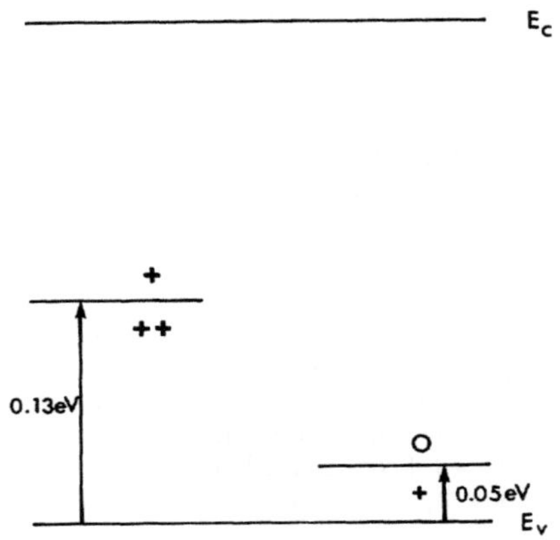

Fig.3.14. Occupancy levels for the vacancy in p-type Si

tion, since the V^0 state can be changed into the V^- state by band-gap light illumination (producing electrons). In n-type material, band-gap light illumination (which produces holes) is necessary to observe the V^- state, indicating that the V^{--} state is then the stable state. Thus, the vacancy seems to possess four different charge states in the gap V^+, V^0, V^- and V^{--}, only two of them being directly observable by EPR.

Later WATKINS [3.32] realized that the V^+ state observed in p-type material is a metastable state. The study of the decay of the V^+ concentration, after the light which produces it ($h\nu < 0.35$ eV) is turned off, shows that it is due to hole emission

$$V^+ \rightarrow V^0 + h \tag{3.129}$$

and the activation energy associated with this process gives the position of the level for the transition $V^0 \rightarrow V^+$ at $\sim E_V + 0.05$ eV. However, the level associated with the vacancy, which is observed by a DLTS study (Sect.6.4.3) of hole emission is at $E_V + 0.13$ eV. Careful examination of the amplitude of the emission signal relative to the vacancy concentration [3.32] indicates that the excitation process involves two holes. It must thus be related to the transition

$$V^{++} \rightarrow V^0 + 2h \quad . \tag{3.130}$$

The reason two holes are emited simultaneously is because the second hole is emitted with a faster rate than the first. Thus the order of the levels associated with the transition $V^0 \rightarrow V^+$ and $V^+ \rightarrow V^{++}$ is inverted (Fig.3.14). This behavior implies that the effective correlation energy for trapping a second electron on V^{++} is negative: the vacancy behaves like an Anderson negative U center (Sect.2.4.4). It is a consequence of a large lattice relaxation, since it overcomes the Coulomb repulsion of two electrons and is thus consistent with the large Jahn-Teller energies observed.

BARAFF et al. [3.31] obtained a similar picture using a quantitative theory in which the electron levels are derived by a self consistent Greens-function technique [Ref.1.1,Sect.4.2.2], and calculating the amplitude of the distortion in a modified valence force model. They argue that the electron-electron repulsion energy U induced when a second electron is added on a V^+ configuration is negative, the energy gained by the distortion being sufficient to compensate the electron-electron repulsion energy. The result is that the transition $V^{++} \rightarrow V^0$ occurs (with two electrons) without intermediate states since the V^+ configuration is unstable.

4. Optical Properties

The interaction of photons with the electrons of a semiconductor provides a large amount of information: the position of an emission or absorption line indicates the energy separating two levels; the lowest photon energy at which a spectral structure begins reveals the threshold for a class of transitions; the shape of a spectrum is a measure of the transition probability or of the distribution of states, etc. In the absorption process, a photon of known energy excites an electron from a lower into a higher energy state. Thus, absorption in a semiconductor occurs due to band-to-band transitions, exciton formation, transitions between localized levels and bands, and between the ground states of localized levels and their excited states. Free-carrier absorption within a band also occurs (this carrier absorption must be removed by compensation of the free carriers, in order to allow observation of other absorption processes). Finally, resonances due to the vibrational states of the lattice and defects also induce absorption as shown in [Ref. 1.1, Chap. 5].

Emission of radiation is the inverse process of absorption. When an electron, occupying a state higher in energy than the state it occupies at equilibrium, makes a transition to an empty, lower, state, the energy difference between the two states can be emitted radiatively, i.e., as photons, or nonradiatively (Chap.6). The emission of photons is called luminescence [4.1]. The emission process requires that the material is not at thermal equilibrium, since the electrons have to be excited in high-energy states. When the excitation is performed by light absorption, the emission process is called photoluminescence; excitation by an electric field is called electroluminescence [4.2], excitation by an electron beam is called cathodoluminescence. When the excitation is made at low temperature in order to freeze the electrons in the high-energy states, from which they can subsequently be released thermally, photon emission is called thermoluminescence [4.3].

Most of the transitions which take place in absorption can also take place in the opposite way to produce emission. The difference between absorption and emission comes from the fact that, if the absorption process couples filled localized states with an empty band, the emission process couples a narrow band of filled states with localized empty states. Thus, usually absorption results in a broad spectrum while emission produces a rather narrow spectrum. Since the aim of this chapter is to describe the information that optical properties can provide on point defects, we shall consider here only absorption and emission due to transitions between the ground state of localized levels and bands or excited states. (For the other types of transitions see, for instance, [4.4-7].)

4.1 Transition Probability

In the semiclassical theory of radiation, the coupling between an electron and an electromagnetic wave leads to a perturbation Hamiltonian

$$\Delta H = \frac{\hbar}{i}\left(\frac{e}{mc}\,\underline{A}\cdot\underline{\nabla}\right) \tag{4.1}$$

where $\underline{A}$ is the vector potential of the radiation

$$\underline{A} = \underline{A}_0 \cos(\underline{q}\cdot\underline{r} - \omega t) \quad , \tag{4.2}$$

$\underline{q}$ being the wave vector of the electromagnetic wave, ω its pulsation, and $\underline{r}$ the position vector. Equation (4.1) is valid to first order in $\underline{A}$. Time-dependent perturbation theory [4.9] shows that such a perturbation induces transitions between the states of the unperturbed system. In the case of absorption, the probability per unit time to make a transition from an initial state ψ_i of energy E_i to a final state ψ_f of energy E_f is given by

$$W = \frac{2\pi}{\hbar}\,|\langle\psi_f|\,\frac{\hbar e}{2imc}\,\underline{A}_0\cdot e^{i\underline{q}\cdot\underline{r}}\underline{\nabla}|\psi_i\rangle|^2\delta(E_f - E_i - h\nu) \tag{4.3}$$

where ν is the photon frequency.

The amplitude A_0 of the wave is related to the photon flux through [4.9]

$$\Phi = \frac{A_0^2\omega}{8\pi\hbar c} \quad . \tag{4.4}$$

When the wavelength $2\pi/q$ of the radiation is many times larger than the extension of the wave function (which is usually the case for localized states

ψ_i), the quantity $\underline{q} \cdot \underline{r}$ is smaller than unity. Then $\exp(i\underline{q} \cdot \underline{r})$ can be replaced by 1 in (4.3). Under these conditions we can write the optical cross section for this transition (i.e., the ratio W/Φ) as

$$\sigma(h\nu) = \frac{\beta}{h\nu} \left|\langle\psi_f|\left(-i\hbar \frac{\partial}{\partial z}\right)|\psi_i\rangle\right|^2 \delta(E_f - E_i - h\nu) \tag{4.5}$$

with

$$\beta = \frac{4\pi^2 e^2 \hbar}{m^2 c} \tag{4.6}$$

where z is the component of $\underline{r}$ along $\underline{A}_0$.

The transitions which are allowed are, of course, those for which the matrix element $\langle\psi_j|(\partial/\partial z)|\psi_i\rangle$ is nonzero. This matrix element can be written in terms of the component of the particle vector position $\underline{r}$. For this we use the fact that

$$[H,z] = -\frac{\hbar^2}{2m} \frac{\partial}{\partial z} \quad . \tag{4.7}$$

The matrix element then reduces to

$$-\frac{2m}{\hbar^2} \langle\psi_f|z|\psi_i\rangle(E_f - E_i) \quad . \tag{4.8}$$

Thus, the allowed transitions are those for which the irreducible representation, to which z belongs, is contained in the product $\Gamma(\psi_i) \times \Gamma(\psi_f)$ of the representations corresponding to the initial and final states (for details see [Ref.1.1, Sect.1.4]).

Consequently, the behavior of transitions observed under uniaxial stress, because it lowers the defect symmetry and splits the associated localized levels, gives information on this symmetry. Because the dipole moment matrix elements depend on polarization, the application of light polarized parallel to the stress (the so-called π component) or perpendicular to the stress (the σ component) can induce different transitions and provide additional information. An example will be described in Sect.4.5.

4.2 The Configuration Coordinate Diagram

To discuss qualitatively the physical properties associated with the electron-lattice interaction it is often interesting to draw what is called the configuration coordinate diagram. For this it is assumed that the defect interacts with only one lattice coordinate Q. The diagram is then simply a

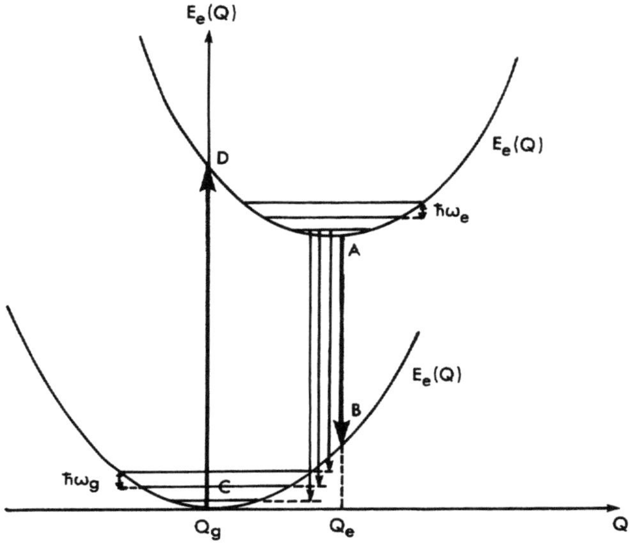

Fig.4.1. Configuration coordinate diagram showing few transitions corresponding to absorption (CD) and emission (AB)

a graphical representation of the electronic energy $E_e(Q)$ for the ground state and the various excited states of interest. This is the case in Fig. 4.1 where a two-level system is represented, $E_g(Q)$ being the ground-state energy and $E_e(Q)$ the energy of an excited state. We call Q_g and Q_e the stable positions for these two states, respectively. Following the Born-Oppenheimer approximation (Ref.1.1, Sect.5.1.1 and Sect.2.1.1) these energies are the potential energies for nuclear motion. Near each minimum we can use the harmonic approximation, so that the vibrational states are harmonic oscillators of energies $(n_g + \frac{1}{2})\hbar\omega_g$ for the ground state and $(n_e + \frac{1}{2})\hbar\omega_e$ for the excited state. Figure 4.1 shows a few low-lying vibrational states for the two energy branches.

In this configuration coordinate diagram, characteristic of the defect associated with the localized state considered, light emission corresponds (according to the Franck-Condon principle [4.10]) to the transition A → B (Fig.4.1): a photon of energy $E_A - E_B$ is emitted, followed by the relaxation of the system towards the minimum C through phonon emission. Actually, as illustrated in Fig.4.1, the emission spectrum is not composed of a single line of energy $E_A - E_B$, but of a band centered around this energy, corresponding to all the possible transitions between the vibrational states of the system. Photon absorption by the localized level corresponds to the transition C→D (Fig.4.1) again with broadening into a band.

In the next section we shall see how the shape of the emission (or of the absorption) spectrum can be derived within the adiabatic approximation, i.e., separating the electronic and vibrational terms of the matrix element. For strong defect-lattice coupling or at high temperature, the vibrational contribution to the matrix element leads to a Gaussian line shape.

This Gaussian function is entirely determined by the knowledge of three parameters which characterize the configuration coordinate diagram: the Franck-Condon shift (Sect.4.3), the pulsation of the vibrational modes and the energy $E_A - E_B$. In the case of a transition between a level and a band the optical cross section is characteristic of the intrinsic properties of the defect: symmetry and extension in space of the wave function, and electronic coupling of the localized level with the band (the variation of the optical cross section with temperature is a way to obtain the temperature dependence of the displacement of the localized level in the forbidden gap). It also reflects the density of states in the band. The analysis of optical cross sections associated with localized states is therefore a powerful means to extract physical information on the corresponding defects.

4.3 Optical Line Shape and the Electron-Lattice Interaction

In this section we treat the origin of band broadening due to the electron-lattice interaction (Chap.2). We consider the case of an optical transition between two electron states of a given defect (the extension to the case of a defect state and a band is trivial). We shall give very simple illustrations of the effect, corresponding to well-defined limiting cases (low- or high-temperature limits) which allow a much simpler analysis than the general formulas. We begin with the coupling to one lattice coordinate at T = 0 K, which is the simplest situation. We then derive some useful results for overlaps between harmonic oscillator wave functions. An application is made to the low-temperature expression of the absorption. We then analyze the strong coupling case and the high-temperature limit where the line shape becomes Gaussian. After that we extend the model to include coupling with a continuum of lattice modes described by an average frequency, for which we can give an exact expression for the line-shape function. Finally, we calculate the moments of the absorption curve for which compact expressions can be derived in the general case.

4.3.1 Coupling to One Lattice Coordinate at T = OK

The situation we consider corresponds to the configuration coordinate diagram of Fig.4.2. On this figure are plotted the electronic energies $\varepsilon_g(Q)$ (ground state) and $\varepsilon_e(Q)$ (excited states), solutions of the electronic equation in the Born-Oppenheimer approximation ([Ref.1.1, Sect.5.1.1 and 2.1.1]). We assume that these energies are the same parabolic functions of only one lattice mode Q which is some combination of the atomic displacements near the defect site: $\varepsilon(Q) = M\omega^2(Q - Q_i)^2/2$ with $Q_i = Q_g$ for the ground state and $Q_i = Q_e$ for the excited state. The vibrational states are then simple harmonic oscillators of energy $(n + \frac{1}{2})\hbar\omega$ (only few of these states are given in Fig.4.2 for the ground and excited states' configurations).

The broadening of the absorption curve is very simple to understand in this case. At T = OK only the ground state of energy $\varepsilon_g(Q_g) + \hbar\omega/2$ is populated. Optical excitation can occur between this state and all phonon states derived from the excited configuration, of energies $\varepsilon_e(Q_e)+(p + \frac{1}{2})\hbar\omega$. In this very simple model the absorption band will then consist of sharp lines of frequencies $h\nu = h\nu_0 + p\hbar\omega$ (Fig.4.2) where $h\nu_0$ is defined by

$$h\nu_0 = \varepsilon_e(Q_e) - \varepsilon_g(Q_g) \tag{4.9}$$

and p is the number of phonons involved in the transition.

Important information can be derived from the relative intensities of these absorption lines. Let us consider the transition involving p phonons. Its intensity is proportional to the quantity I_{op} given by

$$I_{op} = |\langle\psi_g\chi_{go}|\Omega|\psi_e\chi_{ep}\rangle|^2 \quad . \tag{4.10}$$

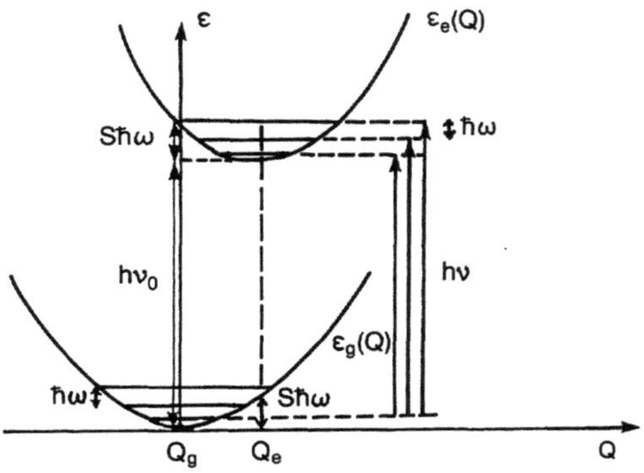

Fig.4.2. Configuration coordinate diagram showing transitions for absorption from the ground state g to phonon states of the excited state e

Here $\psi_g \chi_{go}$ is the total wave function of the ground state which, in the Born-Oppenheimer approximation, is the product of the electronic part ψ_g and of the harmonic oscillator wave function χ_{go}, of quantum number n = 0, centered on $Q = Q_g$. On the other hand, $\psi_e \chi_{ep}$ is a Born-Oppenheimer product of the electronic excited state ψ_e by the pth harmonic oscillator wave function χ_{ep} centered on Q = Qe. Finally Ω is the electronic dipole operator, since we are concerned with an optical transition of electronic origin (Sect. 4.1). To discuss I_{op} in more detail we make the Condon approximation, i.e., we neglect the possible dependence of $\langle\psi_g|\Omega|\psi_e\rangle$ on Q (in the Born-Oppenheimer approximation, ψ_g and ψ_e are functions of Q). Then I_{op} can be factorized and the normalized intensity W_{op} of this transition

$$W_{op} = \frac{I_{op}}{\sum_p I_{op}} \tag{4.11}$$

can be written, using the closure relation $\sum_p |\chi_{ep}\rangle\langle\chi_{ep}| = 1$,

$$W_{op} = |\langle\chi_{go}|\chi_{ep}\rangle|^2 \tag{4.12}$$

which involves overlaps between the harmonic oscillator wave functions only. It will be shown in Sect.4.3.2 that

$$W_{op} = \frac{S^p}{p!} \exp(-S) \quad , \tag{4.13}$$

where S is the Huang-Rhys factor [4.11], defined in Fig.4.2, which measures the relative strength of the electron-phonon interaction. In the limit $S \to 0$ only W_{oo} is nonzero: the transition reduces to a sharp electronic transition, involving no phonons. The line corresponding to p = 0 which is present in the general case with the reduced intensity exp(-S) is called the "zero phonon line".

From (4.13) we can see that the absorption is maximum when $p \sim S$, i.e., when $h\nu = h\nu_o + S\hbar\omega$, the vertical transition on Fig.4.2. S is then the most probable number of phonons involved in the transition (since usually S is not an integer, the most probable number is the integer nearest to S). Three cases are pictured in Fig.4.3, for S = 0, 3 and 10, in order to illustrate the effect of the strength of the interaction. For large values of S, the curve is practically symmetrical with respect to p = S, i.e., around the energy $h\nu_o + S\hbar\omega$. In this large coupling limit the curve looks more and more like a Gaussian function as can be shown by considering the asymptotic behavior of $(S^p/p!)\exp(-S)$. For this we first use Stirling's formula for p! and obtain

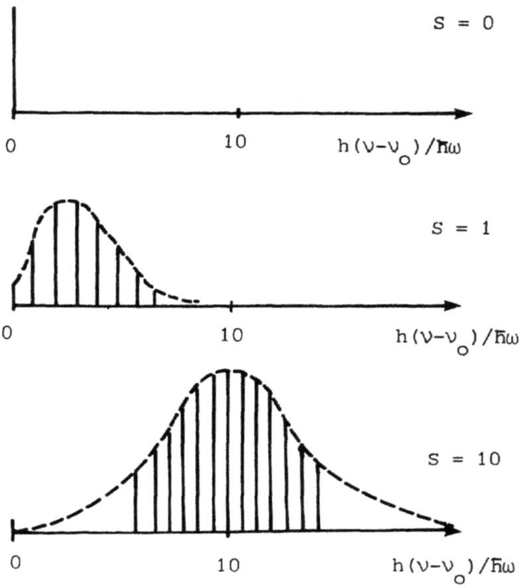

Fig.4.3. The shape of an absorption band as a function of the value of the Huang-Rhys factor S

$$\frac{S^p}{p!}\exp(-S) \sim \exp\left[p\left(\mathrm{Log}\,\frac{S}{p} + 1\right) - S\right] \quad . \tag{4.14}$$

Treating p as a continuous variable, we can search for the maximum of the argument in the exponential (given by p = S) and expand this argument to second order in (p - S) (valid for $|p - S| \ll S$). This leads to

$$\frac{S^p}{p!}\exp(-S) \sim \exp\left[-\frac{(p - S)^2}{2S}\right] \tag{4.15}$$

which shows that for large S the absorption curve is a Gaussian centered on p = S.

4.3.2 Overlap Between Harmonic Oscillators

We calculate now the overlap between harmonic oscillator functions centered on different lattice positions. The expression of this overlap is necessary for the next sections but those who are not interested in its detailed derivation can drop this part and just examine the result.

We aim here at calculating overlaps like $\langle\chi_{g,n}|\chi_{e,n+p}\rangle$ between harmonic oscillator wave functions of quantum numbers n and n + p centered on Q_g and Q_e, respectively. For this, we make use of a very limited number of properties concerning the harmonic oscillator, which we now recall. We use creation and annihilation operators a^+ and a which, for the ground state, are given by

$$a_g^+ = \frac{-\hbar(d/dQ) + M\omega(Q - Q_g)}{\sqrt{2M\hbar\omega}}$$

$$a_g = \frac{\hbar(d/dQ) + M\omega(Q - Q_g)}{\sqrt{2M\hbar\omega}} \tag{4.16}$$

with equivalent definitions for the excited state, replacing g by e. The following properties of these operators will be needed:

$$a_g^+ a_g |\chi_{gn}\rangle = n|\chi_{gn}\rangle \tag{4.17a}$$

$$a_g a_g^+ = 1 + a_g^+ a_g \tag{4.17b}$$

$$|\chi_{g,n}\rangle = n^{-\frac{1}{2}} a_g^+ |\chi_{g,n-1}\rangle = (n!)^{-\frac{1}{2}} (a_g^+)^n |\chi_{g,o}\rangle \tag{4.17c}$$

$$|\chi_{g,n-1}\rangle = n^{-\frac{1}{2}} a_g |\chi_{gn}\rangle \tag{4.17d}$$

which also hold true for the excited state. An important point, derived from (4.16), is that

$$a_e^+ - a_g^+ = a_e - a_g = -\sqrt{\frac{M\omega}{2\hbar}}\,(Q_e - Q_g) \tag{4.18}$$

and, as is apparent on Fig.4.2,

$$\frac{1}{2} M\omega^2 (Q_e - Q_g)^2 = S\hbar\omega \quad . \tag{4.19}$$

Thus (4.18) can be transformed into

$$a_e^+ - a_g^+ = a_e - a_g = -\sqrt{S} \quad . \tag{4.20}$$

We are now in position to calculate the overlap $\langle\chi_{g,n}|\chi_{e,n+p}\rangle$ which we label $F_{n,n+p}$.

Let us begin with the simplest case $F_{o,p}$ (which occurs at zero temperature). We can write, using (4.17) for the excited state,

$$F_{o,p} = \frac{1}{\sqrt{p!}} \langle\chi_{g,o}|(a_e^+)^p|\chi_{e,o}\rangle \quad , \tag{4.21}$$

which can be transformed into

$$F_{o,p} = \frac{1}{\sqrt{p!}} \langle\chi_{g,o}|(a_g^+ - \sqrt{S})^p|\chi_{e,o}\rangle \quad . \tag{4.22}$$

This expression can be expanded in powers of a_g^+. However, any $<\chi_{g,o}|(a_g^+)^m$ is the conjugate of $a_g^m|\chi_{g,o}>$ which is zero, except if m = 0. Then, only the term $(-\sqrt{S})^p$ has to be retained in (4.22) and we can write

$$F_{o,p} = \frac{(-\sqrt{S})^p}{\sqrt{p!}} F_{o,o} \quad . \tag{4.23}$$

The calculation of $F_{o,o}$ can be done very simply by noticing that

$$\sum_{p=0}^{\infty} |F_{o,p}|^2 = \sum_{p=0}^{\infty} <\chi_{g,o}|\chi_{e,p}><\chi_{e,p}|\chi_{g,o}> \tag{4.24}$$

that is:

$$\sum_{p=0}^{\infty} |F_{o,p}|^2 = 1 \tag{4.25}$$

and, on the other hand, by applying (4.23):

$$\sum_{p=0}^{\infty} |F_{o,p}|^2 = |F_{o,o}|^2 \exp(S) \quad . \tag{4.26}$$

Thus, $F_{o,o} = \exp(-S/2)$ and

$$F_{op} = \frac{(-\sqrt{S})^p}{\sqrt{p!}} \exp(-\frac{S}{2}) \quad . \tag{4.27}$$

This expression, when squared, gives the result (4.13) used in Sect.4.3.1.

We can now determine $F_{n,n+p}$ in the general case. For this we proceed in the same manner as for (4.21), writing

$$F_{n,n+p} = \frac{1}{\sqrt{(n+p)!}} <\chi_{gn}|(a_e^+)^{n+p}|\chi_{eo}> \quad . \tag{4.28}$$

As in (4.22) we replace $(a_e^+)^{n+p}$ by $(a_g^+ - \sqrt{S})^{n+p}$ and expand this in power of a_g^+. For the same reason as above only the powers of a_g^+ smaller than n contribute to the expression and we can write

$$F_{n,n+p} = \frac{1}{\sqrt{(n+p)!}} \sum_{k=0}^{n} (-\sqrt{S})^{n+p-k} C_{n+p}^{k} <\chi_{g,n}|(a_g^+)^k|\chi_{e,o}> \quad . \tag{4.29}$$

Now the bra $<\chi_{g,n}|(a_g^+)^k$ is the conjugate of the ket $a_g^k|\chi_{g,n}>$ which, using (4.17) is equal to $\sqrt{n!/(n-k)!}\,|\chi_{g,n-k}>$. This leaves us with the scalar product $<\chi_{g,n-k}|\chi_{eo}>$ simply given by $F_{o,n-k}$ of (4.27) when $-\sqrt{S}$ is replaced by $+\sqrt{S}$. We thus obtain

$$F_{n,n+p} = \sqrt{\frac{n!}{(n+p)!}} \exp\left(-\frac{S}{2}\right) \sum_{k=0}^{n} \frac{(-1)^{n+p-k}}{(n-k)!} (\sqrt{S})^{p+2(n-k)} C_{n+p}^{k} \quad . \qquad (4.30)$$

This expression can be simplified in the limit where $n \ll p$. In such a case C_{n+p}^{k} can be approximated by $(n + p)^k/k!$ and the expression for $F_{n,n+p}$ becomes

$$F_{n,n+p} \sim \exp\left(-\frac{S}{2}\right) \sqrt{\frac{n!}{(n+p)!}} (-\sqrt{S})^{p} \sum_{k=0}^{n} (-1)^{n-k} S^{n-k} \frac{(n+p)^k}{k!(n-k)!} \quad . \qquad (4.31)$$

The term $[k!(n - k)!]^{-1}$ in (4.31) is equal to $C_n^k/n!$ and the sum over k now represents the expansion of $(n + p - S)^n$. Then $F_{n,n+p}$, becomes

$$F_{n,n+p} \sim \exp\left(-\frac{S}{2}\right) \frac{(-\sqrt{S})^{p}}{\sqrt{n!(n+p)!}} (n+p-S)^{n} \quad . \qquad (4.32)$$

This approximate form will be very useful since it corresponds to the low-temperature situation which is considered in the next section. This limit will be also very important in the case of capture by multiphonon emission (Sect.6.3.3).

4.3.3 The Low-Temperature Limit

We now consider the extension of the model treated in Sect.4.3.1 to the general case, when $T \neq 0$ K. There is then thermal excitation and the nth harmonic oscillator state of energy $\varepsilon_g(a_g) + (n + \frac{1}{2})\hbar\omega$ is now populated according to the statistical factor $[1 - \exp(-\hbar\omega/kT)]\exp(- n\hbar\omega/kT)$. The spectrum still consists of sharp lines separated in energy by $\hbar\omega$ and involving p phonons. However, p can now be positive or negative; it is negative in cases where $n\hbar\omega$ becomes greater than the distance $h\nu_0$ separating the bottom of the two potential wells.

The intensity I_p of a transition involving p phonons can be determined by summing over the transitions between an initial state with the quantum number n and a final state with the quantum number n + p. This leads to

$$I_p = \left[1 - \exp\left(-\frac{\hbar\omega}{kT}\right)\right] \sum_n \exp\left(-\frac{n\hbar\omega}{kT}\right) I_{n,n+p} \quad . \qquad (4.33)$$

The corresponding transition occurs at a frequency such that $h\nu$ is equal to $h\nu_0 + p\hbar\omega$ and the intensity $I_{n,n+p}$ is given by an expression analogous to (4.10), i.e., by

$$I_{n,n+p} = |\langle\psi_g\chi_{g,n}|\Omega|\psi_e\chi_{e,n+p}\rangle|^2 \quad . \qquad (4.34)$$

Again, we use the Condon approximation in order to factorize the electronic part. We write the relative intensity W_p of this transition, defined by $I_p/\sum_p I_p$,

$$W_p = \left[1 - \exp(-\frac{\hbar\omega}{kT})\right] \sum_{n=0}^{\infty} \exp(-\frac{n\hbar\omega}{kT})|F_{n,n+p}|^2 \quad , \tag{4.35}$$

where $F_{n,n+p}$ are the overlap terms $\langle\chi_{g,n}|\chi_{g,n+p}\rangle$ calculated in Sect.4.3.2. Expression (4.35) is exact and can quite generally be written in terms of Bessel functions [4.12]. However, we prefer here to calculate it in limiting cases where simple analytic expressions can be derived. We first investigate the low-temperature limit defined by the fact that n is small and we consider frequencies such that $p \gg n$. We have already seen that $F_{n,n+p}$ takes in this case the simple form given by (4.32). Using this expression W_p becomes

$$W_p \sim \left[1 - \exp(-\frac{\hbar\omega}{kT})\right]\exp(-S)S^p \sum_n \exp(-n\frac{\hbar\omega}{kT}) \frac{(n + p - S)^{2n}}{n!(n + p)!} \quad . \tag{4.36}$$

In the limit where $p \gg n$, we can replace $(n + p)!$ by $p!(n + p)^n$. Expression (4.36) then becomes

$$W_p \sim \left[1 - \exp(-\frac{\hbar\omega}{kT})\right]\exp(-S)\frac{S^p}{p!}\sum_n \frac{1}{n!}\left(\frac{\exp(-\hbar\omega/kT)(n + p - S)^2}{n + p}\right)^n \quad . \tag{4.37}$$

This expression can also be simplified in the limit where S is much smaller than p, in which case the term $(n + p - S)^2/(n+p) \sim n+p-2S$. Then W_p is given by

$$W_p \sim \left[1 - \exp(-\frac{\hbar\omega}{kT})\right]\exp(-S)\frac{S^p}{p!}\sum_n \left[\exp(-\frac{\hbar\omega}{kT})(1 - \frac{2S}{p})\right]^n \frac{(n + p)^n}{n!} \quad . \tag{4.38}$$

For $p \gg n$ the term $(n + p)^n$ can be approximated by

$$(n + p)^n \sim \left[\frac{d^n}{du^n}(1 - u)^{-(p+1)}\right]_{u=0} \tag{4.39}$$

and the sum over n corresponds to the power series expansion of this function, which allows us to write W_p in the following way

$$W_p \sim \left[1 - \exp(-\frac{\hbar\omega}{kT})\right]\exp(-S)\frac{S^p}{p!}\left[1 - \exp(-\frac{\hbar\omega}{kT})(1 - \frac{2S}{p})\right]^{-(p+1)} \quad , \tag{4.40}$$

or equivalently

$$W_p \sim W_{op}\left[1 - \exp(-\frac{\hbar\omega}{kT})\right]^{-p}\left(1 + \frac{2S}{p}\ \frac{\exp(-\hbar\omega/kT)}{1 - \exp(-\hbar\omega/kT)}\right)^{-(p+1)} \quad . \tag{4.41}$$

Here W_{op} is the T = 0 K intensity given by (4.13). To obtain the usual form of W_p it is convenient to introduce the average quantum number $\bar{n}$ in the ground state given by

$$\bar{n} = \left[1 - \exp(-\frac{\hbar\omega}{kT})\right] \sum_n n \exp(- n \frac{\hbar\omega}{kT}) \quad , \tag{4.42}$$

that is

$$\bar{n} = \exp(-\frac{\hbar\omega}{kT})\left[1 - \exp(-\frac{\hbar\omega}{kT})\right]^{-1} \quad . \tag{4.43}$$

In (4.41) the term $(1 + 2S\bar{n}/p)^{-(p+1)}$ appears which for large p gives $\exp(-2S\bar{n})$. This leads to a rather simple expression for W_p in the low-temperature limit:

$$W_p \sim W_{op}(\bar{n} + 1)^p \exp(-2S\bar{n}) \quad . \tag{4.44}$$

This formula is valid in the limit $p \gg S$ and $p \gg \bar{n}$, i.e., at high frequencies (since $h\nu$ is equal to $h\nu_o + p\hbar\omega$) and correspondingly low temperatures. This limit will be particularly useful when we treat multiphonon emission capture (Sect.6.3.3).

4.3.4 The Strong Coupling Limit

We have already shown (Sect.4.3.1) that for large values of S and at T = 0K, the absorption curve has a Gaussian shape. Here we generalize this result to nonzero temperatures, using a more direct argument. For this we first write the line shape function at zero temperature under the form

$$W(h\nu) = \sum_p W_{op}\ \delta(h\nu - h\nu_o - p\hbar\omega) \tag{4.45}$$

using W_{op} defined in (4.11). We directly rewrite (4.45) under the form

$$W(h\nu) = \sum_p \langle\chi_{g,o}|\delta(h\nu - h\nu_o - H_e + H_g)|\chi_{e,p}\rangle\langle\chi_{e,p}|\chi_{g,o}\rangle \tag{4.46}$$

where H_e and H_g are the phonon Hamiltonians in the excited and ground states, respectively, with the origin of the energies taken at the bottom of each potential well (Fig.4.2). This transformation is not valid in general but only in the strong coupling limit, as we shall see later. For the moment we use (4.46) as it stands, noticing that the quantity $h\nu_o + H_e - H_g$ is equal to $\varepsilon_e(Q) - \varepsilon_g(Q)$, which quantity varies linearly with Q. We therefore write it under the form

$$\varepsilon_e(Q) - \varepsilon_g(Q) = h\nu_o + S\hbar\omega - I(Q - Q_g) \quad . \tag{4.47}$$

Using the closure relation for the $\chi_{e,p}$ in (4.46), we obtain

$$W(h\nu) = \langle\chi_{g,o}|\delta[h\nu - h\nu_o - S\hbar\omega + I(Q - Q_g)]|\chi_{g,o}\rangle \quad . \tag{4.48}$$

The eigenstate $\chi_{g,o}$ is a Gaussian $\exp[-(M\omega/2\hbar)(Q - Q_g)^2]$. Integrating over Q, with the delta function in (4.48), we find

$$W(h\nu) = \frac{1}{I}\,\frac{\exp\left(-\frac{M\omega}{\hbar}\,\frac{(h\nu - h\nu_o - S\hbar\omega)^2}{I^2}\right)}{\int_{-\infty}^{+\infty} dQ\,\exp\left(-\frac{M\omega}{\hbar}Q^2\right)} \quad . \tag{4.49}$$

We can now express the coefficient I in terms of our fundamental parameter $S\hbar\omega$. For this we first note that $S\hbar\omega$ is given by (4.19). We also notice that the difference between $\varepsilon_e(Q_g) - \varepsilon_g(Q_g)$ and $\varepsilon_e(Q_e) - \varepsilon_g(Q_e)$ is equal to $2S\hbar\omega$ but is also using (4.47), equal to the quantity $I(Q_e - Q_g)$. As a result $S\hbar\omega = I^2(2M\omega^2)^{-1}$. Equation (4.49) then leads to

$$W(h\nu) = \frac{1}{\sqrt{2\pi S(\hbar\omega)^2}}\exp\left(-\frac{(h\nu - h\nu_o - S\hbar\omega)^2}{2S(\hbar\omega)^2}\right) \quad , \tag{4.50}$$

which is strictly equivalent to (4.15) where $p\hbar\omega$ is replaced by $h\nu - h\nu_o$.

We now extend the argument to temperatures different from 0K. As mentioned in Sect.4.3.3 we have to perform a thermal average over the ground state. We obtain for the lineshape

$$W(h\nu) = \mathrm{Av}_n\left[\sum_p |\langle\chi_{g,n}|\chi_{e,n+p}\rangle|^2\delta(h\nu - h\nu_o - p\hbar\omega)\right] \tag{4.51}$$

where $\mathrm{Av}_n\{\ldots\}$ stands for the thermal average over the quantum number n, namely $[1 - \exp(-\hbar\omega/kT)]\sum_n \exp[-n(\hbar\omega/kT)]$. Equation (4.48) is then generalized to

$$W(h\nu) = \mathrm{Av}_n\langle\chi_{g,n}|\delta(h\nu - h\nu_o - S\hbar\omega + I(Q - Q_g)|\chi_{g,n}\rangle \quad . \tag{4.52}$$

To evaluate (4.52) we use Mehler's formula which states that [Ref.4.12, p.973]

$$\mathrm{Av}_n|\chi_{g,n}(Q)|^2 = \sqrt{\frac{M\omega}{\pi\hbar\coth(\hbar\omega/2kT)}}\exp\left(-\frac{M\omega Q^2}{\hbar\coth(\hbar\omega/2kT)}\right) \quad . \tag{4.53}$$

This reduces the determination of $W(h\nu)$ to what was done before and we obtain the generalized expression

$$W(h\nu) = \frac{1}{\sqrt{2\pi S(\hbar\omega)^2 \coth(\hbar\omega/2kT)}} \exp\left(-\frac{(h\nu - h\nu_0 - S\hbar\omega)^2}{2S(\hbar\omega)^2 \coth(\hbar\omega/2kT)}\right) \quad . \qquad (4.54)$$

This expression, which has been derived by many different authors (e.g., [4.10,12]), is obtained here in a very direct way. However, this way is based on the replacement of (4.45) by (4.46) which implies some assumptions regarding H_e and H_g. Let us discuss this point. We have to use in (4.45) the integral representation of the delta function, written

$$\delta(h\nu - h\nu_0 - p\hbar\omega) = \frac{1}{2\pi} \int dt \exp\left(-\frac{i(h\nu - h\nu_0 - p\hbar\omega)t}{\hbar}\right) \quad . \qquad (4.55)$$

We can go from (4.45) to (4.46) replacing the exponential by $\exp[-i(h\nu - h\nu_0 - H_e + H_g)t/\hbar]$ with the condition, which is not valid in general, that H_e and H_g commute. However, in the strong coupling limit, only highly excited states with $p \sim S$ are involved and for such states we can use the classical limit for H_e and H_g. In that limit H_e and H_g can be assumed to commute, which allows us to make the above transformation, i.e., to replace (4.45) by (4.46), justifying our procedure. At $T \neq 0K$ this requires that the average quantum number $\bar{n}$ in the ground state remains much smaller than S.

4.3.5 Classical Treatment for the Lattice

We now make a still more restrictive calculation, treating the lattice vibrations in a completely classical manner. For this we say that at fixed Q, the probability of a transition from $\varepsilon_g(Q)$ to $\varepsilon_e(Q)$ is proportional to

$$|\langle\psi_g|\Omega|\psi_e\rangle|^2 \quad , \qquad (4.56)$$

a quantity which, in the Condon approximation, is taken to be independent of Q. The corresponding contribution to the line-shape function is thus given by

$$\delta[h\nu - \varepsilon_e(Q) + \varepsilon_g(Q)] = \delta[h\nu - h\nu_0 - S\hbar\omega + I(Q - Q_g)] \quad . \qquad (4.57)$$

To obtain the total line shape we have simply to perform an average over the ground state, using classical Boltzmann statistics. We thus write, P being the classical momentum,

$$W(h\nu) = \frac{\int\limits_{-\infty}^{+\infty} dP\, dQ\, \delta[h\nu - h\nu_0 - S\hbar\omega + I(Q-Q_g)]\, \exp\left[-\left(\frac{P^2}{2M} + \frac{M\omega^2 Q^2}{2}\right)/kT\right]}{\int\limits_{-\infty}^{+\infty} dP\, dQ\, \exp\left[-\left(\frac{P^2}{2M} + \frac{M\omega^2 Q^2}{2}\right)/kT\right]} \quad . \tag{4.58}$$

This expression can be readily calculated and gives

$$W(h\nu) = \frac{1}{I}\sqrt{\frac{M\omega^2}{2\pi kT}}\, \exp\left[-\frac{1}{2}\frac{M\omega^2}{I^2 kT}(h\nu - h\nu_0 - S\hbar\omega)^2\right] \tag{4.59}$$

which can be rewritten under the final form

$$W(h\nu) = \frac{1}{\sqrt{4\pi kT\, S\hbar\omega}}\, \exp\left(-\frac{(h\nu - h\nu_0 - S\hbar\omega)^2}{4kT\, S\hbar\omega}\right) \quad . \tag{4.60}$$

It is easy to check that this expression represents the high-temperature limit of (4.54). Indeed, to be valid, this purely classical treatment requires that only states with high quantum numbers are involved, for the ground as well as for the excited state. This is true only in the limit when $kT \gg \hbar\omega$, ensuring that the average quantum number $\bar{n}$ in the ground state is large.

4.3.6 Coupling to a Continuum of Lattice Modes

We now consider the important case where electron-phonon coupling occurs with the continuum of delocalized lattice modes and not, as before, with only one mode Q localized near the defect site. The model we use leads to a result corresponding exactly to that obtained in the case of coupling with one lattice coordinate, so that the limiting formulas developed above still remain valid.

The perfect crystal lattice modes are completely delocalized over the crystal volume, i.e., their amplitude on each lattice site is of the order of $1/\sqrt{N}$, N being the number of atoms in the crystal. The introduction of a defect leads, in the harmonic approximation, to two effects: i) a change in force constants [Ref.1.1, Sect.5.1] corresponding to terms quadratic in the displacements; ii) electron-phonon coupling resulting in terms linear in the atomic displacements. Both perturbations are localized near the defect site. We assume that the changes in force constants do not induce localized vibrational modes, i.e., that the new lattice eigenmodes q_i are still delocalized over the crystal. For each of these coordinates q_i, the linear term leads to diagram similar to Fig.4.2, but where the Huang-Rhys factor

S_i, which is proportional to the square of the electron phonon coupling, is now of the order of 1/N. This fact will introduce a great simplification into the calculation, since only terms of lowest order in S_i will be considered.

The total Born-Oppenheimer eigenstates can be written $|\psi_g>\Pi_i|\chi_{g,n_i}(q_i)>$ for the ground-state configuration and $|\psi_e>\Pi_i|\chi_{e,m_i}(q_i)>$ for the excited one. The products run over all lattice modes q_i, $|\chi_{g,n_i}(q_i)>$ representing an harmonic oscillator eigenstate of quantum number n_i, centered on the equilibrium position $q_{g,i}$.

The same holds true for $|\chi_{e,m_i}(q_i)>$, centered on $q_{e,i}$. The distance between $q_{g,i}$ and $q_{e,i}$ is of the order of $1/\sqrt{N}$. Let us now consider a transition between two such eigenstates $|\psi_g>\Pi_i|\chi_{g,n_i}>$ and $|\psi_e>\Pi_i|\chi_{e,n_i+p_i}>$. It will occur at an energy $h\nu$ given by

$$h\nu = h\nu_0 + \sum_i p_i\hbar\omega_i \quad , \tag{4.61}$$

if $\hbar\omega_i$ is the quantum of vibration for the ith harmonic oscillator. Its relative intensity $W_{\{n_i,p_i\}}$ in the Condon approximation will be given by

$$W_{\{n_i,p_i\}} = \prod_i |<\chi_{g,n_i}|\chi_{e,n_i+p_i}>|^2 \quad . \tag{4.62}$$

Each overlap term can be very easily calculated from the formulas of Sect. 4.3.2, noticing that the corresponding S_i are of the order 1/N. To order $1/N^2$, we find

$$|<\chi_{g,n_i}|\chi_{e,n_i}>|^2 = 1 - (2n_i + 1)S_i$$

$$|<\chi_{g,n_i}|\chi_{e,n_i+1}>|^2 = (n_i + 1)S_i$$

$$|<\chi_{g,n_i}|\chi_{e,n_i-1}>|^2 = n_iS_i \tag{4.63}$$

all other terms being of higher order.

This means that the change in quantum number in the transition can only be equal to 0, +1, or -1. We shall now evaluate the intensity of one transition in which p+s modes are excited by +1, and s modes by -1. We label by the index ℓ the first set of modes, by k the second, and j the modes with no change in quantum number. Using (4.62,63) we can write the intensity of one transition as [4.11]

$$\prod_i^{3N-(p+2s)} [1 - (2n_j + 1)S_j] \prod_\ell^{p+s} (n_\ell + 1)S_\ell \prod_k^{s} n_kS_k \quad , \tag{4.64}$$

where 3N is the total number of phonon modes in the crystal. The finite number p + 2s can be neglected compared to 3N, and the thermal average of (4.64) can be written

$$\prod_{j}^{3N} [1 - (2\bar{n}_j + 1)S_j] \prod_{\ell}^{p+s} (\bar{n}_\ell + 1)S_\ell \prod_{k}^{s} \bar{n}_k S_k \quad , \tag{4.65}$$

where any $\bar{n}_i$ is the average quantum number in the initial state for the ith phonon mode. Its value is clearly given by

$$\bar{n}_i = \left[\exp\left(\frac{\hbar\omega_i}{kT}\right) - 1\right]^{-1} \quad . \tag{4.66}$$

We can now sum the intensities of all transitions corresponding to fixed values of p+s and s. This sum can be expressed in the form

$$W_{p+s,s} = \frac{1}{s!(p+s)!} \prod_{j}^{3N} [1 - (2\bar{n}_j + 1)S_j] \left[\sum_{\ell=1}^{3N} (\bar{n}_\ell + 1)S_\ell\right]^{p+s} \left(\sum_{k=1}^{3N} \bar{n}_k S_k\right)^s \quad . \tag{4.67}$$

The term s!(p + s)! takes account of the fact that the expansion of (4.67) contains groups of identical terms corresponding to permutations of a given set of the p+s ℓ terms between themselves, the same being true for the s k terms. The summation for ℓ and k extends over the 3N phonon modes of the crystal. It is clear that (4.67) contains unphysical terms corresponding to products of terms with equal values of ℓ or k. However their contribution is negligible.

The quantity $W_{p+s,s}$ represents the total intensity of transitions with p+s excitations by +1 and s by -1. However these transitions do not generally occur at the same frequency, and $W_{p+s,s}$ represents the intensity of a band. To simplify these considerations, let us assume that this band is strongly peaked at one value ω so that it is a good approximation to assume that all phonon frequencies can be approximated by ω. In this case, $W_{p+s,s}$ represents the intensity of a sharp line (occuring at $p\hbar\omega$), and the total intensity W_p of all transitions, such that $h\nu$ is equal to $h\nu_0 + p\hbar\omega$, is then:

$$W_p = \sum_{s=0}^{\infty} \frac{1}{s!(p+s)!} [(\bar{n} + 1)S]^{p+s} [\bar{n}S]^s \exp[-(2\bar{n} + 1)S] \quad , \tag{4.68}$$

in which the Huang-Rhys factor S is now defined by

$$S = \sum_{\ell=1}^{3N} S_\ell \tag{4.69}$$

and $\bar{n}$ is equal to (4.66) replacing ω_i by its average value .

This expression is usually rewritten in terms of the Bessel functions I_p with imaginary argument of order p [4.11] as

$$W_p = \left(\frac{\bar{n}+1}{\bar{n}}\right)^{p/2} \exp[-(2\bar{n}+1)S]\; I_p\left\{2S[\bar{n}(\bar{n}+1)]^{\frac{1}{2}}\right\} \quad . \qquad (4.70)$$

To demonstrate (4.68) we have implicitely assumed that p was positive, which may not be true at $T \neq 0$ K. In such a case we interchange the role of the two sets of oscillators, calling $|p| + s$ the number of oscillators with a change in quantum number equal to -1, and s the other one. Then (4.68) remains valid, changing s to $|p| + s$ and $p + s$ to s. This is equivalent to changing $I_{|p|}$ to $I_{-|p|}$ in (4.33) and, because these two Bessel functions are equal, we conclude that (4.70) is valid for all values of p.

The low-temperature limit can be obtained by keeping only the $n = o$ term in (4.68) and clearly we obtain exactly the same expression as (4.44) for the case of coupling with one phonon coordinate. The opposite limit can be determined from an asymptotic expansion of the Bessel function, again leading to the Gaussian shape described in the previous sections.

An important restriction of this calculation and of the model with one coordinate is that both ignore the effects due to the dispersion of the frequencies of the phonon modes (in general we have both localized and extended states). A general formulation of these effects has been given [4.13]. An important conclusion is that, in the weak coupling limit, the absorption spectrum mirrors the phonon density of states weighted by the corresponding coupling function. Thus singularities in the density of states will be present in the optical line shape. We do not derive the whole theory here but, instead, in the next section we derive exact formulas concerning the moments of the line-shape function $W(h\nu)$.

4.3.7 Moments of the Line-Shape Function

Experimentally, the line-shape function $W(h\nu)$ is often characterized by its lower order moments, the moment of order n, M_n, being defined by

$$M_n = \int W(h\nu)(h\nu)^n d(h\nu) \quad . \qquad (4.71)$$

The moment of order zero represents the total intensity. The quantity M_1/M_0 represents the average energy involved in the transition and finally $M_2/M_0 - (M_1/M_0)^2$ is a measure of the width of the absorption band. We shall see that this formulation is ideally suited to a theoretical calculation. To begin, we express $W(h\nu)$ in the form

$$W(h\nu) = Av_{n_i} \sum_{m_i} |\langle \prod_i \chi_{g,n_i} | \prod_i \chi_{e,m_i} \rangle|^2 \delta\left[h\nu - h\nu_o - \sum_i (m_i - n_i)\hbar\omega_i\right] \quad , \tag{4.72}$$

where Av_{n_i} is a thermal average of the initial phonon states of quantum number n_i corresponding to the normal coordinate q_i of equilibrium value $q_{g,i}$. The same notation holds for the excited states, the quantum number being m_i, the equilibrium value $q_{e,i}$. The Huang-Rhys factor for oscillator i is called S_i as above, but is not necessarily of order 1/N when localized states are involved. We label H_g and H_e the phonon Hamiltonians in the ground and excited states. The moment of order n can be written

$$M_n = Av_{n_i} \sum_{m_i} |\langle \prod_i \chi_{g,n_i} | \prod_i \chi_{e,m_i} \rangle|^2 \left[h\nu_o + \sum_i (m_i - n_i)\hbar\omega_i\right]^n \quad . \tag{4.73}$$

The action of H_e on $|\prod_i \chi_{e,m_i}\rangle$ gives $\sum_i (m_i + \frac{1}{2})\hbar\omega_i$. We can introduce H_e into (4.73) and rewrite it as

$$M_n = Av_{n_i} \langle \prod_i \chi_{g,n_i} | \left[h\nu_o + H_e - \sum_i (n_i + \tfrac{1}{2})\hbar\omega_i\right]^n | \prod_i \chi_{g,n_i} \rangle \quad . \tag{4.74}$$

The Hamiltonian H_e can be written as

$$H_e = \sum_i (a^+_{e,i}\, a_{e,i} + \tfrac{1}{2})\hbar\omega_i \quad , \tag{4.75}$$

which, according to (4.20), is equal to

$$H_e = \sum_i \left[(a^+_{g,i} - \sqrt{S_i})(a_{g,i} - \sqrt{S_i}) + \tfrac{1}{2}\right]\hbar\omega_i \quad , \tag{4.76}$$

or equivalently to

$$H_e = H_g + \sum_i \hbar\omega_i \,[S_i - \sqrt{S_i}(a^+_{g,i} + a_{g,i})] \tag{4.77}$$

The moment M_n thus becomes

$$M_n = Av_{n_i} \langle \prod_i \chi_{g,n_i} | \left\{h\nu_o + H_g + \sum_i [(S_i - n_i - \tfrac{1}{2})h\omega_i - \sqrt{S_i}(a^+_{g,i} + a_{g,i})\hbar\omega_i]\right\}^n | \prod_i \chi_{g,n_i} \rangle \quad . \tag{4.78}$$

From (4.78) the evaluation of M_o, M_1 and M_2 is an easy task. M_o is automatically obtained equal to unity. The term $a^+_{g,i} + a_{g,i}$ does not contribute to M_1 and H_g acting on $|\prod_i \chi_{g,n_i}\rangle$ gives $\sum_i (n_i + \frac{1}{2})\hbar\omega_i$. Thus, we find the following general result

$$M_1 = h\nu_o + \sum_i S_i \hbar\omega_i \quad . \tag{4.79}$$

In calculating M_2 care should be taken that H_q does not commute with $a^+_{g,i} + a_{g,i}$. However, there are only two terms of this kind, $H_g(a^+_{g,i} + a_{g,i})$ and $(a^+_{g,i} + a_{g,i})H_g$, for which H_g can again be replaced by $\sum_i(n_i + \frac{1}{2})\hbar\omega_i$. We can thus write

$$M_2 = \mathrm{Av}_{n_i} \langle \prod_i \chi_{g,n_i} | \left[h\nu_o + \sum_i S_i\hbar\omega_i - \sum_i \sqrt{S_i}(a^+_{g,i} + a_{g,i})\hbar\omega_i \right]^2 | \prod_i \chi_{g,n_i} \rangle \quad . \tag{4.80}$$

All terms linear in $a^+_{g,i}$ and $a_{g,i}$ give no contribution, so that M_2 can be rewritten

$$M_2 = \left(h\nu_o + \sum_i S_i\hbar\omega_i \right)^2 + \mathrm{Av}_{n_i} \langle \prod_i \chi_{g,n_i} | \left[\sum_i \sqrt{S_i}(a^+_{g,i} + a_{g,i})\hbar\omega_i \right]^2 | \prod_i \chi_{g,n_i} \rangle \tag{4.81}$$

which can obviously be transformed to

$$M_2 = M_1^2 + \mathrm{Av}_{n_i} \langle \prod_i \chi_{g,n_i} | \sum_i S_i(a^+_{g,i}a_{g,i} + a_{g,i}a^+_{g,i})(\hbar\omega_i)^2 | \prod_i \chi_{g,n_i} \rangle . \tag{4.82}$$

From (4.17a,b) this finally gives

$$M_2 = M_1^2 + \sum_i (2\bar{n}_i + 1)S_i(\hbar\omega_i)^2 \quad . \tag{4.83}$$

If we replace the ω_i by an average value ω as in Sect.4.3.6, we obtain very simple expressions

$$M_1 = S\hbar\omega$$
$$M_2 - M_1^2 = S(\hbar\omega)^2(2\bar{n} + 1) \quad , \tag{4.84}$$

where S is the sum of the S_i. In the strong coupling limit a Gaussian curve adjusted to these two moments gives the result of (4.54) and at high temperature the result of (4.54). In the same limit this result can be generalized to a Gaussian adjusted to the exact moments given by (4.79,83).

In the general case we can compare the experimental values of the lower order moments to the theoretical formulas and see if they satisfy, for instance, a simple model with an average frequency. The comparison can also serve as a test of the internal consistency of the model. The method of moments is also interesting in cases of orbital degeneracy as encountered in the Jahn-Teller effect, to which case (4.79) and (4.83) have been generalized [4.13-14].

4.4 Optical Cross Section

4.4.1 Theoretical Models

The cross section for the optical transition between a localized state and a band can be written from (4.5) as

$$\sigma_e^o(h\nu) = \frac{\beta}{h\nu} |\langle\psi_{b,\underline{k}}| (-i\hbar \tfrac{\partial}{\partial z})|\psi_T\rangle|^2 \rho_b(E_{\underline{k}}) \quad , \tag{4.85}$$

where $\psi_{b,\underline{k}}$ is the state in band b for wave vector $\underline{k}$, $\rho_b(E_{\underline{k}})$ the corresponding density of states, ψ_T and E_T are the wave function and energy of the localized state. Conservation of energy requires that

$$h\nu = E_{\underline{k}} - E_T \quad . \tag{4.86}$$

When the wave function ψ_T is written in terms of the Bloch functions of the perfect crystal, we have

$$\psi_T = \sum_{b',\underline{k}'} A_{b',\underline{k}'}\psi_{b',\underline{k}'} \quad . \tag{4.87}$$

To simplify, let us assume that only one band b' contributes to the expansion of ψ_T. The cross section then becomes

$$\sigma_e^o(h\nu) = \frac{\beta}{h\nu} |A_{b',\underline{k}}|^2 |\langle\psi_{b,\underline{k}}|(-i\hbar \tfrac{\partial}{\partial z})|\psi_{b',\underline{k}}\rangle|^2 \rho_b(E_{\underline{k}}) \tag{4.88}$$

since there is conservation of the wave vector $\underline{k}$ ($\partial/\partial z$ cannot mix states for which $\underline{k} \neq \underline{k}'$).

In the literature, the coefficients $A_{b',\underline{k}}$ have been calculated by substituting plane waves to Bloch functions. Then $A_{b',\underline{k}}$ is the Fourier transform of ψ_T. In many of the models used, e.g. [4.15], ψ_T is taken to be of the form $\exp(-\alpha r)$ or $r^{-1}\exp(-\alpha r)$. If we choose the second form, the calculation of $A_{b',\underline{k}}$ reduces to the computation of the integral

$$\int r^{-1} \exp(-\alpha r) \frac{\exp(i\underline{k}\cdot\underline{r})}{\sqrt{V}} dV = \frac{1}{\sqrt{V}} \frac{4\pi}{\alpha^2 + k^2} \tag{4.89}$$

where V is the volume of the crystal. This gives

$$|A_{b',\underline{k}}|^2 = \frac{1}{V} \frac{(4\pi)^2}{(\alpha^2 + k^2)^2} \quad . \tag{4.90}$$

The matrix element is taken to be constant or proportional to k. The reason is that

$$<\psi_{b,k}|(-i\hbar \frac{\partial}{\partial z})|\psi_{b',\underline{k}}> = -i\hbar \int u^*_{b,\underline{k}} \frac{\partial}{\partial z} u_{b',\underline{k}}\, dV + \hbar k_z \int u^*_{b,\underline{k}} u_{b',\underline{k}}\, dV \quad , \tag{4.91}$$

where $u_{b,\underline{k}}$ is the periodic part of the Bloch function $[\psi_{b,\underline{k}} = u_{b,\underline{k}} \exp(i\underline{k}\cdot\underline{r})]$. Thus, when the transition occurs with the band from which ψ_T is built (i.e., b = b'), the first integral is equal to zero because of symmetry. When the transition occurs with a different band (b ≠ b') the second integral is zero since the initial and final states have different symmetries. Such assumptions are, of course, severe, especially for the transitions involving states far from the bottom of the band, i.e., in the region of the spectrum far from the optical threshold. However they give an indication about the behavior of the transition matrix element and they are the most currently used hypotheses in analytical models.

The result is an optical cross section having the following form [with $\rho(E_{\underline{k}}) \propto V(m^*)^{3/2}(E_{\underline{k}} - E_o)^{1/2}$]

$$\sigma_e^o \propto \frac{1}{h\nu} \frac{1}{(\alpha^2 + k^2)^2} (m^*)^{3/2}(h\nu - E_o)^{1/2} \tag{4.92}$$

for a constant matrix element, and

$$\sigma_e^o \propto \frac{1}{h\nu} \frac{k_z^2}{(\alpha^2 + k^2)^2} (m^*)^{3/2}(h\nu - E_o)^{1/2} \tag{4.93}$$

for a matrix element proportional to k_z. In these expressions m^* is the effective mass of the electron in the band. When the transition occurs with several bands, σ_e^o must be summed over these i bands, assuming the spectrum is the superposition of the spectra associated with each band

$$\sigma_e^o = \sum_i \beta_i \sigma_e^i(m_i^*, E_o^i) \quad . \tag{4.94}$$

The coefficient β_i is there to account for the corresponding oscillator strengths of the various transitions.

Since E_o^i can be obtained independently, an optical spectrum can be fitted to this expression, once corrected to account for the electron-phonon coupling (by convolution with the line-shape function of Sect.4.3, usually taken

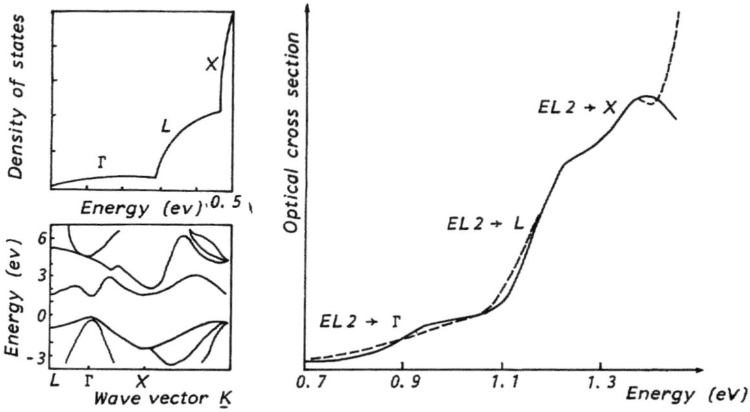

Fig.4.4. Band structure of GaAs, density of states after [4.19] and optical spectrum of the so-called "EL2" center (E_0 = 0.86 eV) in GaAs obtained [4.24] by the DLOS technique (Sect.4.4.3). The values of the parameters used to obtain a good theoretical fit (full line) with experimental data (dashed line) are: α^{-1} = 5Å, d_{FC} = 0.12 eV, β_L/β_Γ = 0.25, β_X/β_Γ = 0.1

as a Gaussian) using three independent parameters: 1) the respective oscillator strengths β_i of the transitions towards the different extrema of the i bands; 2) the Franck-Condon parameter $S\hbar\omega$; 3) the extension in space α of the wave function associated with the localized level. Figure 4.4 gives a good example of such fit for an electron trap situated at 0.8 eV below the conduction band in n-type GaAs. The defect associated with this level, labelled EL2 [4.16] or sometimes O [4.17] or A [4.18], commonly found in materials grown by different methods was generally attributed to the presence of oxygen [4.19]. It has been the subject of various studies (pressure effect [4.20], electrical field effect [4.21], alloying effect [4.22], in addition to optical studies [4.23]). It can be seen [4.24] from Fig.4.4 that the relative weights β_i of the oscillator strengths of the transition from level (EL2) towards the minima X, L and Γ of the conduction band can be easily determined since the contributions of these three transitions are well separated at low temperature. Moreover, because the width of the spectrum is sensitive only to the parameter α, this parameter can be obtained by the adjustment of the slope of the spectrum. Finally, because of the general shape of the spectrum, the Franck-Condon parameter d_{FC} is also obtained independently from the energy position of the threshold of the transition. The values derived for d_{FC} (120 meV) and E_0 are in agreement with photoluminescence studies. The luminescence spectrum is a structureless broadband [4.20] which peaks at 0.645 eV. The width at half height W_0 at 4 K is 130 meV.

From the expression of W_o and d_{FC}

$$W_o = \hbar\omega\sqrt{8S \ln 2} \tag{4.95}$$

and

$$d_{FC} = S\hbar\omega \tag{4.96}$$

we deduce $S \sim 5.5$ and $\hbar\omega \sim 20$ meV in agreement with phonon-assisted tunneling observations (Sect.6.5.3). The energy position E_ℓ agrees with a recombination occuring with the valence band since $E_\ell \sim E_G - E_o$ (the band gap is 1.49 eV).

4.4.2 Exact Expression for the Case of a Delta-Function Potential

It is instructive to know the analytical behavior of $\sigma_e^o(h\nu)$ in simple cases, as was done by LUCOVSKY [4.25]. In his work he compared some experimental results with theory for the hydrogenic model and the delta-function potential. In this last case, simulating the conduction-band states by free-electron states of effective mass m^*, the normalized ground-state wave function is

$$\psi_T(r) = \left(\frac{\alpha}{2\pi}\right)^{\frac{1}{2}} \frac{\exp(-\alpha r)}{r} \quad , \tag{4.97}$$

where α is related to the binding energy E_I of the impurity by

$$E_I = \frac{\hbar^2}{2m^*} \alpha^2 \quad . \tag{4.98}$$

We calculate $\sigma_e^o(h\nu)$ from (4.85). We consider here that the conduction-band states are plane waves and calculate the matrix element $\langle\psi_T|(-i\hbar \frac{\partial}{\partial z})|\psi_{\underline{k}}\rangle$ as

$$\langle\psi_T|(-i\hbar \frac{\partial}{\partial z})|\psi_{\underline{k}}\rangle = \sqrt{\frac{\alpha}{2\pi}} \int \frac{\exp(-\alpha r)}{r} (\hbar k_z)\, e^{i\underline{k}\cdot\underline{r}}\, d^3\underline{r} \quad , \tag{4.99}$$

which can be transformed to

$$\langle\psi_T|(-i\hbar \frac{\partial}{\partial z})|\psi_{\underline{k}}\rangle = \frac{2\sqrt{2\pi\alpha}}{\alpha^2 + k^2} \hbar k_z \quad . \tag{4.100}$$

This can be injected into (4.85) where $\rho(E_k)$ can be replaced by its expression $(2\pi^2)^{-1}(2m^*/\hbar^2)^{3/2}E_k^{1/2}$ (including spin degeneracy). Noticing that $E_{\underline{k}}$ is equal to $h\nu - E_I$, we get

$$\sigma_e^o(h\nu) = \frac{16\pi}{3}\frac{e^2\hbar}{m^*c}\frac{E_I^{1/2}(h\nu - E_I)^{3/2}}{(h\nu)^3} \tag{4.101}$$

which is identical to the exact formula derived in [4.26]. This differs from the hydrogenic case in that it reaches a maximum for $h\nu = 2E_I$ (instead of $h\nu = E_I$) and falls off approximately as $(h\nu)^{-3/2}$ for large ν, instead of $(h\nu)^{-3}$. This is in agreement with some experimental results concerning deep centers but not for more loosely bound centers [4.25].

4.4.3 Measurement

The optical cross section σ^o is usually derived from absorption measurements. The coefficient of absorption μ, defined by the change dI of the photon intensity I due to the absorption in a thickness de of the sample

$$dI = -\mu\, I\, de \quad , \tag{4.102}$$

directly gives σ^o, once the concentration of centers N_T, which give rise to the absorption, is known

$$\mu = N_T\sigma^o \quad . \tag{4.103}$$

At a given photon energy the absorption coefficient is measured by transmission or reflection. The transmittance T, ratio of the transmitted energy I to the incident energy I_o for a sample of reflectivity R is

$$T = \frac{I}{I_o} = \frac{(1 - R)^2 \exp(-\mu e)}{1 - R^2 \exp(-2\mu e)} \quad . \tag{4.104}$$

This expression is readily derived: the radiation transmitted through the first face of the sample is $(1 - R)I_o$; it is $(1 - R)I_o \exp(-\mu e)$ when it reaches the opposite face and the fraction which emerges is $(1 - R)^2 \exp(-\mu e)$. The denominator in (4.104) comes from the fact that there are multiple internal reflections. As soon as μe is large enough, μ is given by the measurement of T, since (4.104) reduces to

$$T \sim (1 - R)^2 \exp(-\mu e) \tag{4.105}$$

(for $\mu e \gtrsim 1.2$ the error is less than 1%). When the reflectivity R is unknown μ is simply obtained by measuring the transmittances T_1 and T_2 of two samples of different thicknesses e_1 and e_2

$$\frac{T_1}{T_2} = \exp[\mu(e_2 - e_1)] \quad . \tag{4.106}$$

The transmittance

$$R = \left(\frac{n-1}{n+1}\right)^2 \qquad (4.107)$$

allows the determination of the refractive index n [Ref.1.1, Sect.5.3] from which μ is determined using the Kramers-Kronig relations (the extinction coefficient k is zero since the measurements are performed in samples containing practically no free carriers, i.e., of zero conductivity).

For reflectance or absorption measurements it is best to place the sample at the exit of the slits of the monochromator (which selects the desired photon energy) in order to avoid simultaneous excitation of other transitions, at energies larger than the energy of interest. (For technical details on monochromators, spectrometers, radiation sources and detectors see [4.4].) The main difficulty encountered in the determination of an optical cross section using absorption techniques is that, usually, the semiconductor contains more than one type of defect and the cross section which is obtained is the sum of the cross sections due to all centers whose energy levels lie in the gap, at a depth smaller or equal to the energy of the photon excitation. However, there is another type of experiment which is selective, i.e., able to give the optical cross section of each individual defect.

This technique is based on the observation of the kinetics associated with the return to steady state (after an external perturbation) of the electronic population of the defect level, under optical excitation. The evolution of the carrier population b(t) [Ref.1.1, Sect.6.2.3] of the defect (in total concentration N_T), after the establishment of nonequilibrium initial conditions, is described by

$$\frac{db(t)}{dt} = -G_e b(t) + G_h[N_T - b(t)] \quad , \qquad (4.108)$$

where G_e and G_h are the emission rates of electrons and holes from the level to the conduction and valence bands, respectively [Ref.1.1, Sect.7.3.2] and Chap.6. This equation does not contain any capture terms because the measurement of b(t) requires that the defects be placed in the space charge region of a junction region, which do not contain any free carriers (Sect.6.2). The solution of (4.108) is

$$b(t) = \left[b(0) - \frac{G_h}{G_e + G_h} N_T\right] \exp\left[-(G_e + G_h)t\right] + \frac{G_h}{G_e + G_h} N_T \quad , \qquad (4.109)$$

where b(0) is the concentration of the centers populated at t = 0. We shall show in Sect.6.2 that b(t) can be measured directly using capacitance techniques and that, consequently, it is possible to determine N_T, G_e and G_h. In Sect.6.2 the emission rates considered are thermal emission rates g_e and g_h, i.e., rates associated with the return to equilibrium under thermal excitation. Under optical excitation by a flux $\phi(h\nu)$, the optical emission rates $\sigma_e^o\phi$ and $\sigma_h^o\phi$ must be added to the thermal rates g_e and g_h, and b(t) is given by (4.109) where

$$G_e = g_e + \sigma_e^o\phi \tag{4.110}$$

$$G_h = g_h + \sigma_h^o\phi \quad . \tag{4.111}$$

Thus, the determination of an optical cross section consists in extracting σ_e^o or σ_h^o for varying photon energies from the kinetics b(t). The experiment is performed at a low enough temperature so that g_e and g_h are negligible. Depending on the initial condition b(0), the kinetics correspond to filling or emptying by the illumination (Fig.4.5), with a time constant $(\sigma_e^o + \sigma_h^o)^{-1}\phi^{-1}$ and a final value

$$b(\infty) = \frac{\sigma_h^o}{\sigma_e^o + \sigma_h^o} N_T \quad . \tag{4.112}$$

The method of extraction consists in the analysis of the transient itself or of the quasi-equilibrium situation $b(\infty)$. Measurement of the change in the population immediately after the application of the optical excitation, i.e., of the quantity

$$\left(\frac{db}{dt}\right)_{t=0} = -(g_e + \sigma_e^o\phi)b(0) + (g_h + \sigma_h^o\phi)[N_T - b(0)] \quad , \tag{4.113}$$

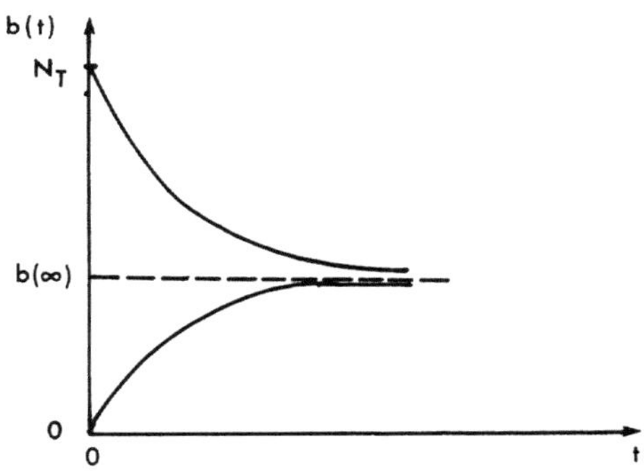

Fig.4.5. Evolution of a defect occupancy under the application of an optical excitation, starting from empty or filled levels

is particularly powerful. Indeed, it is possible, using initial conditions such that b(0) = 0 and b(0) = N_T, to eliminate the transition to the conduction and valence bands respectively, and to obtain independently σ_e^o and σ_h^o since, at low temperature where g_e and g_h can be neglected,

$$\left(\frac{db}{dt}\right)_{t=0} = \sigma_h^o \phi N_T \qquad (4.114)$$

when

$$b(0) = 0 \qquad (4.115)$$

and

$$\left(\frac{db}{dt}\right)_{t=0} = -\sigma_e^o \phi N_T \qquad (4.116)$$

when

$$b(0) = N_T \quad . \qquad (4.117)$$

The measurement and analysis of the transient b(t), from which (db/dt)t=0 is deduced, is described in Sect.6.2.5. The experimental difficulty proper to the case of optically excited transitions lies in the fact that it is necessary, in order to obtain reasonable emission rates, to have a large flux of monochromatric photons.

The main interest of this method over conventional methods such as absorption, photocapacitance, photoconductivity (Chap.7) comes from the fact that it allows the determination of σ_e^o and σ_h^o for a localized level, independently of any others. The reason is that each level has its own emission rates and it is possible to separate various emission rates using correlation techniques (Sect.6.2.5).

4.5 An Example. The GR Absorption Band in Diamond

4.5.1 Experimental Situation

The shape of the absorption or emission spectrum associated with a defect depends on the character of the associated localized level. Since we have already briefly discussed the case of impurities which introduce shallow levels and the case of donor-acceptor pairs [Ref.1.1, Sect.2.4], we describe here an example corresponding to a defect which gives rise to a deep level.

The so-called GR absorption spectrum, associated with a defect introduced by electron irradiation in diamond, is attributed to the vacancy. There is no direct evidence for this identification because this defect has not (yet) been detected by electron paramagnetic resonance (which would be the only technique able to provide a firm identification, see Chap.3).

The absorption spectrum associated with the GR center (Fig.4.6) is composed of several groups of bands called GR1 to GR8. Because the relative intensities of these groups of lines are always the same, whatever the dose and energy of irradiation, i.e., they do not depend on defect concentration, it is concluded [4.27] that the GR1-8 lines are all associated with transitions at the same GR defect. From the study of the variation of the different lines (labeled A,B,C etc.) which compose the GR1 system versus temperature, it is easy to see [4.28] that this system corresponds to two zero phonon transitions at 1.665 and 1.673 eV. (The A and B lines, respectively, Fig. 4.7.) The temperature dependence of the intensity of the 1.665 eV line indicates that the population of the corresponding states depends on the thermal promotion of electrons from a lower state lying precisely at 1.673 eV - 1.665 eV = 0,08 eV above a ground state which thus should be the 1.673 eV state. This leads to the energy diagram of Fig.4.8.

The possibility that there are other transitions allowed in this energy range, which could not be seen in absorption because the corresponding levels are unoccupied, is ruled out by the fact that no zero phonon lines, other than the A and B lines, are observed in photoluminescence [4.29]. The symmetry of the GR defect has been studied using optical absorption with polarized light, under uniaxial stress. The stress-splitting patterns [4.30] for the A and B lines are schematized in Fig.4.9 (which in fact gives the theoretical predictions, the experimental results being very close to them).

A stress-splitting pattern gives the energy position of the π and σ polarized components (Sect.4.1) with their relative intensities. The line intensities depend on the applied stress as well as on the temperature at which the measurements are recorded. A normalization to given stress and temperature is thus necessary, before such pattern is drawn, if the results obtained in various conditions are to be compared.

These two stress-splitting patterns correspond respectively to A → T (the A line) and E → T (the B line) transitions in Td symmetry. Indeed, consider a defect of Td symmetry. Electronic transitions are allowed between states having the T,A and E symmetries [Ref.1.1, Sect.1.4.4].

The fact that only one π and one σ components of the A line under <100> or <111> stresses are observed and the fact that they appear at different energies implies that an A → T transition is involved. This can easily be checked using [Ref.1.1, Table 1.7], which gives the matrix elements and the

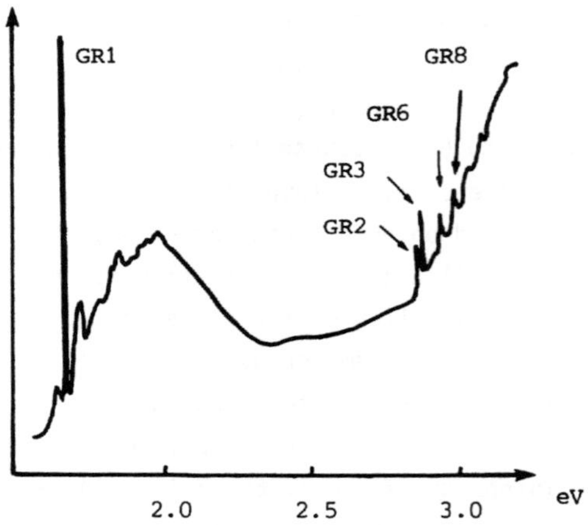

◄ Fig.4.6. The absorption spectrum of the GR center in diamond

Fig.4.7. Absorption and luminescence spectra of the GR1 system

Fig.4.8. Energy diagram of the GR1 system, showing the transitions involved in absorption

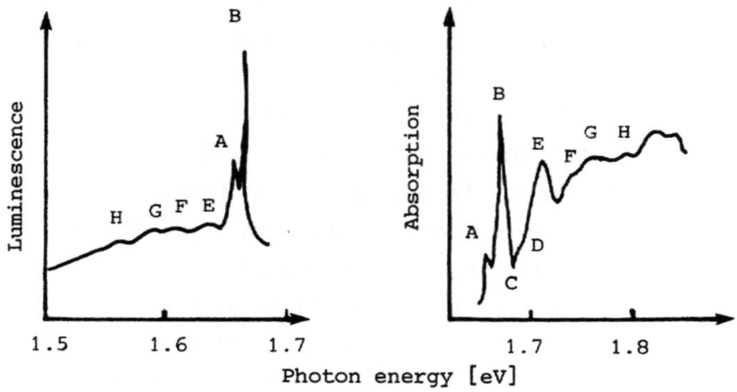

Fig.4.7

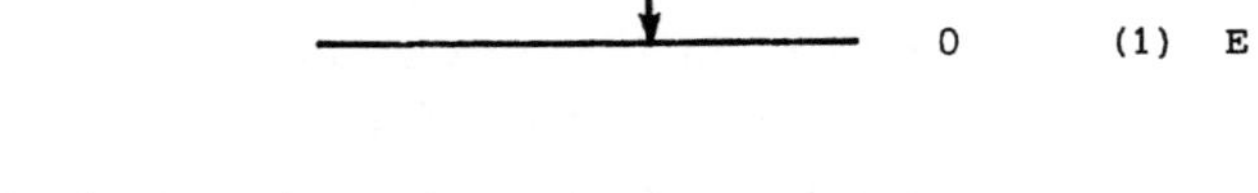

Fig.4.8

basis functions of a T_2 state under <100>, <111> and <110> stresses. Consider the <100> stress; the T_2 level is then split into two components (Fig.4.9), one singly degenerate at energy A + 2B (see notations in [Ref.1.1,Sect.1.5]) of x symmetry, and one doubly degenerate at energy A - B, of y,z symmetry. Transition from a (nondegenerate) A state to one of the split T_2 states

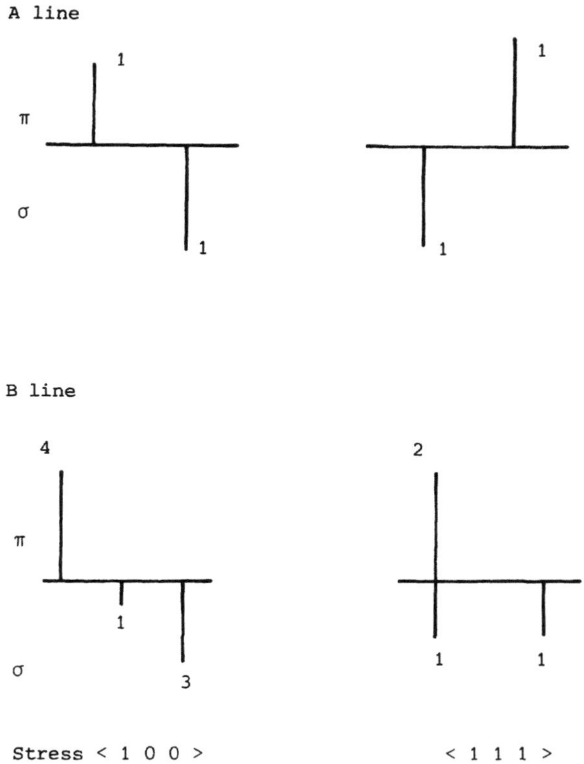

Fig.4.9. Stress pattern obtained for the A and B lines of the GR1 system

occurs i) with the state of x symmetry for π polarized light E = E(1,0,0); ii) with a state of y or z symmetry for σ polarized light (E = E(0,1,0) or E(0,0,1)). Thus, the stress-splitting pattern is, as shown in Fig.4.9, composed of one σ and one π line separated by an energy 3B.

It can be readily seen that the stress-splitting pattern is also the same under <111> stress since the T_2 splitting is the same (the energy difference between the π and σ lines is now C). Since the defect is of T_d symmetry, the ground state corresponding to level 1 in Fig.4.8 is of A, E or T symmetry. It cannot be of A symmetry (which is nondegenerate) since this could not explain the more complex stress-splitting pattern of the B line (Fig.4.9). This leaves only E or T symmetry (the same being true for level 3 of Fig.4.8). A first conclusion from this is that since the A line is due to a A → T transition, then it is level 2 which must be of A symmetry.

To have more information on levels 1 and 3 we must consider in greater detail the splitting pattern for a <100> stress. In this case E and T states reduce to [Ref.1.1, Sect.1.4.4]

$$E \rightarrow A_1 + B_1$$

$$T_1 \rightarrow A_2 + E$$

$$T_2 \rightarrow B_2 + E \quad . \tag{4.118}$$

The energy of the line is found to have a nonlinear variation with the amplitude of the applied stress. This means that there is an interaction between the stress-split components of levels 1 and 2. Such interaction can occur between states having the same symmetry only. Since level 2 which is of A symmetry (A_1 or A_2), transforms either to A_1 or B_1 under a <100> stress then this interaction is possible only if level 2 is of E symmetry (which also gives A_1 or B_1). This finally leaves T symmetry for level 3.

The experimental results leave an uncertainty since we can obtain for levels 1,2,3 the following sequences: either E, A_1, T_2 or E, A_2, T_1, both sets leading to similar conclusions. We shall now see that theory can provide an explanation for such sequences and give some arguments in favor of E, A_1, T_2.

4.5.2 Theoretical Interpretation

We base our analysis on the assumption that the GR1 band could be due to an optical transition at the neutral vacancy V^o (there are a variety of arguments in favor of such an assumption [4.31]). As discussed in Ref.1.1, Sect.4.2.2 , the three lower many-electron states of V^o should be ^1E, $^3T_1$ and $^1T_2$ (this was the original finding by Coulson and Kearsley confirmed recently by calculations for V^o in silicon [4.32]). Following this result, the GR1 should originate from the electronic transition $^1E \rightarrow {}^1T_2$. This finding corresponds to Coulson and Kearsley's original proposal but can explain the situation only for the B line, giving an E state for level 1 and a T_2 state for level 3. It does not provide any explanation for the A line (which should then correspond to an $A_1 \rightarrow T_2$ transition, the origin of the A_1 level being unexplained).

An explanation for this was proposed a long time ago by LANNOO and STONEHAM [4.14] in terms of a dynamic Jahn-Teller effect in the ground state. We have seen in Sects.2.3.1,2 that a dynamic Jahn-Teller effect corresponding to coupling of an E state with tetragonal distortion modes effectively leads to an E vibronic ground state and an A_1 first excited state. In the static limit we have seen that the E-A_1 splitting was equal to 3γ with γ given by (2.85), i.e.,

$$\gamma = \frac{\hbar\omega_\theta}{4\pi} \exp\left(-\frac{4E^\theta_{JT}}{\hbar\omega_\theta}\right) \tag{4.119}$$

where E^θ_{JT} is half the barrier for rotational motion and ω_θ the frequency of the soft mode associated with the motion in one of the equivalent minima in θ. Thus the E-A_1 splitting is a unique function of the two parameters E^θ_{JT} and ω_θ. The small experimental value of $3\gamma \sim 8$ meV is a guarantee that we are effectively in the strong coupling limit.

Confirmation of this model has been recently given by STONEHAM and others (see [4.33] for a recent review). From the experimental values of the stress-splitting coefficients it is possible to get an idea of all the parameters characterizing the Jahn-Teller effect in the E ground state. We do not present the detailed arguments [4.31] here but simply reproduce the essential numbers:

a) The barrier height for motion in θ is 0.026 eV
b) The energy $\hbar\omega_\theta$ is 0.036 eV
c) The Jahn-Teller energy (Sect.2.3.2) E_{JT} is 0.63 eV
d) The main phonon frequency $\hbar\omega$ is 0.131 eV.

These numerical values show the complete internal consistency of the model. First, the barrier height for rotation is much smaller than the Jahn-Teller energy. Second, $\hbar\omega_\theta$ is also much smaller than $\hbar\omega$, showing that the motion in θ is effectively softer. Finally the phonon energy is found to be smaller than the Raman frequency 0.165 eV, which is in agreement with the theoretical estimations [4.34]. Other conclusions also emerge: i) the large value found for $E_{JT}/\hbar\omega$ leads to a Ham reduction factor, given by (2.98), of the order of 10^{-3} ruling out the possibility that distinct electronic states are involved; ii) the next higher vibronic levels are only 12 meV higher, i.e., $\sim$20 meV above the ground state, probably enough to explain why they are not observed.

It is thus reasonable to think that GR1 is associated with the neutral vacancy V^0, the two lowest E and A_1 states being derived from the ^1E electronic ground state, the A and B lines resulting from transitions between these and the $^1T_2$ excited state. Other arguments based on line-shape theory [4.14] tend to confirm this view, the theoretical predictions for the Huang-Rhys factor and the effective frequency being in good agreement with experimental data.

5. Electrical Properties

In this and the following chapter, we consider all aspects of the interaction between charged carriers (electrons and holes) which are in the conduction and valence bands and those localized on defect levels when only phonons are involved. Interaction with photons is treated in Chap.4. As in the other chapters, this interaction is studied in order to investigate information on defects obtained from the behavior of the free carriers.

The carriers available at a given temperature in a semiconductor are distributed in the bands and on the defect sites. At equilibrium, the carrier distribution is reached through trapping and emission of carriers between defect levels and bands. The equilibrium distribution results from a temporal average: a defect level which is occupied by a carrier a fraction α of the time, because of successive emissions and captures, has an occupancy α. As a result, the free-carrier concentration, i.e., the concentration of carriers in a band, reflects the concentration of defects. Its evolution with temperature gives (in simple cases) the defect concentrations and the associated ionization energies, i.e., the positions of the energy levels of the defects in the forbidden gap. Emission and capture processes, which can be directly observed by studying the return to equilibrium of a perturbed distribution and, therefore, allow the measurement of energy levels, cross sections and entropies of ionization, will be considered in the next chapter. In the present chapter we consider only the distribution of carriers between bands and defect levels at equilibrium and free-carrier mobility, which involves defects since they play the role of scattering centers.

5.1 Carrier Distribution Between Bands and Defect Levels

5.1.1 Intrinsic Semiconductor

We first briefly recall the way the concentration of carriers in a band is derived in the case of a semiconductor not containing any defect or impur y [5.1]. At temperature T = 0 K, the valence band is filled and the conduction band empty. For T above 0 K, electrons are excited above the band gap: there are n electrons in the conduction band and p holes in the valence band, in equal concentration n_i

$$n = p = n_i \quad . \tag{5.1}$$

The probability f of finding an electron at an energy E in the conduction band is given by the Fermi-Dirac statistics

$$f(E) = \left[g + \exp\left(\frac{E - E_F}{kT}\right)\right]^{-1} \tag{5.2}$$

where g is the so-called degeneracy factor (Sect.5.1.3) and E_F the Fermi level. Thus, the electron concentration given by

$$n = 2\int_{E_c}^{\infty} N_c(E)f(E)\,dE \quad , \tag{5.3}$$

where $N_c(E)$ is the density of states in the conduction band (the factor two expresses the fact that a level is spin degenerated) can be written

$$n = 2\exp\left[-\frac{(E_c - E_F)}{kT}\right]\int_{E_c}^{\infty} N_c(E)\exp\left[-\frac{(E - E_c)}{kT}\right]dE \quad , \tag{5.4}$$

when f(E) reduces to

$$f(E) = \exp -\left(\frac{E - E_F}{kT}\right) \quad , \tag{5.5}$$

valid for a nondegenerate semiconductor for which $(E_c - E_F)/kT \gg 1$. Calculation of (5.4) leads to

$$n = N_c \exp -\left(\frac{E_c - E_F}{kT}\right) \tag{5.6}$$

with

$$N_c = \frac{1}{\sqrt{2}}\left(\frac{m_e^* kT}{\pi\hbar^2}\right)^{3/2} \quad , \tag{5.7}$$

because the density of states per unit volume $N_c(E)$, for energies above the bottom E_c of the conduction band, reduces to the density of states of free electrons (replacing the electron mass by the effective mass m_e^*) in the region of energy where the exponential term in the integral has a nonnegligible value (i.e., a few kT above E_c), and E can be approximated by

$$E = E_c + \frac{\hbar^2 |\underline{k}|^2}{2m_e^*} . \tag{5.8}$$

Similarly the hole concentration in the valence band is

$$p = N_v \exp\left(\frac{E_v - E_F}{kT}\right) \tag{5.9}$$

with

$$N_v = \frac{1}{\sqrt{2}} \left(\frac{m_h^* \, kT}{\pi\hbar^2}\right)^{3/2} \tag{5.10}$$

where E_v is the energy of the top of the valence band. The Fermi level E_F is obtained using (5.1) which expresses the fact that the system is electrically neutral

$$E_F = \frac{E_c + E_v}{2} + \frac{3}{4} kT \ln \frac{m_h^*}{m_e^*} . \tag{5.11}$$

Usually, the ratio of the effective masses is in the range 0.1-10 and E_F remains close to the middle of the gap.

From the point of view of carrier concentration, we are dealing with a two-level system, one level at E_c containing N_c sites and one level at E_v containing N_v sites. The carrier concentration in these levels is

$$n_i = \sqrt{np} \tag{5.12}$$

that is

$$n_i = \sqrt{N_c N_v} \exp - \left(\frac{E_c - E_v}{2kT}\right) . \tag{5.13}$$

Typically, at T = 300 K, $n_i \sim 10^{10}$ cm^{-3} in silicon and $n_i \sim 10^{13}$ cm^2 in germanium.

In a semiconductor with a single isotropic band, the effective mass has the usual definition (Ref.1.1, Sect.2.2.2). In case the semiconductor contains two isotropic bands (for instance the valence band for silicon, germanium and gallium arsenide), two effective masses m_1^* and m_2^* are defined;

the two bands give additive contributions to the density of states and $m^{*3/2}$ must be replaced by $m_\ell^{*3/2} + m_2^{*3/2}$. In the case of anisotropic bands ($m_\ell^* \neq m_t^*$) it is easy to show that the above simple formalism remains valid when m^* is replaced by $m_\ell m_t^2)^{1/3}$.

5.1.2 Extrinsic Semiconductor

We now consider the case of a semiconductor containing electrically active defects, which introduce energy levels E_T in the forbidden gap. If a defect is neutral when the Fermi level E_F lies above E_T, then it is said to be a donor when, once ionized (i.e., when E_F is below E_T), it becomes positively charged by giving an electron to the conduction band. It is said to be an acceptor when it gives a hole to the valence band (i.e., when it traps an electron).

We treat here the case of a semiconductor containing one donor level at E_D, and one acceptor level at E_A. This is the only interesting case in practice because the shallow level whose concentration is the largest (say, for instance, the donor level) plays the role of the doping level, i.e., provides free carriers to the corresponding band, while all the other levels do not play a role unless they compensate the doping level. Compensating levels are levels lying deeper than the doping level, which trap free carriers. Thus an acceptor level E_A is the compensating level in the case of a donor-doping level.

The populations of carriers in the bands (n and p) and in the donor and acceptor levels (n_D and p_A) are obtained by the application of Fermi-Dirac statistics. The expressions of n and p have been derived in Sect.5.1.1, (5.2,3):

$$n = 2\int_{E_c}^{\infty} \frac{N_c(E)\,dE}{1 + \exp[(E - E_F)/kT]} \tag{5.14}$$

$$p = 2\int_{-\infty}^{E_v} \frac{N_v(E)\,dE}{1 + \exp[(E_F - E)/kT]} \quad . \tag{5.15}$$

For the donor and acceptor, whose concentrations are respectively N_D and N_A, we have similarly:

$$n_D = \frac{g_D N_D}{g_D + \exp[(E_c - E_D - E_F)/kT]} \tag{5.16}$$

and

$$p_A = \frac{g_A N_A}{g_A + \exp[(E_F - E_v - E_A)/kT]} \tag{5.17}$$

where g_D and g_A are the degeneracy factors of the donor and acceptor respectively (see Sect.5.1.3 for a more detailed discussion).

In order to get n, p, n_D and p_A, it is necessary to solve a system of five equations since the concentrations are expressed in terms of another unknown quantity, the Fermi level E_F. The fifth equation is the relation expressing the electrical neutrality (valid only for homogeneous systems with no space charge region):

$$n + n_D - p - p_A = N_D - N_A \quad . \tag{5.18}$$

The solution of this system for a given temperature is not simple and must be performed numerically. However, there are temperature ranges in which approximations can be made, allowing us to obtain simple solutions. We derive these solutions in the particular case of a n-type semiconductor (i.e., $N_D > N_A$) containing single donors and acceptors, in which case: $g_D = 2$ and $g_A = 2$ (Sect.5.1.3). We also assume that $\exp[+(E_c - E_F)/kT]$ and $\exp[-(E_v - E_F)/kT]$ are very large compared to unity so that the semiconductor is non-degenerate (we shall see later the kind of conditions this hypothesis implies). This assumption allows the simplification of (5.14,15). The system to be solved is then

$$n = N_c \exp -\left(\frac{E_c - E_F}{kT}\right) \quad ; \tag{5.19}$$

$$p = N_v \exp\left(\frac{E_v - E_F}{kT}\right) \quad ; \tag{5.20}$$

$$n_D = \frac{N_D}{1 + \frac{1}{2}\exp[(E_c - E_D - E_F)/kT]} \quad ; \tag{5.21}$$

$$p_A = \frac{N_A}{1 + \frac{1}{2}\exp[(E_F - E_v - E_A)/kT]} \quad ; \tag{5.22}$$

$$n + n_D - p - p_A = N_D - N_A \quad . \tag{5.23}$$

We now look for the solution of this system in different temperature ranges:

i) Solution at 0 K. It is obvious that

$$n = p = p_A = 0 \tag{5.24}$$

$$n_D = N_D - N_A \tag{5.25}$$

and

$$E_F = E_c - E_D \quad . \tag{5.26}$$

The holes of the acceptor levels are compensated by the electrons provided by the donors and only N_D-N_A electrons remain on the donors.

ii) Solution at Low Temperature. When the temperature increases slightly from 0 K, the Fermi level remains close to the donor level. As a result p and p_A are negligible compared to n and n_D and we can write:

$$p_A \sim p \sim 0 \quad . \tag{5.27}$$

Eliminating E_F between (5.19) and (5.21), we obtain:

$$n_D = \frac{N_D}{1 + \frac{1}{2} N_c/n \exp - \left(\frac{E_D}{kT}\right)} \tag{5.28}$$

$$\frac{2n(N_D - n_D)}{n_D N_c} = \exp - \left(\frac{E_D}{kT}\right) \quad . \tag{5.29}$$

Then, n_D can then be eliminated using the neutrality condition (with $p = p_A = 0$).

The result is

$$\frac{2n(n + N_A)}{N_c(N_D - N_A - n)} = \exp - \left(\frac{E_D}{kT}\right) \quad . \tag{5.30}$$

This expression can be simplified again, when:

a) the temperature is low enough so that $n \ll N_A$ and $n \ll N_D$-N_A. Then

$$n = \frac{N_c(N_D-N_A)}{2N_A} \exp - \left(\frac{E_D}{kT}\right) \quad ; \tag{5.31}$$

b) the temperature is such that $N_A \ll n$ and $n \ll N_D$-N_A. This happens when the doping concentration is large compared to the concentration of compensating defects ($N_D \gg 2N_A$). Then

$$n = \left[\frac{N_c(N_D-N_A)}{2}\right]^{\frac{1}{2}} \exp - \left(\frac{E_D}{2kT}\right) \quad . \tag{5.32}$$

iii) Solution at High Temperature. When kT becomes of the order of E_D, $\exp(-E_D/kT) \sim 1$ and (5.28) indicate that $n_D \ll N_D$, since we consider only nondegenerate semiconductors for which $n \ll N_c$. All the donors and acceptors are ionized. The system of equations (5.19) to (5.23) reduces to:

$$np = n_i^2$$

that is

$$np = N_c N_v \exp - \left(\frac{E_c - E_v}{kT}\right) \quad , \tag{5.33}$$

and

$$n - p = N_D - N_A \tag{5.34}$$

whose solution is

$$n = \frac{1}{2}\left\{N_D - N_A + [(N_D - N_A)^2 + 4n_i^2]^{\frac{1}{2}}\right\} \quad . \tag{5.35}$$

Usually, the concentration N_D is chosen so that, around room temperature, the semiconductor is effectively doped, i.e., $N_D - N_A$ is large compared to the intrinsic concentration n_i. In that case (5.35) reduces to

$$n \sim N_D - N_A \quad , \tag{5.36}$$

and p, given by

$$p \sim \frac{n_i^2}{N_D - N_A} \tag{5.37}$$

is small compared to $N_D - N_A$.

5.1.3 The Degeneracy Factor

We have used in (5.16,17) statistical expressions for the populations of the simple donors and acceptors, including degeneracy factors g_D and g_A. Such expressions go beyond simple Fermi-Dirac statistics and can be understood from the thermodynamical considerations of [Ref.1.1, Sect.6.2.3]. Let us then consider the single donor impurity. The ratio N_D^+/N_D^0 of the number of ionized donors (D^+) to the number of neutral donors (D^0) is obtained by considering the reaction $D^0 \rightleftharpoons D^+ + e^-$ at thermal equilibrium. This gives

$$\frac{N_D^+}{N_D^0} = \frac{Z(D^+)}{Z(D^0)} \exp\left[-\frac{E(D^+) - E(D^0) + E_F}{kT}\right] \quad . \tag{5.38}$$

In this expression $E(D^0) - E(D^+)$ is equal to $E_c - E_D$, where E_D is the binding energy of the donor electron with respect to the bottom of the conduction band. The quantities $Z(D^+)$ and $Z(D^0)$ are electronic degeneracy factors, which correspond here to spin degeneracy and are $Z(D^+) = 1$ and $Z(D^0) = 2$. The total density N_D of donors being the sum of N_D^+ and N_D^0, we can rewrite

(5.38) under the usual form taken by Fermi-Dirac statistics

$$N_D^0 = \frac{2N_D}{2 + \exp[(E_c - E_D - E_F)/kT]} \tag{5.39}$$

which is completely equivalent to (5.16).

The case of acceptor impurities can be treated exactly in the same manner. Here, we label N_A^- the density of ionized acceptors A^- (which have lost a hole and thus bear a negative charge) and $N_{A_0}^0$ the density of neutral defects A^0. Again, we can write from the reaction $A^0 \rightleftharpoons A^- + h^+$

$$\frac{N_A^-}{N_A^0} = \frac{Z(A^-)}{Z(A^0)} \exp\left[-\frac{E(A^-) - E(A_0) - E_F}{kT}\right] , \tag{5.40}$$

Here $E(A^0) - E(A^-)$ is equal to $- E_v - E_A$ where E_A is the binding energy of the hole with respect to the top of the valence band. The degeneracy factors are again due to spin, and $Z(A^-) = 1$, $Z(A^0) = 2$. This leads to

$$N_A^0 = \frac{2N_A}{2 + \exp[(-E_v - E_A + E_F)/kT]} \tag{5.41}$$

where N_A is the total concentration of acceptors. The generalization [5.2] to orbitally as well as spin-degenerate states can be done along the same lines.

5.1.4 Experimental Determination of Defect Concentration

Knowledge of how the free carrier concentration varies with temperature provides a means to evaluate the concentration of the doping impurity and eventually the concentration of the compensating centers. Consider for instance the example developed in Sect.5.1.2 of an n-type semiconductor ($N_D > N_A$). The relations (5.31,36), which provide n(T) in the various temperature ranges, indicate that the plot of ln(n) versus T^{-1} is composed of straight lines whose slopes are E_D, $E_D/2$, 0 and $(E_c - E_v)/2$ with increasing temperatures (Fig.5.1). In the temperature range where n is constant (when all the uncompensated donor impurities are ionized), its value is just

$$n = N_D - N_A \quad . \tag{5.42}$$

The regime where n is independent of T occurs usually around room temperature or below for most semiconductors since i) the binding energies of the donor and acceptor doping impurities are then of the order of kT [Ref.1.1,

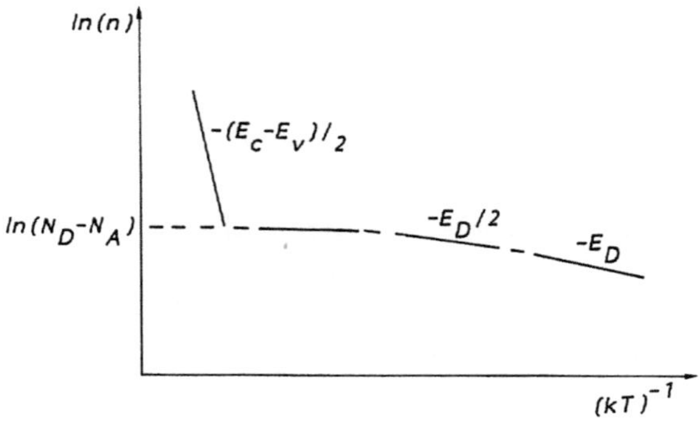

Fig.5.1. Arrhenius plot of the free-carrier concentration in a partially compensated semiconductor. The slopes of the lines indicated give the associated activation energies

Sect.2.1], and ii) the intrinsic carrier concentration ($n_i = 10^{10}$ cm^{-3} and 10^{13} cm^{-3} in Si and Ge, respectively) remains negligible compared to N_D for the lowest doping concentrations used ($\sim 10^{14}$ cm^{-3}) when the material is not intentionally counterdoped. The regime corresponding to the ionization of the donor impurity occurs below ~200 K in silicon [5.3] and below ~100 K in germanium [5.4].

The quantities $(N_D - N_A)$ and E_D can be determined quite accurately. However the evaluation of N_A and N_D separately, which is done by the fit of n(T) with (5.30) is far less accurate. As we shall see in Sect.5.3, it is also possible to get an order of magnitude of the concentration N_A from the carrier mobility. Moreover, the binding energy associated with the compensating defect cannot be obtained. Only the total concentration N_A of the defects, when there are several types of compensating centers, can be obtained.

The measurement of the free-carrier concentration n(T) is made basically using two techniques: capacitance-voltage and Hall effect plus conductivity measurements. The capacitance technique consists in using the variation of the capacitance C of an abrupt junction (or Schottky barrier) versus reverse bias V_b to extract n using the following relation

$$\frac{d(C^{-2})}{dV_b} = \frac{2}{e\varepsilon(N_D - N_A)} \quad . \tag{5.43}$$

Indeed, the application of a reverse bias depletes a region of thickness W (the depleted zone) of the free charges it contains, i.e., of the concentration $(N_D - N_A)$ of ionized impurities which are not compensated. Integration of Poisson's equation

$$\frac{d^2V(x)}{dx^2} = -\frac{\rho(x)}{\varepsilon} \tag{5.44}$$

with

$$\rho(x) = e(N_D - N_A) \tag{5.45}$$

gives the variation of the potential V versus depth x

$$V(x) = \frac{e(N_D - N_A)}{\varepsilon}\left(Wx - \frac{x^2}{2}\right) \quad . \tag{5.46}$$

The voltage at x = W being the reverse bias V_b corrected for the built-in voltage V_i characteristic of the junction, we have

$$V_b + V_i = \frac{e(N_D - N_A)W^2}{2\varepsilon} \tag{5.47}$$

and the capacitance (per unit area) associated with this depleted region

$$C = \frac{\varepsilon}{W} \tag{5.48}$$

is given by

$$C = \left[\frac{e(N_D - N_A)\varepsilon}{2(V_b + V_i)}\right]^{\frac{1}{2}} \quad . \tag{5.49}$$

The only experimental difficulty is to make a reasonably good junction or Schottky barrier. When the junction is not abrupt (i.e., when $N_D - N_A$ varies with x), this can be detected since the C^{-2} versus V plot is not linear.

Conductivity and Hall effect techniques provide, respectively, the quantities (in the simple case of a material containing a single type of carrier)

$$\sigma = ne\mu \tag{5.50}$$

$$R_H = \frac{1}{en} \tag{5.51}$$

from which the free-carrier concentration n and mobility μ are deduced. To obtain the conductivity σ, a known current flux J_x flows through the material in an x direction and the electric field ε_x parallel to x is measured

$$\sigma = \frac{\varepsilon_x}{J_x} \quad . \tag{5.52}$$

To obtain the Hall constant R_H, a magnetic field B_y is applied in a direction y perpendicular to x and the electric field ε_z is measured in the direction perpendicular to both x and y:

$$R_H = \frac{\varepsilon_z}{J_x B_y} \quad . \tag{5.53}$$

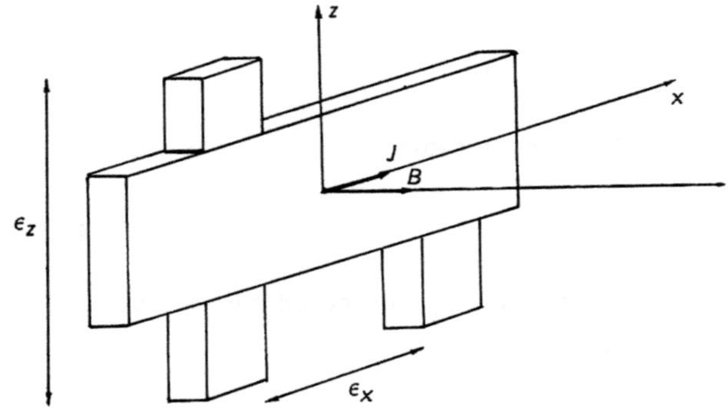

Fig.5.2. Sample geometries for Hall effect measurements

The Hall voltage is due to the action of the Lorentz force $e\underline{v}\wedge\underline{B}$ induced by the magnetic field on the free carrier moving with a velocity $\underline{v}$. Since the sign of this force depends on the sign of the electric charge e, the sign of the electric field ε_z produced indicates the nature, p or n, of the carriers. This electric field is given by

$$\varepsilon_z = -v_x B_y \tag{5.54}$$

i.e.,

$$\varepsilon_z = \frac{J_x}{ne} B_y \quad . \tag{5.55}$$

The above relations are more complicated when two types of carriers (n and p) coexist in the material. They assume that the carrier relaxation time τ (Sect. 5.3.1) is independent of the carrier energy. Otherwise R_H is given by [5.5]

$$R_H = \frac{r}{en} \tag{5.56}$$

with

$$r = \frac{\langle\tau^2\rangle}{\langle\tau\rangle^2} \quad . \tag{5.57}$$

The practical difficulty lies in the realization of electrical contacts, which must be ohmic in the whole temperature range of study. The effect of non-ohmic, i.e., of rectifying contacts, can be reduced by using a method of zero current for voltage measurements (equivalent to the use of an apparatus having an infinite impedance) and a geometry with long arms such as the one depicted in Fig.5.2, to avoid the effect of minority carrier injection.

5.2 Conduction in Case of Defect Interaction

When defects interact, the energy level associated with a particular defect is modified by an amount corresponding to the interaction energy with the other neighboring defects. Because the spatial distribution of defects is not perfectly uniform, the interactions between defects present a certain variation which reflects their distribution. The result is that, instead of being degenerate, the defect levels form a band. An illustration of such a phenomenon has been described in [Ref.1.1, Sect.2.3] for the case of a Coulomb interaction. We now describe the consequences of such interaction on carrier conduction.

5.2.1 Metallic Conduction

In the case of shallow defects, such as donor impurities whose individual localized level (E_D) is close to the conduction band (E_c), the impurity band thus formed can be large enough to overlap with the conduction band. The conduction becomes metallic since the band, impurity band plus conduction band, is now only partially filled. The impurity concentration N_D necessary to reach such metallic conduction can be evaluated by stating that the overlap of the impurity band with the conduction band occurs when the binding energy E_D of the isolated donor is of the order of the average interaction energy.

For a Coulomb interaction, this gives

$$E_D \sim \frac{e^2}{\varepsilon R} \tag{5.58}$$

where R, the mean distance between impurities, is given by

$$\frac{4}{3}\pi R^3 N_D = 1 \quad . \tag{5.59}$$

For a donor impurity in germanium ($\varepsilon = 16$, $E_D \sim 10$ meV), $R \sim 10^{-6}$ cm and the minimum impurity concentration necessary to induce a metallic conduction is $N_D \sim 10^{18}$ cm^{-3}. In a material in which the impurity lies deeper, such as boron in diamond ($E_A \sim 370$ meV), the concentration which would induce a metallic conduction should be considerably higher ($N_A \sim 10^{22}$ cm^{-3}!).

5.2.2 Hopping Conduction

At low enough temperature the carriers cannot be excited in the valence or conduction band and remain frozen on their original sites. The doping impurity band is completely filled unless some impurities are compensated

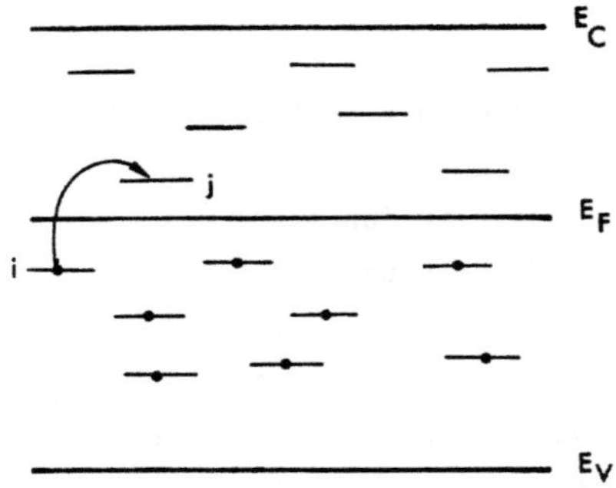

Fig.5.3. Schematic representation of carrier hopping in a partially filled band

by deep defects. In that case, the impurity band is partially filled (up to Fermi level E_F). Conduction is then possible since there are empty sites in which the carriers can hop. As illustrated in Fig.5.3, a carrier located on a site i at energy ε_i just below E_F can hop, due to thermal agitation, in an empty site j at energy ε_j just above E_F. The energies ε_i and ε_j are slightly different since they are the energies of the isolated impurity modified by (variable) interactions with other impurities.

We shall now consider this kind of conduction because it provides information on the concentrations N_D and N_A of impurities and of compensating centers, as well as on the extension in space of the wave function associated with the centers between which hopping takes place. These centers are usually the doping impurities but hopping conduction can, and has been observed to take place between any kind of defects in concentration large enough to give rise to a band [5.6].

a) *Jump Probability*

When the density of defects is low enough, the interaction between defects can be treated in the tight-binding approximation [Ref.1.1, Chap.3]. We can then write the interaction which allows the jump of an electron from site i to site j as

$$V = \langle\varphi_i|H|\varphi_j\rangle \quad . \tag{5.60}$$

The jump probability is obtained using the method of time-dependent perturbation theory. We write for this the wave function of the system as a linear combination of the wave functions φ_i and φ_j corresponding to the electron on sites i and j

$$\psi(t) = c_i(t)\varphi_i + c_j(t)\varphi_j \quad . \tag{5.61}$$

Thus, at time t, $|c_j(t)|^2$ and $|c_i(t)|^2$ are the probabilities to find the electron on sites i and j respectively. The standard perturbation method, applied to the Hamiltonian

$$H = \begin{bmatrix} \varepsilon_i & V \\ V & \varepsilon_j \end{bmatrix} \tag{5.62}$$

with the initial conditions

$$\begin{aligned} c_i(t=0) &= 1 \\ c_j(t=0) &= 0 \end{aligned} \tag{5.63}$$

gives:

$$|c_j(t)|^2 = V^2 \left\{ \frac{\sin[(\varepsilon_i - \varepsilon_j)t/2\hbar]}{\varepsilon_i - \varepsilon_j} \right\}^2 \quad . \tag{5.64}$$

From (5.64) we deduce:

$$\frac{d}{dt} |c_j(t)|^2 = \frac{V^2}{2\hbar} \; \frac{\sin[(\varepsilon_i - \varepsilon_j)t/\hbar]}{\varepsilon_i - \varepsilon_j} \quad . \tag{5.65}$$

For long times it is easy to see, using the equality $\int_0^\infty [\sin(u)/u]du = \pi/2$, that the quantity $d|c_j(E)|^2/dt$ reduces to a Dirac function:

$$\frac{d}{dt} |c_j(t)|^2 = \frac{V^2}{\pi\hbar} \delta(\varepsilon_i - \varepsilon_j) \quad . \tag{5.66}$$

A purely electronic jump is impossible since ε_i is different from ε_j; it occurs only when the difference in energy $w = \varepsilon_j - \varepsilon_i (\varepsilon_j > \varepsilon_i)$ is provided by phonons. Consequently, the jump probability ν_{ij} is proportional to the Boltzman factor exp(-w/kT) and can be written [5.7]

$$\nu_{ij} = CV^2 \exp(-w/kT) \tag{5.67}$$

where C is a constant.

We now have to evaluate V in order to get the complete expression of ν_{ij}. This is done assuming that the defect concentration is low enough so that the carrier interacts only with its initial i and final j sites and not with more distant defects. In that case, the tight-binding matrix element V defined by (5.60) can be rewritten

$$V = \langle \varphi_i | H_i + V_j | \varphi_j \rangle \quad , \tag{5.68}$$

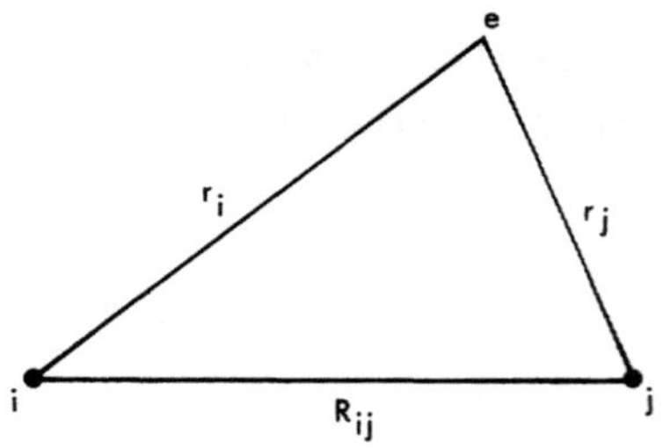

Fig.5.4. Local situation for an electron jumping from site i to j

where H_i is the Hamiltonian of the isolated site i (of eigenstate φ_i and eigenvalue ε_i) and V_j is the attractive potential due to site j equal to $-e^2/r_j$ (using the notations of Fig.5.4). This allows us to rewrite V as

$$V = \varepsilon_i \langle \varphi_i | \varphi_j \rangle + \langle \varphi_i | V_j | \varphi_j \rangle \quad . \tag{5.69}$$

For large distances R_{ij}, the behavior of V is dominated by the product $\varphi_i \varphi_j$ which is exponentially decreasing. Taking φ_i and φ_j as proportional to $\exp(-\alpha r_i)$ and $\exp(-\alpha r_j)$, then this product behaves as $\exp[-\alpha(r_i + r_j)]$ which for very large distances gives

$$\varphi_i \varphi_j \propto \exp(-\alpha R_{ij}) \quad . \tag{5.70}$$

The integrals giving V are products of smooth functions of r_i, r_j so that, in the limit of large R_{ij}, we have

$$V = V_0 \exp(-\alpha R_{ij}) \quad . \tag{5.71}$$

We can thus write ν_{ij} in the form

$$\nu_{ij} = C V_0^2 \exp\left(-\ 2\alpha R - \frac{w}{kT}\right) \tag{5.72}$$

where R is an average value of the distance between defects and V_0 another constant whose value can be calculated.

b) *Hopping Conductivity*

Carriers jumping from one defect site to another induce a net conductivity under the application of an electric field. This type of conductivity is treated in the framework of percolation theory [5.8]. The main contribution to the conductivity is due to carriers which follow paths for which ν_{ij} is maximum, i.e., for which the quantity $2\alpha R + w/kT$ is minimum. There are thus two cases to be considered for the behavior of conductivity versus temperature, depending on the magnitude of w/kT compared to $2\alpha R$. When:

$$\frac{w}{kT} \ll 2\alpha R \quad , \tag{5.73}$$

it is obvious that ν_{ij} is maximum when R is minimum. A carrier jumps from one site to the next nearest one and the conductivity, proportional to exp(-w/kT), is thermally activated. The activation energy w is the energy necessary for a carrier to surmount the (coulombic) potential between the occupied state and the neighboring unoccupied one. This energy, which depends on the compensation, i.e., on the relative concentration of unoccupied sites, has been calculated by MILLER and ABRAHAMS [5.9] and later by SHKLOWSKI et al. [5.10].

When w/kT is not negligible in front of $2\alpha R$, then ν_{ij} is maximum when carriers jump on sites situated at a distance R_m and at an energy w_m such that the quantity $2\alpha R + w/kT$ is minimum for this couple of values. Because carriers jump at various distances R in order to find sites for which w is such that the couple (R,w) is minimum, this hopping mechanism is usually called variable range hopping. This situation is equivalent to a population of carriers moving from site to site in a medium containing a density N_m of sites

$$N_m = \frac{3}{4\pi R_m^3} \quad . \tag{5.74}$$

Since the energy difference w is roughly proportional to the density of the interacting sites, we can write

$$w_m = N_F N_m = \frac{N_F}{4\pi R_m^3/3} \quad . \tag{5.75}$$

The condition that ν_{ij} is maximum can then be written

$$\frac{d}{dR_m}\left(2\alpha R_m + \frac{3N_F}{4\pi R_m^3 kT}\right) = 0 \tag{5.76}$$

which gives

$$R_m = \left(\frac{9N_F}{8\pi\alpha kT}\right)^{1/4} \quad . \tag{5.77}$$

The maximum jump probability takes the simple form

$$\nu = CV_0^2 \exp\left(-\frac{A}{T^{1/4}}\right) \tag{5.78}$$

with

$$A = 2.063\left(\frac{\alpha^3}{kN_F}\right)^{\frac{1}{4}} \quad . \tag{5.79}$$

We now derive the conductivity σ, following a simple argument which is equivalent to the one given by MOTT [5.7]. For this we start from the fact that, with an applied electric field $\underline{E}$, the induced current density $\underline{J}$ is given by

$$\underline{J} = -ne\delta\underline{v} \tag{5.80}$$

where n the carrier density and $\delta\underline{v}$ their change in velocity due to the application of the electric field. For a jump between sites i and j, the velocity $\underline{v}$ is given by the product of the jump frequency ν_{ij} times the vector $\underline{R}_{ij}$ joining the two sites. Thus, from (5.72) we have

$$\underline{v} = \underline{R}_{ij}CV_0^2 \exp\left(-2\alpha R_{ij} - \frac{w}{kT}\right) \quad . \tag{5.81}$$

The application of the electric field changes the quantity w by an amount equal to $-e\underline{E}\cdot\underline{R}_{ij}$. The first-order change $\delta\underline{v}$ in velocity is thus given by

$$\delta\underline{v} = \nu_{ij}\underline{R}_{ij}\left(-\frac{e}{kT}\,\underline{E}\cdot\underline{R}_{ij}\right) \quad . \tag{5.82}$$

We can now estimate $\underline{J}$ by taking an average value for $|\underline{R}_{ij}|$ equal to R_m, performing a spherical average over angles. This leads to

$$\delta\underline{v} = -\frac{e}{kT}\,\frac{\nu_{ij}R_m^2}{3}\,\underline{E} \quad , \tag{5.83}$$

from which we obtain the conductivity σ under the form

$$\sigma = \frac{ne^2}{kT}\,\frac{R_m^2}{3}\,\nu_0 \tag{5.84}$$

where ν_0 takes the form given by (5.72). Finally σ can be written

$$\sigma = \sigma_0 \exp\left(-\frac{A}{T^{\frac{1}{4}}}\right) \tag{5.85}$$

which is the form of MOTT's law [5.7].

5.2.3 Observation of Hopping Conductivity

Hopping conduction can be observed: i) when the temperature is low enough to prevent the carriers from being excited in a band; ii) when the defect (or impurity) concentration is large enough to allow a nonnegligible hopping probability; iii) when the defects are partially compensated so that there are empty sites in which carriers can hop.

Usually, the concentration of impurities cannot be very large before the conduction becomes metallic. In silicon and germanium for instance, as shown in Sect.5.2.1, the conduction is metallic for a donor or acceptor impurity concentration typically of 10^{18} cm^{-3}. Consider for instance the case of a shallow doping impurity in germanium, whose level is at $E_T \sim 10$ meV from a band (i.e., corresponds to a Bohr radius $\alpha^{-1} \sim 50$ Å). For a concentration of 10^{16} cm^{-3}, i.e., for impurities separated by a distance $R \sim 300$ Å, the quantity w/kT (where $w \leq E_T$) always remains small compared to $2\alpha R$ to temperatures as low as 1 K. Consequently, only hopping through nearest neighbors can be observed (Fig.5.5). Hopping conduction has indeed been observed in germanium [5.11] and silicon [5.12], in the range 10-1 K (Fig.5.6). Unfortunately, it is difficult to judge the validity of the hopping mechanism involved because the temperature range in which it is observed is too small. However, one possible way to verify the existence of a given hopping mechanism consists in using irradiation (Chap.8) as a tool to vary the concentration N of the compensating centers [5.14]: a change of N induces a variation of the activation energy w associated with the conductivity which can be calculated once the expression of w (N) has been derived [5.9,10]. For instance the linear variation of w with a dose of irradiation, proportional to N, (Fig.5.7) in cadmium telluride [5.14] indicates that the hopping regime which takes place between 10-4 K is indeed due to hopping between nearest neighbors.

Observation of variable range hopping can be made only in the following case:

a) For shallow impurity states, very low temperatures are necessary so that w/kT does not remain negligible compared to $2\alpha R$. A study of conductivity in an ion-implanted layer in silicon [5.15] shows that the conductivity is due to variable range hopping below ~0.3 K as witnessed by the fact that the conductivity follows Mott's law (Fig.5.8) given by (5.85). In this case, the doping impurities are introduced by implantation, and the compensating centers are the defects produced by the implantation which have not been annealed completely. When the temperature is decreased (below 50 mK) the hopping range R increases (~600 Å in that case), the conduction changes from a three-dimensional regime, characterized by a $T^{-1/4}$ dependence, to a two-dimensional regime, chararacterized by a $T^{-1/3}$ dependence (Fig.5.8). Indeed, (5.75) which relates w_m, R_m and N_F becomes in a two-dimensional regime:

$$w_m = \frac{N_F}{\pi R^2} \quad . \tag{5.86}$$

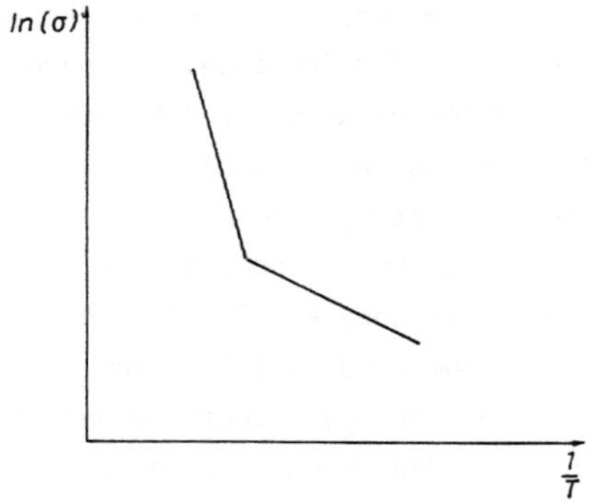

Fig.5.5. Arrhenius plot of conductivity showing the two ranges of conduction in the band and through hopping between nearest neighbors (low-temperature range)

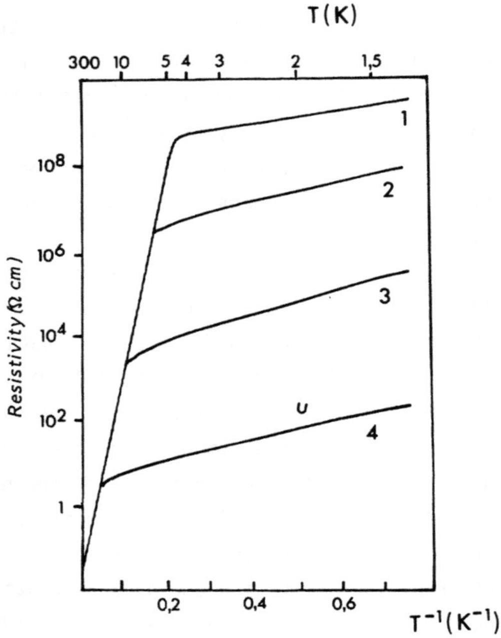

Fig.5.6. Resistivity versus T in transmutation doped germanium. The doping concentration (Ga) is varied by the dose of irradiation (with slow neutrons).
1. 7.5×10^{14} cm^{-3}
2. 1.5×10^{15} cm^{-3}
3. 4.9×10^{15} cm^{-3}
4. 3.5×10^{16} cm^{-3}

Because the amount of deep defects produced by neutrons for one transmuted Ge is a constant, the compensation ratio remains the same when the concentration of Ga varies (after [5.13])

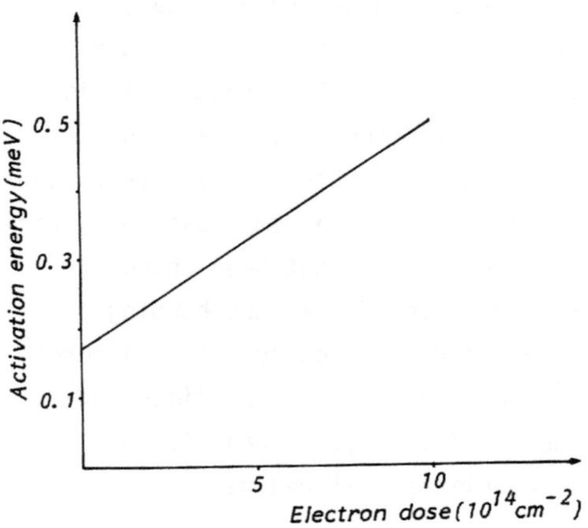

Fig.5.7. Variation of the activation energy of hopping conduction in CdTe versus the dose of irradiation with 1.5 MeV electrons, after [5.14]

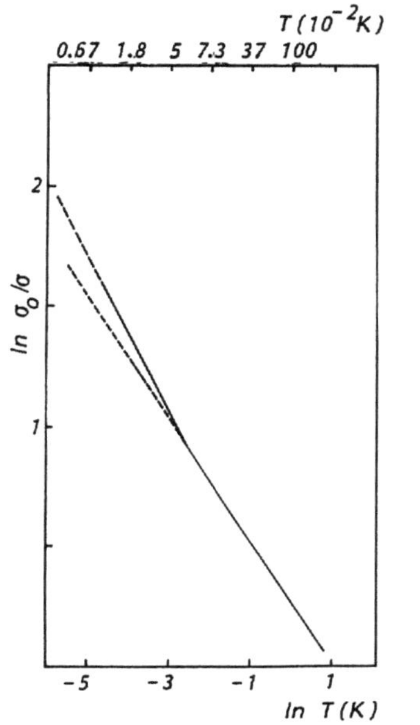

Fig.5.8. Variation of conductivity versus temperature in ion-implanted and annealed Si, after [5.15]. The slope of the curve indicates the temperature power law

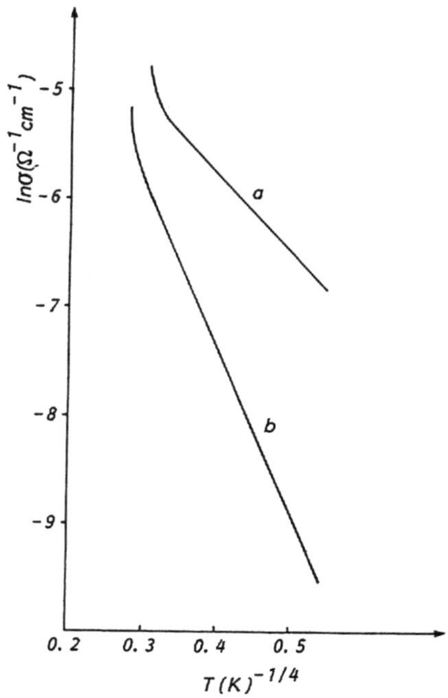

Fig.5.9. Variation of conductivity versus temperature in boron-doped diamond, before (a) and after (b) irradiation with 6.10^{16} 0.7 MeV electrons cm^{-2}, after [5.14]

b) In the case of deep states, since the electronic states are more localized, they can be present in large concentration before they induce metallic conductivity. Thus, w/kT can be comparable to $2\alpha R$ even at reasonably high temperature and conduction through variable range hopping can be observed easily. This is, for instance, the case below 200 K in Gallium arsenide heavily irradiated by fast neutrons [5.6] and in boron-doped diamond [5.16]. Diamond can be heavily doped ($\sim 10^{19}/cm^3$) with boron before the conduction becomes metallic because the boron level lies at 0.37 eV above the valence band (Sect.5.2.1). Variable range hopping takes place (Fig.5.9) below 150 K because the boron impurities which are not ionized in this temperature range are partially compensated by the numerous deep defects this material contains naturally.

5.3 Carrier Scattering

The interaction of free carriers with point defects is the origin of two different phenomena: i) scattering, whereby a carrier is diffused from one propagating Bloch state to another one; ii) capture, whereby the initial state is propagating but the final one is localized at the defect site, with exponentially decreasing wave function. Since both processes can be analyzed with the same general concepts we introduce them here although the capture process (and the ionization process which is the inverse of the capture) are described in the next chapter.

5.3.1 Scattering Cross Section

Let us bombard a finite volume V containing scattering (or capture) centers of density N, with a constant flux (number of particles per unit area and per unit time) J of incident carriers. The number n of scattered (or captured) carriers per unit time being proportional both to J and to the product NV, we can write

$$n = \sigma NVJ \quad . \tag{5.87}$$

The coefficient σ has the dimension of an area and is defined as a scattering (or capture) cross section; this is the area in which each center scatters (or traps) the free carriers.

Consider now the one-dimensional situation depicted in Fig.5.10. In the distance dx, the flux of particules changes from J to J + dJ; -dJ is equal to the number of diffused (or captured) particles per unit time and per unit surface. Dividing (5.87) by the transverse section S we have

$$-dJ = \sigma NJ \frac{V}{S} \quad , \tag{5.88}$$

that is

$$-dJ = \sigma NJ\, dx \quad . \tag{5.89}$$

Integrating this relation we obtain

$$J = J_0 \exp(-\sigma Nx) \quad . \tag{5.90}$$

It is clear that the fraction of carriers which propagate over a distance x is proportional to $\exp(-\sigma Nx)$. The mean free path ℓ is the average distance over which such a carrier propagates; it is thus given by

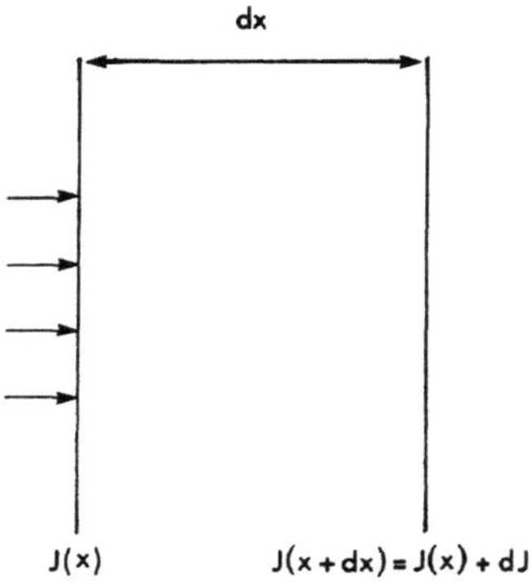

Fig.5.10. Variation of the flux of particles between two planes dx distance apart

$$\ell = \frac{\int_0^\infty x \exp(-\sigma N x)\, dx}{\int_0^\infty \exp(-\sigma N x)\, dx} \quad . \tag{5.91}$$

With this mean free path is associated a characteristic time, the relaxation time

$$\tau = \frac{\ell}{v} = \frac{1}{N\sigma v} \quad , \tag{5.92}$$

where v is the carrier velocity given by

$$kT = \frac{1}{2} m^* v^2 \quad . \tag{5.93}$$

In a diffusion process, we can make an angular analysis of the scattered carriers, defining by $n(\theta,\varphi)d\Omega$ the number of scattered carriers in the element of solid angle $d\Omega$ in the direction θ,φ, and hence write

$$n(\theta,\varphi)d\Omega = NVJ\sigma(\theta,\varphi)d\Omega \quad . \tag{5.94}$$

The quantity $\sigma(\theta,\varphi)$ is the differential scattering cross section. The total cross section is

$$\sigma = \int \sigma(\theta,\varphi)d\Omega \quad . \tag{5.95}$$

Relaxation times and differential scattering cross sections are quantities necessary to express the conductivity of solids. Let us summarize the results. In cubic systems, the conductivity is a scalar equal to $n\mu e$ where n is the concentration of free carriers and e the electron charge. The mobility μ is given by

$$\mu = \frac{e\langle\tau\rangle}{m^*} \quad , \tag{5.96}$$

$<\tau>$ being some average performed over $\tau(\underline{k})$, the time between collisions for a carrier of wave vector $\underline{k}$. It can be shown [5.17] that $\tau(\underline{k})$ is given by an expression similar to (5.92), i.e.,

$$\tau(\underline{k}) = [N\sigma_c v(\underline{k})]^{-1} \tag{5.97}$$

where σ_c, the cross section for conduction, is given by (Sect.5.3.2a)

$$\sigma_c = \int \sigma(\theta,\varphi)(1 - \cos\theta)d\Omega \quad . \tag{5.98}$$

When there are different kinds of scattering centers, the quantities $1/\tau(\underline{k})$ clearly have to be added to give the total inverse relaxation time.

The average relaxation time is a weighted average over all the carriers which have a certain distribution of energy

$$<\tau> = \frac{<\tau v(\underline{k})^2>}{<v(\underline{k})^2>} \quad . \tag{5.99}$$

Replacing an integration in the $\underline{k}$ space by an integration of energy, the expression of $<\tau>$ becomes

$$<\tau> = \frac{\int \tau\, E^{3/2} f(E)dE}{\int E^{3/2} f(E)dE} \quad , \tag{5.100}$$

where f is the function describing the occupancy of the allowed energy states in the band (for a nondegenerate semiconductor this Fermi function f reduces to the Boltzman distribution exp(-E/kT).

The physical meaning of the mobility μ is the following: the equation of motion of a charged carrier in an electric field contains a friction term

$$\underline{v} = \mu\underline{E} \tag{5.101}$$

necessary to account for the constant drift velocity $\underline{v}$ of the carrier in the electric field $\underline{E}$. The friction coefficient μ is due to the fact that the carriers are scattered by all kinds of imperfections, i.e., point defects, dislocations, lattice vibrations (phonons), optical-mode lattice waves, etc. The scattering due to dislocations, calculated by DEXTER and SEITZ [5.18], is not considered here. It results in a mobility which varies proportionally with temperature and is inversely proportional to the dislocation concentration.

At high temperature phonons are the dominant scattering centers and mobility varies as $T^{-3/2}$ [5.19]. At low temperature, when the phonon contribution decreases, the dominant scattering centers are impurities and defects. Of course charged defects, because the coulombic interaction is long range at low temperature [Ref.1.1, Sect.6.2.5], are more effective than neutral defects to scatter the carriers. A crude way to estimate the scattering cross section due to charged defects consists in using the notion of critical radius, developed in [Ref.1.1, Sect.6.2.5]. A carrier is scattered when its interaction energy with the scattering center is sufficiently larger than the thermal energy kT. For a Coulomb interaction with a center of charge Ze, an order of magnitude for the scattering radius r_s is thus given by

$$\frac{Ze^2}{\varepsilon r_s} = bkT \quad , \tag{5.102}$$

(where b is a constant of the order of 1 to 10) and the scattering cross-section

$$\sigma = \pi r_s^2 \tag{5.103}$$

is [5.20]

$$\sigma = \frac{\pi}{b^2}\left(\frac{Ze^2}{\varepsilon kT}\right)^2 \quad . \tag{5.104}$$

Typically, for a singly charged center, the magnitude of σ is

$$\sigma[\mathrm{cm}^2] = aT^{-2}[\mathrm{K}^{-2}] \tag{5.105}$$

with $a \sim 10^{-7}$ for $\varepsilon = 10$.

5.3.2 Mobility

The mobility μ is derived from the relaxation time $\langle\tau\rangle$ by (5.96). The relaxation time itself is an average over $\tau(\underline{k})$ given by (5.100) in terms of the velocity $v(\underline{k})$. For an order of magnitude estimation, we assume that $\langle\tau\rangle$ is related to the average thermal velocity v by (5.92), so that we can write

$$\frac{1}{\langle\tau\rangle} = aNvT^{-2} \quad . \tag{5.106}$$

As the average thermal velocity is such that

$$\frac{1}{2}m^*v^2 = \frac{3}{2}kT \quad , \tag{5.107}$$

we obtain for the mobility μ the final result

$$\mu = \frac{eT^{3/2}}{aN(3km^*)^{1/2}} \quad . \tag{5.108}$$

As is shown in the following, such an approximate derivation describes correctly the scattering by charged defects.

a) *Scattering by a Charged Center*

More generally, as recalled in Sect.5.3.1, the cross section σ must be obtained by integrating the differential cross section $\sigma(\theta)$ over the solid angle

$$d\Omega = 2\pi \sin\theta \, d\theta \quad , \tag{5.109}$$

i.e., by calculating the following expression

$$\sigma = 2\pi \int_0^\pi \sigma(\theta) \sin\theta \, d\theta \quad . \tag{5.110}$$

The differential cross section $\sigma(\theta)$, and the probability that the carrier is scattered between the angles θ and $\theta + d\theta$, can be expressed in terms of the impact parameter p which characterizes the separation between the initial carrier path and the scattering center (Fig.5.11). The relation between p and θ depends on the type of interaction. We have

$$2\pi p \, dp = \sigma(\theta) \, d\Omega \quad . \tag{5.111}$$

Since the carrier is scattered in the direction θ from its original direction of motion (the Z axis), the change of carrier velocity in this direction is

$$v_z = v(1 - \cos\theta) \tag{5.112}$$

and

$$\frac{1}{\tau} = 2\pi N v \int_0^\pi \sigma(\theta) \sin\theta (1 - \cos\theta) \, d\theta \quad . \tag{5.113}$$

This shows why the cross section for conduction σ_c (5.98), to be used in the calculation of τ, includes the weighting factor $1 - \cos\theta$.

Since the interaction is coulombic the relation between p and θ is given by Rutherford's formula [5.17] which we write in the form

$$p = \frac{K}{\tan(\theta/2)} \tag{5.114}$$

with

$$K = \frac{Ze^2}{\varepsilon m^* v^2} \quad . \tag{5.115}$$

The calculation of

$$\sigma(\theta) = \frac{p \, dp}{\sin\theta \, d\theta} \tag{5.116}$$

is straightforward. The result is

$$\sigma(\theta) = \left(\frac{K}{2 \sin^2\theta/2}\right)^2 \quad . \tag{5.117}$$

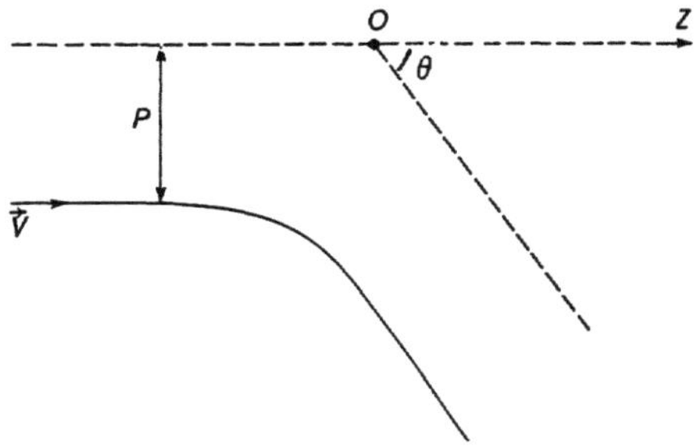

Fig.5.11. Schematic representation of the scattering of an electron of velocity $\vec{V}$ by a defect center O

The derivation of σ from this expression runs into a difficulty: the integral diverges for $\theta = 0$ since $\sigma(0)$ is infinite. This divergence corresponds to an infinite impact parameter p, a situation which cannot physically occur. The reason is that the Coulomb field due to the defect is modified by the neighboring charged defects and is screened by free electrons and holes. To account for the first limitation, the Coulomb field is assumed to be effective up to a distance p_m equal to half the average distance between two impurities (this means that beyond that distance the carrier is scattered by the neighboring impurity). Then the integration in (5.110) is now taken between a minimum value θ_m and π [5.21], θ_m being related to p_m through

$$p_m = \frac{K}{\tan(\theta_m/2)} \quad . \tag{5.118}$$

The maximum impact parameter p_m is related to the impurity density N by the relation

$$(2p_m)^3 = \frac{1}{N} \quad . \tag{5.119}$$

The second limitation corresponds to the screening of the Coulomb field by free carriers. Using a method similar to the one described in [Ref.1.1, Sect.6.2.5], it can be shown that a potential $-e^2/r$ is screened in such a way to give

$$V(r) = -\frac{e^2}{r}\exp\left(-\frac{r}{L}\right) \quad , \tag{5.120}$$

resulting in a cross section σ(θ) of the form [5.21]

$$\sigma(\theta) = \left[\frac{K/2}{\sin^2(\theta/2) + \beta^2}\right]^2 \tag{5.121}$$

which does not diverge any more.

Let us then calculate μ in the first case, where the integration over θ is truncated to θ_m. Expression (5.113) combined with $\sigma(\theta)$, given by (5.117), leads to

$$\frac{1}{\tau} = 2\pi Nv\left(\frac{K}{2}\right)^2 \int_{\theta_m}^{\pi} \frac{(1 - \cos\theta)\sin\theta}{\sin^4(\theta/2)}\, d\theta \qquad (5.122)$$

which can also be rewritten

$$\frac{1}{\tau} = 4\pi Nv\left(\frac{K}{2}\right)^2 \int_{\theta_m}^{\pi} \frac{(d/d\theta \sin^4\theta/2)d\theta}{\sin^4(\theta/2)} \quad . \qquad (5.123)$$

Integration of this expression leads to

$$\frac{1}{\tau} = -2\pi NvK^2 \ln\left(\sin^2 \frac{\theta_m}{2}\right) \quad . \qquad (5.124)$$

Since p_m is simply related to N (5.119) we now introduce p_m instead of θ_m in (5.124). For this we use (5.118) to write

$$\left(\frac{p_m}{K}\right)^2 = \frac{\cos^2(\theta_m/2)}{\sin^2(\theta_m/2)} \quad . \qquad (5.125)$$

Adding unity on both sides of this relation, we get

$$\frac{1}{\sin^2(\theta_m/2)} = 1 + \left(\frac{p_m}{K}\right)^2 \quad . \qquad (5.126)$$

Introduction of this relation in (5.124) gives

$$\frac{1}{\tau} = 2\pi NvK^2 \ln\left[1 + \left(\frac{p_m}{K}\right)^2\right] \quad . \qquad (5.127)$$

Replacing p_m by its value, deduced from (5.119), in terms of N leads to

$$\frac{1}{\tau} = 2\pi NvK^2 \ln\left[1 + \frac{1}{4K^2}\left(\frac{1}{N}\right)^{2/3}\right] \quad . \qquad (5.128)$$

Usually the expression of $1/\tau$ is given in terms of the thermal energy $E = m^*v^2/2$ and of the effective mass m^* of the carrier, i.e., in the form

$$\frac{1}{\tau} = \frac{\pi NZ^2e^4}{\varepsilon^2(2m^*)^{\frac{1}{2}}E^{3/2}} \ln\left[1 + \frac{\varepsilon^2E^2}{Z^2e^4}\left(\frac{1}{N}\right)^{2/3}\right] \quad . \qquad (5.129)$$

As recalled in Sect.5.3.1, because of the energy distribution of the carriers, it is necessary to use an averaging procedure on τ in order to get the mobility μ. When τ can be expressed in the form of a simple power law of energy, i.e., when the dependence on E of the logarithm in (5.129) can be neglected

$$\tau = AE^{-s} \tag{5.130}$$

where A is a constant, then $\langle\tau\rangle$, given by (5.100), can be calculated as a function of s. The result is

$$\langle\tau\rangle = \frac{A}{(kT)^s}\frac{\Gamma(5/2 - s)}{\Gamma(5/2)} \tag{5.131}$$

where Γ is the factorial function

$$\Gamma(z) = \int_0^\infty e^{-t}\, t^{z-1}\, dt \quad . \tag{5.132}$$

In the case we consider here, (5.129) shows that τ is proportional to $E^{3/2}$ divided by a logarithmic function of E^2. As the logarithm is a slowly varying function of E, it is a good approximation to take it outside the integral, replacing E in the logarithm by the value 3 kT, which corresponds to the maximum of the remainder of the integrand $E^3 \exp(-E/kT)$. With this approximation, τ can be written in the form (5.130) with s = -3/2 and

$$A = \frac{(2m^*)^{1/2}\varepsilon^2}{\pi N Z^2 e^4}\left\{\ln\left[1 + \left(\frac{3\varepsilon kT}{Ze^2N^{1/3}}\right)^2\right]\right\}^{-1} \quad . \tag{5.133}$$

The average $\langle\tau\rangle$ is thus given by

$$\langle\tau\rangle = \frac{8A}{\sqrt{\pi}}(kT)^{3/2} \tag{5.134}$$

since $\Gamma(3) = 6$ and $\Gamma(5/2) = 3\sqrt{\pi}/4$. The final expression for the mobility $e\langle\tau\rangle/m^*$ is thus

$$\mu = \frac{8\sqrt{2}}{(m^*)^{1/2}}\frac{\varepsilon^2}{NZ^2e^3}\left(\frac{kT}{\pi}\right)^{3/2}\left\{\ln\left[1 + \left(\frac{3\varepsilon kT}{Ze^2N^{1/3}}\right)^2\right]\right\}^{-1} \quad . \tag{5.135}$$

This is exactly the same result as (5.108), which was obtained by an approximate derivation. The only improvement is that we obtain the numerical value for the constant b (which in fact varies slowly with T since it is proportional to a logarithm).

b) *Scattering by Pairs*

In case of paired defects, the Coulomb field is greatly reduced. The pairs are then less efficient than isolated charged defects in scattering free carriers and the mobility is higher. The derivation of the expression of the mobility in this case can be made simply by replacing the Coulomb field by a dipole field [Ref.1.1, Sect.2.3]. Scattering by dipoles has been treated by BOARDMAN [5.23]. The result, for a dipole of distance R, is

$$\mu = \frac{2^{\frac{1}{2}}\hbar^2(4\pi\varepsilon)^2(kT)^{\frac{1}{2}}}{\pi^{3/2}\, e^2 m^{*3/2} NR^2} f(x) \quad , \tag{5.136}$$

where f(x) is the function

$$f(x) = \left[2\left(\frac{1+x}{1+2x}\right) - \frac{1}{x}\ln(1+2x)\right]^{-1} \tag{5.137}$$

with

$$x = \frac{8m^*\varepsilon(kT)^2}{\hbar^2 e^2 n} \quad , \tag{5.138}$$

n being the free-carrier concentration.

The $T^{\frac{1}{2}}$ dependence of the mobility is obvious. Indeed when the Coulomb field (in r^{-1}) is replaced by a dipole field (in r^{-2}) in the approximate derivation of the preceding section, the scattering cross section is found to vary as T^{-1} and thus μ as $T^{\frac{1}{2}}$.

c) *Scattering by Neutral Defects*

In the case of neutral defects, the problem is the choice of the interaction potential between the free carrier and the defect. It is similar to the problem of the scattering of low-energy electrons by a gas which is treated quantum mechanically [5.24]. Since the theory of such scattering, based on the method of partial waves, is developed in textbooks on quantum mechanics [5.25], it will not be given here. This treatment can be readily extended to the case of scattering in crystals [5.26]. The calculation can include both the effect of electron exchange and the effect of polarization of the defect by the carrier. A good approximation of the result for the scattering cross section is

$$\sigma_c = \frac{20a}{k} \tag{5.139}$$

where $a = (m_0/m^*)(\varepsilon/\varepsilon_0)a_0$ is the Bohr radius of the defect, m_0 the electron mass and k the magnitude of the wave vector of the carrier.

Since

$$k = \frac{m^* v}{\hbar} \quad , \tag{5.140}$$

we obtain for τ, using (5.97)

$$\frac{1}{\tau} = \frac{20a\hbar N}{m^*} \quad , \tag{5.141}$$

which is both temperature and energy independent. From this, we finally get for the mobility

$$\mu = \frac{e}{20a\hbar N} \quad . \tag{5.142}$$

The magnitude of this mobility is, for germanium for instance, ($\varepsilon = 16$, $m^*/m_0 = 0.12$)

$$\mu[\mathrm{cm}^2\ \mathrm{V}^{-1}\ \mathrm{s}^{-1}] = \frac{10^{20}}{N[\mathrm{cm}^{-3}]} \quad . \tag{5.143}$$

The reason a quantum-mechanical treatment is necessary in the case of scattering by a neutral defect, while a classical treatment is sufficient in the case of a charged defect, is due to the nature and range of the potential involved. For long-range potential, slowly variable with distance, such as a coulomb potential, the characteristic length over which the perturbation varies is large compared to the wavelength of the carrier. This is not the case for the potential of a neutral defect, short range compared to the wavelength, so scattering must be treated using quantum theory.

5.3.3 Experimental Results

Resistivity combined with Hall data provide Hall mobility data (Sect.5.1-4). Typical variations of μ with temperature are represented in Fig.5.12 for n-type germanium doped with various impurity concentrations. A wide range of behavior is observed as the impurity concentration varies [5.27]. For a pure sample, μ practically does not depart from the $T^{-3/2}$ behavior, characteristics of lattice scattering (Sect.5.3.2), even at low temperature, indicating that lattice scattering dominates at all temperatures. With increasing impurity concentration, departure from $T^{-3/2}$ behavior appears: the larger the impurity concentration, the higher the temperature at which this departure takes place because of the increasing importance of ionized impurity scattering compared to lattice scattering.

The difference between the observed temperature dependence of μ at low temperature and the $T^{-3/2}$ dependence is a means to obtain the contribution of impurity scattering. The mobility never reaches the value predicted for impurity scattering because of the existence of neutral impurity (or defect) scattering. However, data in the temperature range where neutral scattering is negligible (80-300 K in Ge) and in which the concentration of ionized impurities is known, has been used to check the validity of the ionized impurity scattering theory. In this temperature range, the concentration of

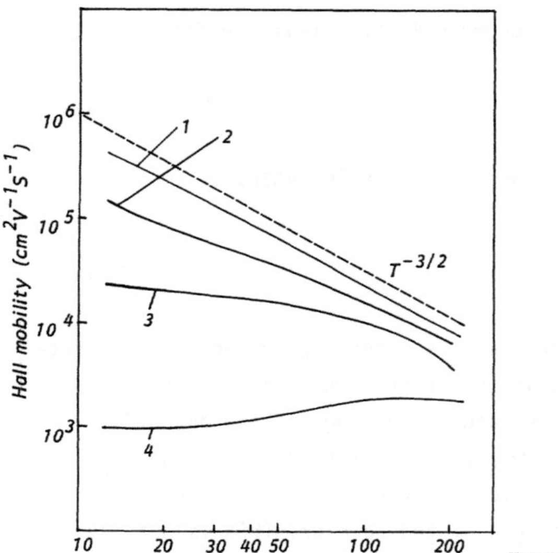

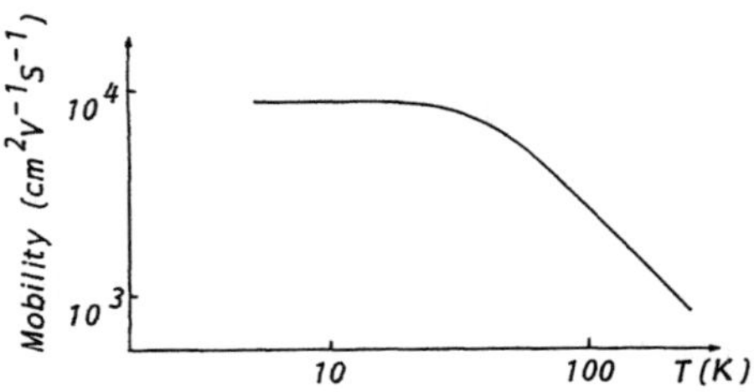

Fig.5.14. Mobility versus temperature in n-type CdTe doped with 3.5×10^{13} cm^{-3} obtained by the so-called "travelling heater method" annealed in Cd atmosphere for 2 hours at 650°C

Fig.5.12. Variation of Hall mobility versus temperature in n-type germanium doped with various impurity concentrations: 10^{13} cm^{-3} (1), 1.7×10^{15} cm^{-3} (2), 7.5×10^{15} cm^{-3} (3), 5.5×10^{16} cm^{-3} (4), from [5.4]

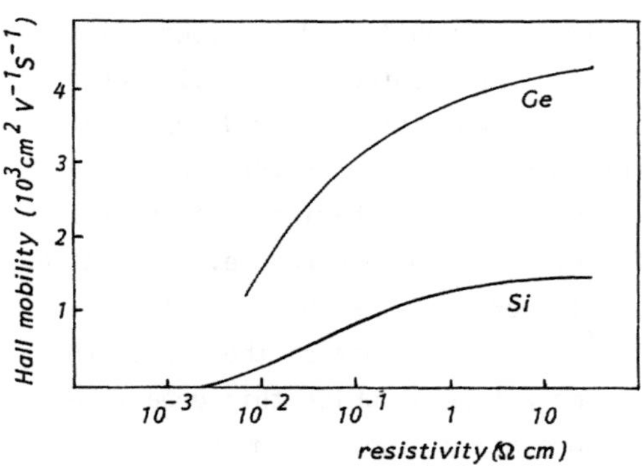

Fig.5.13. Variation of Hall mobility at room temperature in Si (after [5.3] and Ge (after [5.4]) versus the resistivity of the material at room temperature

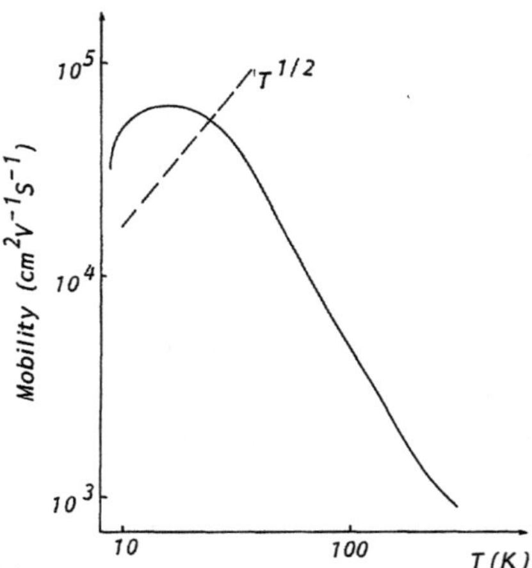

Fig.5.15. Mobility versus temperature in n CdTe doped with 10^{14} cm^{-3} obtained by the Bridgeman method and purified by vertical zone melting

the ionized impurities is practically the doping concentration; then the mobility is a unique function of this concentration, i.e., of the resistivity of the material, which can be calculated [5.3] and compares favorably with experiment (Fig.5.13). Data in a lower temperature range (below 80 K in Ge), in which μ is due to ionized impurity as well as neutral impurity scattering, can be then used to evaluate the concentration of neutral defects once the ionized impurity concentration has been determined.

Cases, such as the one described for CdTe ($\varepsilon = 10.2$) in Fig.5.14 are rare, in which a mobility governed by scattering on neutral impurities is observed independently. Only in such a case can the concentration of neutral centers be determined unambiguously. Application of (5.142) in Fig.5.14 ($\mu = 10^4$ cm^2 V^{-1} s^{-1}) gives a concentration of 10^{16} cm^{-3} centers. In cadmium telluride the high mobilities sometimes observed at low temperature can be interpreted in terms of pair scattering (Fig.5.15). The temperature law of μ cannot be verified. However, once the concentration of pairs N_p has been determined, the average distance R between the elements of the pair can be estimated. In Fig.5.14, $N_p \sim 2 \times 10^{14}$ cm^{-3}, gives $R \sim 8$ Å.

6. Carrier Emission and Recombination

A defect whose associated energy level lies in the forbidden gap exchanges carriers with the conduction and valence bands through the emission and recombination of electrons and holes. As we discussed in [Ref.1.1, Sect. 7.3.2], the electronic transitions between level E_T and the bands allow the determination of the average time a defect is occupied by a carrier, i.e., the occupancy of the level. In this chapter we consider the rates of emission of electrons or holes from the defect level to the bands and the rates of recombination. Because these rates depend on the free energy of ionization and on the cross sections for electrons and holes trapping on the defect, their study provides information from which practically all the electrical characteristics of the defect (enthalpy and entropy of ionization, trapping cross section etc.) can be deduced. Moreover, variation of the cross section versus temperature appears to be a powerful tool to get an insight into the defect-phonon interaction and consequently the correlated lattice distortion around the defect.

6.1 Emission and Capture Rates

We use the notations introduced in [Ref.1.1, Sect.7.3.2]: S is the $(n + 1)^{th}$ charge state of the defect as defined in Sect.2.1.3, and B its nth charge state which contains one more electron (Fig.6.1). Equations [Ref.1.1, Eqs. (7.44,45)] describe the electronic transitions between the bands and level E_T, corresponding to $E(n,n + 1)$ defined in Sect.2.1.3. At equilibrium, the occupancy α of the level by an electron is given by [Ref.1.1,Eq. (7.50)] in which g_e, g_h, k_e, k_h are the probabilities for emission (g) and capture (k) of electrons (e) and holes (h).

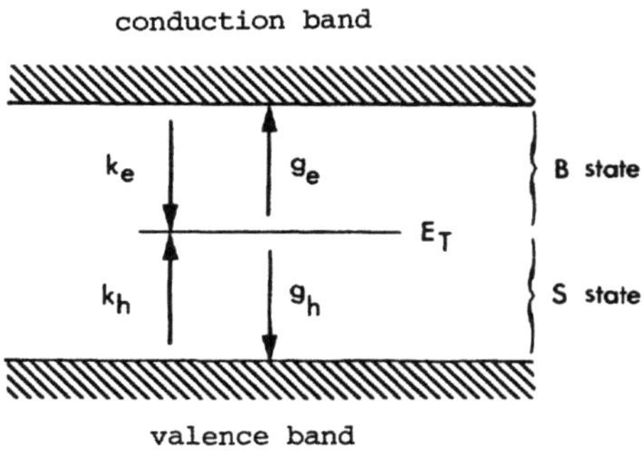

Fig.6.1. Representation of electron (e), hole (h), emission (g) and capture (k) processes between level E_T and the conduction and valence bands

6.1.1 The Principle of Detailed Balance

The probability per unit time k_e (introduced in [Ref.1.1, Sect.7.3.2] that a defect captures an electron from the conduction band (and changes its charge state from S to B) is proportional to the concentration n of electrons in this band

$$k_e = c_e n \tag{6.1}$$

where c_e is the capture coefficient for electrons. Similarly, for hole capture (transition from B to S) the probability per unit time is

$$k_h = c_h p \tag{6.2}$$

where p is the hole concentration in the valence band. The coefficients c_e and c_h, which have the dimension of a volume per unit time, can be expressed as the product of a cross section for capture, σ_e or σ_h, times the carrier velocity v_e or v_h

$$c_e = \sigma_e v_e \tag{6.3}$$

$$c_h = \sigma_h v_h \quad . \tag{6.4}$$

The velocities correspond to thermal averages and can be obtained from

$$\frac{1}{2} m^*_{\substack{e\\h}} v^2_{\substack{e\\h}} = \frac{3}{2} kT \quad . \tag{6.5}$$

Since the effective masses m_e^* or m_h^* depend on the type of the carrier, this thermal velocity is slightly different for electrons and holes.

Carrier emission from the defect to a band occurs with a probability per unit time g_e for electron emission (B→S) and g_h for hole emission (S→B). Both quantities depend on the number of empty states in the bands, equal to the total number of states minus n or p. However, n or p are usually small compared to the number of empty states so that they can be neglected.

Thus, g_e and g_h can be considered as independent of the free-carrier concentration.

Now, the total rates, i.e., the changes in concentration per unit time, for the different processes are the products of the probability per unit time (k_e, k_h, g_e, g_h) times the concentration s or b of the defect in the corresponding initial charge state. These rates are given by:

$$\begin{aligned} &k_e s = c_e ns && \text{for electron capture}\\ &g_e b && \text{for electron emission}\\ &k_h b = c_h pb && \text{for hole capture}\\ &g_h s && \text{for hole emission} \quad . \end{aligned} \tag{6.6}$$

The coefficients c_e, g_e, c_h and g_h are not independent since, at thermal equilibrium, each capture rate $k_e^0 s^0$ or $k_h^0 b^0$ (the index 0 indicates thermal equilibrium) is equal to the corresponding emission rate $g_e^0 b^0$ or $g_h^0 s^0$. This allows us to express g_e and g_h in terms of c_e and c_h [by coupling (6.1,2,6)] which leads to

$$g_e = c_e n^0 \frac{s^0}{b^0} \tag{6.7}$$

and

$$g_h = c_h p^0 \frac{b^0}{s^0} \quad . \tag{6.8}$$

From [Ref.1.1, Sect.6.2.3], we know that the ratio s^0/b^0 is given by

$$\frac{s^0}{b^0} = \gamma \exp\left(\frac{E_T - E_F}{kT}\right) \quad , \tag{6.9}$$

where E_T is the negative free energy of ionization of the defect corresponding to the change in charge state from B to S. The factor γ is the degeneracy factor, equal to the ratio Z(S)/Z(B), as shown in [Ref. 1.1, Eq. (6.33)]. To obtain symmetrical expressions for g_e and g_h, we write

$$n^0 = n_i \exp\left(\frac{E_F - E_i}{kT}\right) \tag{6.10}$$

$$p^0 = n_i \exp\left(\frac{E_i - E_F}{kT}\right) \tag{6.11}$$

where n_i is the intrinsic concentration and E_i the intrinsic Fermi-level position. We thus get

$$g_e = c_e \gamma n_i \exp\left(\frac{E_T - E_i}{kT}\right) \tag{6.12}$$

$$g_h = \frac{c_h}{\gamma} n_i \exp\left(\frac{E_i - E_T}{kT}\right) \quad . \tag{6.13}$$

6.1.2 Enthalpy and Entropy of Ionization

Relations (6.12,13) can be rewritten by replacing c_e and c_h by their expressions (6.3,4) and also by using the equalities

$$n_i \exp\left(\frac{E_F - E_i}{kT}\right) = N_c \exp\left(\frac{E_F - E_c}{kT}\right) \tag{6.14}$$

$$n_i \exp\left(\frac{E_i - E_F}{kT}\right) = N_v \exp\left(\frac{E_v - E_F}{kT}\right) \quad , \tag{6.15}$$

where E_c and E_v are the conduction and valence band limits, N_c and N_v being the effective number of states in these bands in the nondegenerate situation (see Sect.5.1.1 for details). This gives

$$g_e = \sigma_e v_e \gamma N_c \exp\left(-\frac{E_c - E_T}{kT}\right) \tag{6.16}$$

$$g_h = \frac{\sigma_h v_h}{\gamma} N_v \exp\left(\frac{E_v - E_T}{kT}\right) \quad . \tag{6.17}$$

These expressions result from thermodynamical considerations and the quantities $E_c - E_T$ as well as $E_v - E_T$ are in fact free energies of ionization [Ref.1.1, Sect.6.2.3] of an electron to the conduction band in the first case, of a hole to the valence band in the second one. They can thus formally be rewritten as

$$g_{\substack{e\\h}} = v_{\substack{e\\h}} \sigma_{\substack{e\\h}} N_{\substack{c\\v}} \exp\left(\frac{\Delta S}{k}\right) \exp\left(-\frac{\Delta H}{kT}\right) \tag{6.18}$$

where we have split the free energy into its entropy part ΔS and its enthalpy part ΔH (the degeneracy factors γ or $1/\gamma$ have been included in the entropy). This defines the entropy and enthalpy of ionization of electrons or holes from a given defect. In the following, we shall be interested in the temperature dependence of g_e and g_h. We thus recall that N_c and N_v behave as $T^{3/2}$, while v_e and v_h are proportional to $T^{1/2}$, which allows us to write

$$g_e = \nu_e T^2 \sigma_e(T) \exp\left[\frac{\Delta S_e(T)}{k}\right] \exp\left[-\frac{\Delta H_e(T)}{kT}\right] \tag{6.19}$$

$$g_h = \nu_h T^2 \sigma_h(T) \exp\left[\frac{\Delta S_h(T)}{k}\right] \exp\left[-\frac{\Delta H_h(T)}{kT}\right] \tag{6.20}$$

where ν_e and ν_h are temperature independent coefficients. As discussed in Sect.6.3.3, the capture cross sections $\sigma_e(T)$ and $\sigma_h(T)$ can be temperature dependent, i.e., can behave as $\exp(-\Delta E/kT)$ where ΔE corresponds to some energy barrier.

As noticed above the entropy parts $\Delta S_e(T)$ or $\Delta S_h(T)$ contain two terms. The first one is due to the degeneracy factor γ (or $1/\gamma$) (Sect.5.1.3)

$$\Delta S_{1e} = k \ln\gamma \tag{6.21}$$

or

$$\Delta S_{1h} = -k \ln\gamma \quad . \tag{6.22}$$

The second term is due to the fact that when the defect is ionized, its bonding with the host lattice is modified. As we discussed in [Ref.1.1, Sect.5.4.4], the change in force constants between the defect and its neighbors, caused by the modification of the bonds, results in a vibrational entropy term (this term can be calculated using the methods in [Ref.1.1, Sect.5.4].

6.1.3 Trapping and Recombination Centers

Emission and capture rates determine the defect occupancy at thermal equilibrium and the way this equilibrium occupancy is reached. The kinetics of return to equilibrium are therefore governed by quantities such as cross sections, carrier concentrations and temperature. Depending on the relative values of k_e, k_h, g_e and g_h, which are functions of the doping concentrations, temperature, and compensation, a defect is said to be a trapping center, a recombination center, a generation center, etc.

Consider a carrier captured at a defect site; if this carrier stays there until it is reemitted back into the band it comes from, then the center is classified as a trap (a minority or majority one depending on the type of captured carrier). When a carrier of opposite sign gets trapped on the same site, resulting in electron-hole recombination (before the first carrier is reemitted), then the defect is classified as a recombination center. The conditions on the relative values of k_e, k_h, g_e, g_h which correspond to this classification differ with authors. An electron trap, for instance, is sometimes defined as a center for which $k_e \gg k_h$, by the condition $g_e \gg g_h$. Here we choose to use conditions also adopted by SAH [6.1] which directly result from the above definitions for traps and recombination centers.

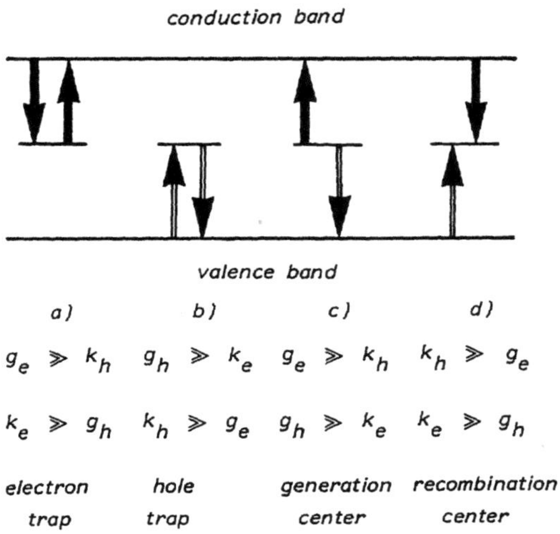

Fig.6.2a-d. Definition and illustration of traps, generation and recombination centers

A defect acts as an electron trap if, when empty (i.e., in the S state), it captures an electron from the conduction band, and when filled (i.e., in the B state), it tends to reemit this electron to the conduction band. This implies that, when filled, the probability per unit time g_e for emission of an electron to the conduction band is much larger than k_h for the capture of a hole from the valence band. Similarly, in the empty state, k_e must be much larger than g_h. This is the situation in Fig.6.2a. The other possibilities are given in Figs.2b-d. Figure 6.2b corresponds obviously to the case of a hole trap. In the case of Fig.6.2c the defect, when filled, emits an electron and, when empty, emits a hole. This favors the transfer of electrons from the valence band to the conduction band and the defect acts as a generation center. The case of Fig.6.2d where electrons are transferred from the conduction to the valence band corresponds to a recombination center.

The mathematical conditions to determine the actual behavior of a given defect are thus obtained by calculating the ratios g_e/k_h and g_h/k_e. For this, we use (6.1,2) for k_e, k_h, (6.12,13) for g_e, g_h and finally the standard expressions of n and p in terms of the pseudo-Fermi levels E_{Fe} for electrons and E_{Fh} for holes:

$$n = n_i \exp\left(\frac{E_{Fe} - E_i}{kT}\right) \tag{6.23}$$

$$p = n_i \exp\left(\frac{E_i - E_{Fh}}{kT}\right) \quad . \tag{6.24}$$

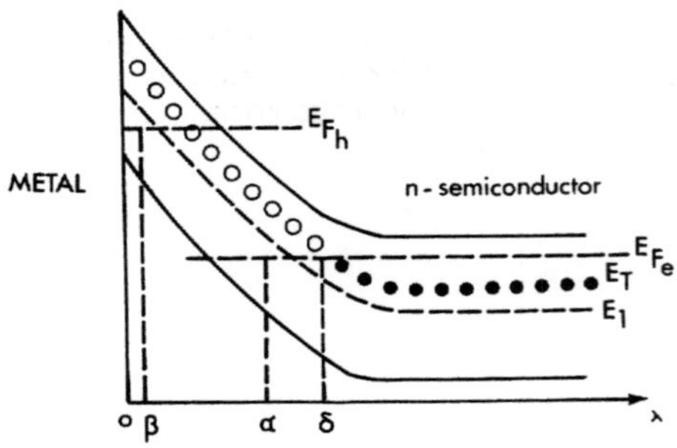

Fig.6.3. Energy diagram of a Schottky barrier under reverse bias showing the distribution of filled (•) and empty (o) traps when the trap level E_T is situated above E_1

This gives

$$\frac{g_e}{k_h} = \gamma \frac{c_e}{c_h} \exp\left(\frac{E_T + E_{Fh} - 2E_i}{kT}\right) \tag{6.25}$$

and

$$\frac{k_e}{g_h} = \gamma \frac{c_e}{c_h} \exp\left(\frac{E_T + E_{Fe} - 2E_i}{kT}\right) \quad . \tag{6.26}$$

Thus, the condition $g_e/k_h \gg 1$ is equivalent to

$$\frac{E_T + E_{Fh}}{2} \gg E_1 \tag{6.27}$$

with

$$E_1 = E_i + \frac{kT}{2} \ln\left(\frac{c_h}{\gamma c_e}\right) \tag{6.28}$$

and the condition $k_e/g_h \gg 1$, to

$$\frac{E_T + E_{Fe}}{2} \gg E_1 \quad . \tag{6.29}$$

This means that, when the trap level E_T is high enough in the gap, for both (6.27,29) to be satisfied, then electron processes dominate over hole processes. To examine these conditions, we recall the following important points: i) E_i is near the middle of the gap, ii) the sign of $(kT/2)\ln(c_h/\gamma c_e)$ depends upon the ratio of the capture coefficients c_e and c_h for electrons and holes. To get a rough idea, let us investigate the single donor defect in which the filled B state is the neutral charge state and the empty one S is the positively charged state [in other words E_T is equal to $E(0,+)$]. Before capture of an electron, the defect is in the S state of charge +1, leading to a long-range attractive potential, while before capture of a hole, it is in the neutral B state with no long-range potential. It is thus likely that $c_e > c_h$ and the quantity $(kT/2)\ln(c_h/\gamma c_e)$ is presumably negative.

An interesting example is the Schottky barrier for n-type material under reverse bias. We consider the situation where E_T is above E_1, which corresponds to the case of Fig.6.3 [6.2]. We define: $x = \delta$ the point where $E_T = E_{Fe}$, $x = \alpha$ where (6.29) is an equality, and $x = \beta$ where (6.27) is an equality. When $x > \alpha$, both inequalities (6.27,29) are satisfied so that electron processes dominate. In the steady state the rate of change of the S state, given by [Ref.1.1, Eq. (7.47)], vanishes. This means that the rates of electron emission and capture from the level must be equal. Thus, from (6.6) we have

$$c_e ns = g_e b \quad . \tag{6.30}$$

If we replace g_e by its expression (6.12) and n by (6.23), we can determine the nonequilibrium ratio s/b, which is given by

$$\frac{s}{b} = \gamma \exp\left(\frac{E_T - E_{Fe}}{kT}\right) \quad . \tag{6.31}$$

This expression is just the same as the equilibrium one (6.9) with E_F replaced by E_{Fe}. This comes from the fact that the defect only exchanges electrons with the conduction band, so that it is in thermal equilibrium with this band, and Fermi-Dirac statistics can be used with the electron pseudo-Fermi level.

If we come to the left of $x = \alpha$, the first condition to be broken is (6.29) so that, in such a case, there is a region $\beta < x < \alpha$ where g_e and g_h dominate respectively over k_e and k_h: the defect is a generation center. In the steady state regime we have

$$g_e b = g_h s \tag{6.32}$$

which, using (6.12,13), leads to

$$\frac{s}{b} = \gamma \exp\left(\frac{E_T - E_1}{kT/2}\right) \quad . \tag{6.33}$$

In this region, the occupancy of the center no longer depends upon the position of the Fermi level E_{Fe}. However, as $E_T > E_1$, $s/b \gg 1$ and the level is empty (i.e., all defects are in the S state).

A third situation occurs if $x < \beta$. Then, g_h and k_h dominate and we recover Fermi-Dirac statistics but with the hole quasi-Fermi level

$$\frac{s}{b} = \gamma \exp\left(\frac{E_T - E_{Fh}}{kT}\right) \quad . \tag{6.34}$$

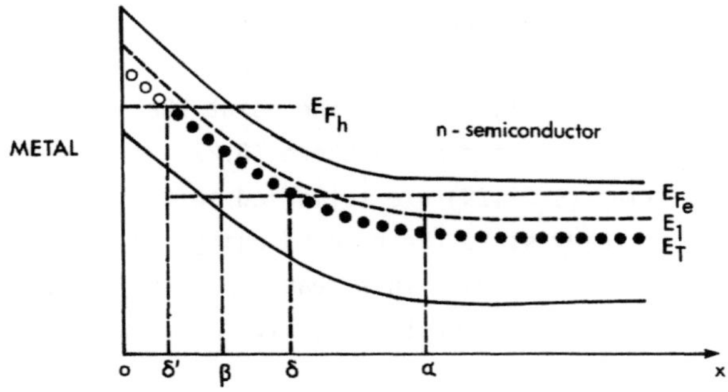

Fig.6.4. Energy diagram of a Schottky barrier under reverse bias showing the distribution of filled (●) and empty (○) traps when the trap level E_T is situated below E_1

In the configuration of Fig.6.3, $E_T > E_{Fh}$ when $x < \beta$ so that again the level is empty.

In conclusion, the physical situation is the following: when $x > \delta$ the level is filled (the defects are in state B), while for $x < \delta$ it is empty (the defects are in state S). We thus recover the results which are intuitively obtained by applying Fermi-Dirac statistics (6.31) over the whole range of values for x.

However, this simple picture does not remain valid when $E_T < E_1$ (Fig.6.4). In that case the point α comes to the right of δ. Similar considerations show that: i) For $x > \alpha$, as $E_T < E_{Fe}$, (6.31) shows that the level is filled; ii) For $\beta < x < \alpha$, application of (6.33) with $E_T < E_1$ shows that, again, the level is filled; iii) For $x < \beta$ Fermi-Dirac statistics with the holes are valid (6.34). Defining as $x = \delta'$ the point where $E_{Fh} = E_T$, when $x > \delta'$ the level is filled, and when $x < \delta'$ the level is empty. In this case we again obtain a simple picture, but with the hole Fermi level replacing the electron one.

It is obvious that all the above discussion advanced for an n-type material remains valid for p-type material at the condition of exchanging the roles of electrons and holes.

It now appears, from the examples given above, that the same defect can be either a trap or a recombination center, depending on the position of E_{Fe} and E_{Fh}, i.e., on the magnitude of n and p. For the situation in Fig.6.3, according to the definitions given on Fig.6.2, the defect acts for $x > \alpha$ as an electron trap. For $\beta < x < \alpha$ the defect behaves as a generation center. It is only for $x < \beta$ that it becomes a hole trap. For the same situation in Fig.6.3, where the semiconductor is n-type, the electrons are the majority carriers, and the holes, the minority carriers. Thus one often speaks of majority carrier traps and minority carrier traps.

6.2 Experimental Observation of Emission Rates

6.2.1 Principle

The way to observe the emission of a carrier from a defect level E_T to a band consists in monitoring, for instance, the variation of the space charge of the depletion layer of a diode, induced by sudden application of an external perturbation. The rate of change of the carrier population on the defect is given by ds/dt (equal to -db/dt) and is reflected by a corresponding change -e(db/dt) of the stored space charge density (this is due to the fact that the emitted carriers leave the space charge region instantaneously [6.3]). The equations for this rate of change in population can be written as [Ref. 1.1, Eq. (7.47)]

$$\frac{ds}{dt} = -\frac{db}{dt} = (g_e + k_h)b - (g_h + k_e)s \quad . \tag{6.35}$$

From the kinetics of b(t) it is possible to get the rate of emission of the defect, for instance. To demonstrate this, let us choose a simple case where the defect acts as an electron trap (i.e., $g_e, k_e \gg g_h, k_h$) in an n-type material. In such a case we have

$$\frac{db}{dt} = g_e b - k_e s \quad . \tag{6.36}$$

The steady state corresponds to db/dt = 0, when the emission ($g_e b$) and capture rates ($k_e s$) are equal. Let us now assume that an external perturbation (described later) creates, at time t = 0, a new situation where $g_e b$ and $k_e s$ are different. We want to study the transient associated with the return to steady state, characterized at $t = \infty$ by

$$g_e b(\infty) = k_e s(\infty) \quad . \tag{6.37}$$

For this we define $\Delta b = -\Delta s$ as the change in concentrations with respect to the steady state. We rewrite (6.36) in the form

$$\frac{d\Delta b}{dt} = -(g_e + k_e)\Delta b \quad . \tag{6.38}$$

To solve this equation we assume that $g_e + k_e$ remains constant in time at a given position. This is obvious for g_e, but not for k_e, equal to $c_e n$. However, the latter is usually true since the excess emitted carriers (corresponding to db/dt) immediately disappear so that n (and thus k_e) can be considered to be constant [equal to (∞)] in time. Under this assumption, the solution of (6.38) is

$$\Delta b(t) = \Delta b(0) \exp(- g_e - k_e)t \quad . \tag{6.39}$$

This solution can be transformed by noting that k_e and g_e are related through (6.37) and that for an electron trap, as shown in (6.31), $s(\infty)$ and $b(\infty)$ are related through Fermi-Dirac statistics, with E_F replaced by E_{Fe}. We thus obtain

$$\Delta b(t) = \Delta b(0) \exp\left\{- g_e\left[1 + \frac{1}{\gamma} \exp \frac{(E_{Fe} - E_T)_\infty}{kT}\right]t\right\} \quad . \tag{6.40}$$

If we define the occupancy factor as

$$f = \frac{b}{b + s} \quad , \tag{6.41}$$

(6.40) can be rewritten

$$f(t) - f(\infty) = [f(0) - f(\infty)] \exp\left(- \frac{g_e t}{1 - f(\infty)}\right) \quad . \tag{6.42}$$

This expresses the probability that the defect is occupied (i.e., is in the B state) at time t. Expression (6.42), derived in [6.4], will prove to be very convenient in the following.

In cases where the final steady-state situation (for $t = \infty$) corresponds to empty defects (in the S state), then $f(\infty)$ is zero and (6.42) reduces to the simple form

$$f(t) - f(\infty) = [f(0) - f(\infty)] \exp(-g_e t) \quad . \tag{6.43}$$

As $\Delta b(t)$ has the same behavior, we see that the emission rate can be obtained by measuring the time constant associated with the transient behavior of $\Delta b(t)$. According to (6.18), the measurement of g_e at various temperatures enables us to calculate the ionization enthalpy ΔH and the preexponential factor $\sigma\exp(\Delta S/k)$. Once σ is measured (6.4.1), then the entropy term can also be deduced.

Usually, the logarithm of the quantity $g_e T^{-2}$ is plotted versus T^{-1} (Fig. 6.5) in order to account for the temperature dependence of the carrier velocity and of the density of states in the band (6.19,20). The quantity ΔH is the slope of the straight line obtained, provided that the temperature dependence of $\sigma(T)$ is negligible.

6.2.2 Observation Techniques

One way the occupancy of a level can be modified consists in using a junction or a Schottky barrier (other devices such as p-i-n transitors, MOS and MIS structures, MESFET's and MOSFET's can and have been used as well).

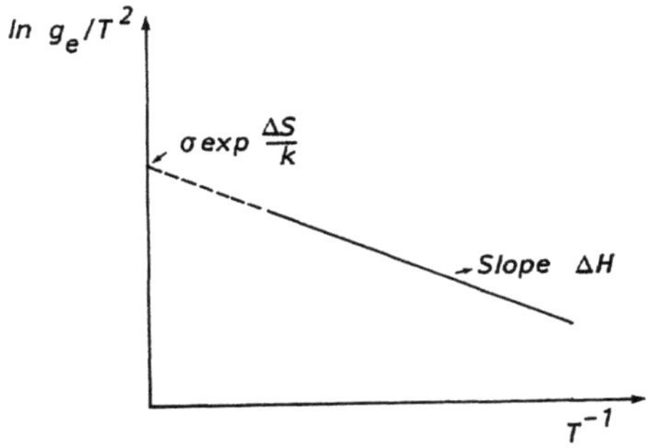

Fig.6.5. Variation of an emission rate with inverse temperature, from which the characteristics of the trap can be extracted

Consider, for instance, the n side of a p^+ - n junction (Fig.6.6) under reverse bias. Due to the formation of a space-charge region, all energy levels exhibit a strong curvature, except for the quasi-Fermi levels E_{Fe} and E_{Fh} which tend to remain constant in this region [6.2]. If the defect behaves as an electron trap, then the steady-state situation is similar to that in Fig.6.3 and the defects are filled for $x > W - \lambda$ and empty for $x < W - \lambda$. On Fig.6.6, the abscissa $x = W$ represents the limit of the depleted zone, i.e., the zone which has been deserted by the free carriers, due to the electric field in the junction; and λ is the abscissa at which E_{Fe} crosses E_T.

The process used to observe electron emission from the defect level starts initially with a reversed biased junction, such as shown on Fig. 6.7a. In that initial situation, the levels are filled for $x > W_0 - \lambda_0$. If we suddenly increase the reverse bias, then the new steady-state situation will correspond to Fig.6.7b with $W_1 > W_0$ and $W_1 - \lambda_1 > W_0 - \lambda_0$. There will be a transient regime between these two steady-state situations during which the filled defects located in the region $W_0 - \lambda_0 < x < W_1 - \lambda_1$ will emit their electrons to the conduction band. This emission can be obtained by measuring the current which flows through the junction or the change in capacitance of this junction.

Let us examine the second technique. The depleted zone contains a space charge of density $\rho(x)$. If we assume an abrupt transition at $x = W$ between the depleted zone and the neutrality region, then the potential drop V across this region is quite generally given by

$$V = \frac{1}{\varepsilon} \int_0^W x\rho(x)\, dx \quad , \tag{6.44}$$

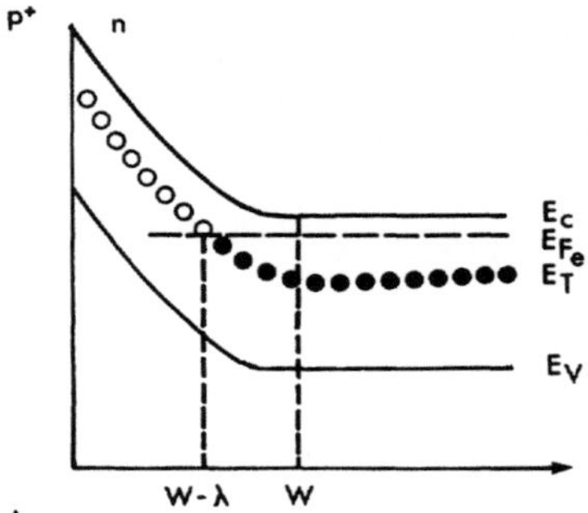

Fig.6.6. Energy diagram of a p^+ n junction under reverse bias showing the distribution of filled and empty traps

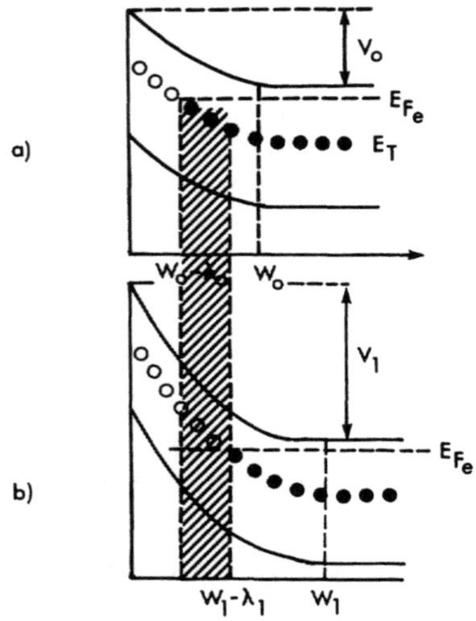

Fig.6.7. Distribution of filled and empty traps at steady state in a p^+n junction under two different reverse biases V_0(a) and $V_1 > V_0$(b)

while the stored charge in that region is given by

$$Q = S \int_0^W \rho(x)\, dx \tag{6.45}$$

where S is the junction area. The capacitance C

$$C = \frac{dQ}{dV} \tag{6.46}$$

induced by the applied voltage increment dV, corresponds to a change dW of the depletion layer width which, in turn, induces a modification dQ of the stored charge. Differentiating (6.44,45) and taking their ratio leads to

$$C = \frac{\varepsilon S}{W} \tag{6.47}$$

which is the usual expression for a plane capacitor. This expression, of course, assumes that ρ is not modified by the increment dV, i.e., that the traps have no time to respond.

We have seen that the transient regime corresponding to the transition depicted in Figs.6.7a,b is characterized by a continuous change of W(t) between its two limits W_0 and W_1. We define

$$\Delta W(t) = W(t) - W_1 \tag{6.48}$$

and assume this quantity remains small, so that we can write

$$\frac{\Delta C(t)}{C} = - \frac{\Delta W(t)}{W_1} \quad . \tag{6.49}$$

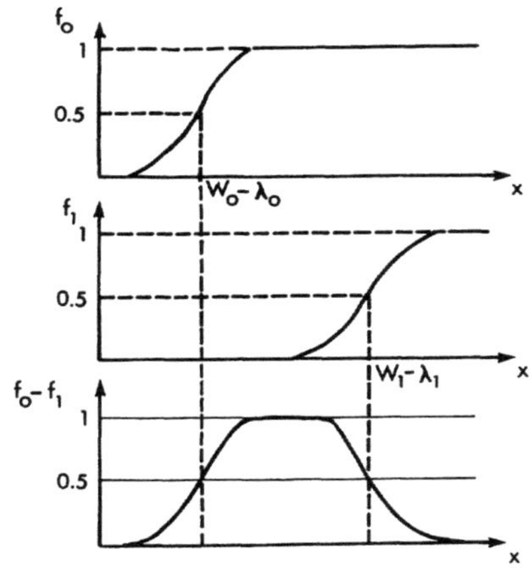

Fig.6.8. Occupancy factors f_0 and f_1 of the trap level as a function of the distance x from the junction under reverse biases V_0(a) and V_1(b) respectively, and f_0-f_1 versus x(c) when the quantity γ of (6.31) is taken to be 1

To obtain $\Delta C/C$ in terms of the emission rate we have still to relate $\Delta W/W_1$ to the emission process which is occuring. For this we notice that during all the transient regime, the potential V is a constant, equal to V_1.

There is thus a relation between W(t) and $\rho(x,t)$ which can be obtained by differentiation of (6.44)

$$0 = W_1\rho(W_1)\Delta W + \int_0^{W_1} x\Delta\rho(x,t)\,dx \qquad (6.50)$$

which, coupled to (6.49), gives

$$\frac{\Delta C}{C} = \frac{1}{W_1^2\rho(W_1)} \int_0^{W_1} x\Delta\rho(x,t)\,dx \quad . \qquad (6.51)$$

The change $\Delta\rho(x,t)$, equal to $-e\Delta b(t)$, is induced by the emission of carriers in a region roughly situated between $W_0 - \lambda_0$ and $W_1 - \lambda_1$. Figure 6.8 shows the Fermi-Dirac function f_0 corresponding to the initial situation, f_1 to the final one, and f_0-f_1 which represents the initial perturbation with respect to f_1. If $(W_1 - \lambda_1) - (W_0 - \lambda_0)$ is sufficiently large compared to the width of the Fermi distributions, then f_0-f_1 can be approximated by 1 in this interval and zero outside. Using (6.42) with $f(\infty) = f_1$, we can finally write

$$\frac{\Delta C}{C} = -\frac{N_T}{2N_d(W_1)} \frac{(W_1 - \lambda_1)^2 - (W_0 - \lambda_0)^2}{W_1^2} \exp(-g_e t) \quad , \qquad (6.52)$$

where N_T is the defect density in the interval $[(W_0 - \lambda_0), (W_1 - \lambda_1)]$ and

$N_d(W_1)$ the donor density at $x = W_1$. This expression can be simplified further when N_d is a constant in the region of interest. Then, integration of Poisson's equation is extremely easy and we get

$$W_0 = \left(\frac{2\varepsilon}{eN_D} V_0\right)^{\frac{1}{2}} \tag{6.53}$$

$$W_1 = \left(\frac{2\varepsilon}{eN_D} V_1\right)^{\frac{1}{2}} \tag{6.54}$$

and

$$\lambda_0 = \lambda_1 = \left[\frac{2\varepsilon}{e^2 N_D} (E_{Fe} - E_T)_{x=\infty}\right]^{\frac{1}{2}} \quad . \tag{6.55}$$

These last quantities usually being much smaller than W_0 and W_1, we get

$$\frac{\Delta C}{C} \simeq - \frac{N_T}{2N_D} \; \frac{W_1^2 - W_0^2}{W_1^2} \exp(-g_e t) \tag{6.56}$$

which we can simplify still further when $W_1 \gg W_0$ (i.e., $V_1 \gg V_0$)

$$\frac{\Delta C}{C} \simeq - \frac{N_T}{2N_D} \exp(-g_e t) \quad . \tag{6.57}$$

The equations (6.44-47) which are the basis of our treatment have been derived using the depletion approximation. It will be shown in Sect.6.2.5 how they can be generalized to provide an exact treatment of the problem.

6.2.3 Emission from Minority and Majority Carrier Traps

Let us first consider the case of a majority carrier trap, i.e., in an n-type material, a defect for which both k_e and g_e dominate (this is the example detailed in Sect.6.2.2). We have derived in that case the change in capacitance given by (6.52) which we found to be negative. This result can be obtained qualitatively quite readily. After the sudden increase of the reverse bias at $t = 0$, the depletion region does not extend to W_1 but to $W_1 + \Delta W(t = 0)$. This is because, between $t = 0$ and ∞, the electron traps are ionized, increasing positively the space charge ρ from eN_D to $e(N_D + N_T)$. Since $W(t)$ behaves as $\rho^{-\frac{1}{2}}$, W_1 is smaller than $W(t)$, and thus ΔW is positive. The capacitance C is given by (6.47) and $\Delta C/C$ by $- \Delta W/W$, i.e., by $-(N_T/2N_D) \exp(-g_e t)$ which is the result given by (6.57) under conditions specified in the previous section. These conditions of observation for majority carrier traps, are schematized in Fig.6.9a, which shows the negative sign of the transient capacitance.

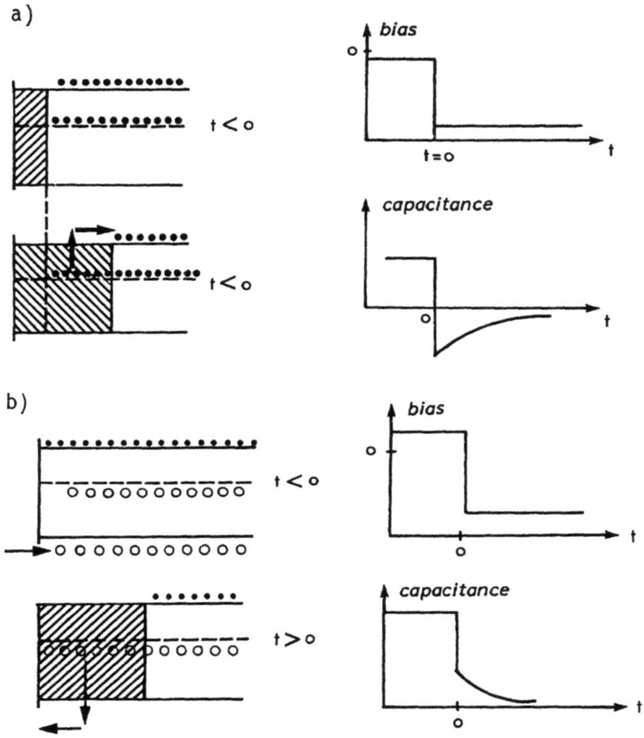

Fig.6.9a,b. Energy diagram (band bending has been omitted for simplicity) of the depleted region (shaded area) of a p^+n junction showing the emission processes occurring after the application of a reverse bias when the junction is initially reverse (a) and forward (b) biased and the resulting capacitance change

Minority carrier traps can be observed as well. Minority carriers, i.e., holes in the case of the n-type material considered here, can be injected in the depletion region of a junction by application of a forward bias. Once trapped on the defects of the depleted region, these holes can be reemitted after application of a reverse bias. As illustrated in Fig.6.9b, this emission of minority carriers, corresponding to positive charges leaving the depletion region, induces a positive change of capacitance ΔC. Thus majority and minority carrier traps are readily distinguishable since they induce capacitance changes of opposite signs.

6.2.4 Capture and Reemission from Majority Carrier Traps

Let us consider the case of a filling impulsion of width t_p and amplitude ΔV, represented in Fig.6.10. The polarization V_0 corresponds to a reverse bias (including the equilibrium barrier height of the junction) while $V_1 = V_0 - \Delta V$ corresponds to a reduced reverse bias. During the pulse the situation is different from that described in Sect.6.2.2 because at time t_p when the polarization returns to V_0, only part of the traps are filled and can reemit their electrons.

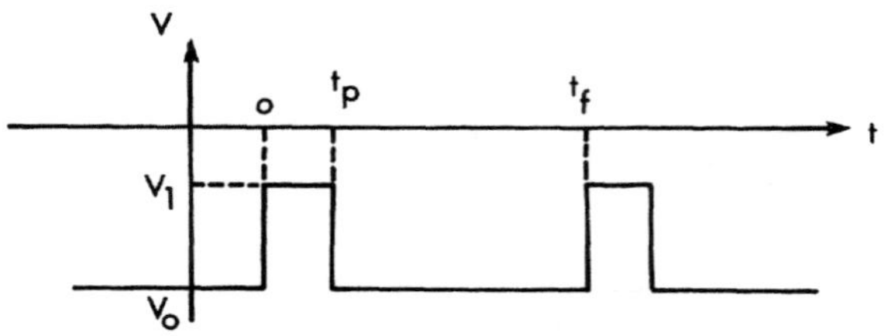

Fig.6.10. Periodic reverse bias applied on a junction to fill traps and study their reemission

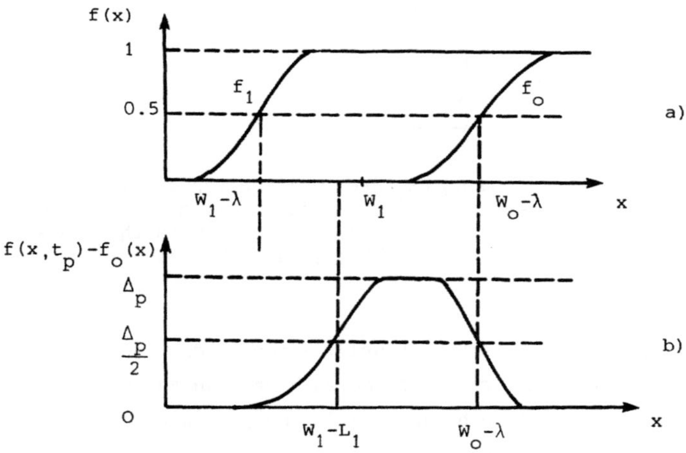

Fig.6.11a,b. Occupancy factor of the trap level: a) f_0 and f_1 corresponding to the potentials V_0 and V_1; b) the difference $f(x,t_p) - f_o$, where $f(x,t_p)$ is the occupancy factor at the end of the pulse

We discuss this situation in terms of the Fermi-Dirac functions f which give the probability of occupation of the trap. Firstly there are two steady-state situations, corresponding to the polarizations V_0 and V_1, which are of importance. We call W_0 and W_1 the respective limits of the depleted zone, and f_0 and f_1 the corresponding Fermi-Dirac functions. These are drawn on

Fig.6.11a, where the notations are the same as in Sect.6.2.2 and where we assume that λ has the same value for both polarizations.

To begin, let us consider the case of one filling pulse alone. We have to calculate $f(t_p)$, which serves as the initial condition for the transient regime we want to discuss.

The equation to solve is (6.42) which can also be written

$$f(x,t_p) - f_1(x) = [f_0(x) - f_1(x)] \exp\left[-\frac{c_e n_1(x) t_p}{f_1(x)}\right] \quad . \tag{6.58}$$

The quantities f_0 and f_1 are drawn on Fig.6.11a. At time t_p the filling of the traps is not uniform since $n_1(x)$ is a rapidly varying function of x. There are essentially two regions: $x > W_1$ where $n_1(x) \sim N_D$ and the filling of the traps is faster; $x < W_1$ where $n_1(x)$ varies rapidly from $N_D(W_1)$ to zero and the filling becomes correspondingly slow. The quantity $f(x,t_p) - f_0$ has thus the shape of Fig.6.11b with a maximum value called Δ_p corresponding to fast filling of the traps.

The transient regime for $t > t_p$ is given by

$$f(x,t) - f_0(x) = [f(x,t_p) - f_0(x)] \exp\left[-\frac{g_e(t - t_p)}{1 - f_0(x)}\right] \quad , \tag{6.59}$$

which can significantly differ from zero, but only if the difference $(W_0 - \lambda) - (W_1 - L_1)$ is large enough compared to kT.

Up to now, we have discussed the case of only one pulse. Let us then consider a sequence of pulses of periodicity t_f, such as the one represented in Fig.6.10. The situation has been analyzed in detail by PONS [6.4,5]. We have to solve the two equations

$$f(x,t_p) - f_1(x) = [f(x,0) - f_1(x)]A_1(x) \tag{6.60}$$

$$f(x,t) - f_0(x) = [f(x,t_p) - f_0(x)]A_0(x,t) \tag{6.61}$$

with

$$A_1(x) = \exp\left[-\frac{c_e n_1(x) t_p}{f_1(x)}\right] \tag{6.62}$$

$$A_0(x,t) = \exp\left[-\frac{g_e(t - t_p)}{1 - f_0(x)}\right] \quad . \tag{6.63}$$

Now $f(x,0)$, the value of $f(x,t)$ at time $t = 0$, is unknown, contrary to the case of a single pulse. If energy losses can be neglected we can use the relation

$$f(x,0) = f(x,t_f) \tag{6.64}$$

where $f(x,t_f)$ can be obtained from (6.61) at time $t = t_f$

$$f(x,t_f) - f_0(x) = [f(x,t_p) - f_0(x)]A_0(x,t_f) \quad . \tag{6.65}$$

We can replace $f(x,t_f)$ by $f(x,0)$ in (6.65) and inject this value of $f(x,0)$ into (6.60). This allows us to calculate $f(x,t_p) - f_0(x)$, giving

$$f(x,t_p) - f_0(x) = [f_1(x) - f_0(x)]\left[\frac{1 - A_1(x)}{1 - A_0(x,t_f)A_1(x)}\right] \quad . \tag{6.66}$$

PONS has derived a numerical solution to this equation (see [6.5] and Sect.6.2.5). Here we begin by giving an approximate analytic solution based on replacing $f(x,t_p) - f_0(x)$ in Fig.6.11b by a rectangle of height Δ_p between $x = W_1 - L_1$ and $W_0 - \lambda$. The maximum value of $f(x,t_p) - f_0(x)$ is Δ_p, corresponding to the region where the filling of the trap is faster. This is the region where $n_1(x) \sim N_D(W_1)$ and $f_1(x) - f_0(x)$ is unity. The application of (6.66) under these conditions gives

$$\Delta_p = \frac{1 - \alpha_1}{1 - \alpha_0(t_f)\alpha_1} \ , \tag{6.67}$$

where α_1 and α_0 are the corresponding values of A_1 and A_0, i.e.,

$$\alpha_1 = \exp[-c_e N_D(W_1)t_p)] \tag{6.68}$$

$$\alpha_0(t_f) = \exp[- g_e(t_f - t_p)] \quad . \tag{6.69}$$

From (6.66) the upper limit of the rectangle is $W_0 - \lambda$, the value at which $f_0(x)$ is equal to 0.5. The lower limit $W_1 - L_1$ is the value at which $f(x,t_p) - f_0(x)$ becomes equal to $\Delta_p/2$ in a region where $n_1(x)$ is rapidly varying but where $f_0(x)$ is equal to zero. The point $W_1 - L_1$ is thus the solution of (6.66) where $f(W_1 - L_1,t_p) - f_0(W_1 - L_1)$ becomes equal to $\Delta_p/2$, given by the equation

$$A_1(W_1 - L_1) = \frac{f_1(W_1 - L_1) - \Delta_p/2}{f_1(W_1 - L_1) - \alpha_0(t_f)\Delta_p/2} \quad . \tag{6.70}$$

This equation can be solved by successive iterations with the initial value $f_1(W_1 - L_1) = 1$ on the right-hand side. We obtain

$$\frac{n_1(W_1 - L_1)}{N_D(W_1)} \sim \frac{\ln\{[1 - \alpha_0(t_f)\Delta_p/2]/[1 - \Delta_p/2]\}}{c_e N_D(W_1)t_p} \quad . \tag{6.71}$$

To extract L_1 from this equation requires knowledge of the potential $V(x)$ in the region near W_1. This can be done exactly but numerically (see [6.5] and Sect.6.2.5). Here, to consolidate these ideas, we make use of the depletion approximation. First we express $n_1(W_1 - L_1)$ as

$$n_1(W_1 - L_1) = N_D(W_1) \exp\left[\frac{e\Delta V(W_1 - L_1)}{kT}\right] \quad , \tag{6.72}$$

where $V(x)$ is the change in potential with respect to the neutral region [(6.72) implicitly assumes $N_D(x)$ to be constant over a few Debye lengthes near W_1]. We then use for $\Delta V(x)$ the parabolic form characteristic of the depletion approximation and obtain

$$n_1(W_1 - \lambda_1) = N_D(W_1) \exp(-L_1^2/2L_D^2) \tag{6.73}$$

where L_D is the Debye length $[\varepsilon kT/e^2 N_D(W_1)]$.

Combining (6.73) and (6.71) we get

$$L_1^2 = 2L_D^2 \ln\left[\frac{c_e N_D(W_1) t_p}{\ln\{[1 - \alpha_0(t_f)\Delta_p/2][1 - \Delta_p/2]\}}\right]^{\frac{1}{2}} \quad . \tag{6.74}$$

Once L_1 is known, we can calculate the expression of the relative change in capacitance, using (6.51) and taking for Δ_p the value $-e(f - f_0)N_T(x)$ in the interval $\{W_1 - L_1, W_0 - \lambda\}$:

$$\frac{\Delta C}{C} = -\frac{N_T\Delta_p}{2N_D(W_0)}\left[\left(1 - \frac{\lambda}{W_0}\right)^2 - \left(\frac{W_1}{W_0} - \frac{L_1}{W_0}\right)^2\right] \exp[-g_e(t - t_p)] \quad . \tag{6.75}$$

This expression is valid for a uniform density of traps and, in order that the above approximations be meaningful, for ΔV larger than a few kT/e. In such a case we also have $W_1 - L_1 > W_0 - \lambda$. When the amplitude of $\Delta C/C$ at $t = t_p$ is extrapolated towards zero, we obtain an apparent threshold ΔV_0 such that

$$W_1 - L_1 = W_0 - \lambda \quad . \tag{6.76}$$

Equations (6.53-55) give W_1, W_0 and λ so that (6.76) is equivalent to

$$\frac{\Delta V_0}{V_0} = 1 - \left(1 + \frac{L_1 - \lambda}{W_0}\right)^2 \tag{6.77}$$

with

$$\frac{\lambda}{W_0} = \left[\frac{(E_{Fe} - E_T)_\infty}{eV_0}\right]^{\frac{1}{2}} \quad . \tag{6.78}$$

Important information extracted from these expressions concerns the capture cross section σ_e contained in c_e, i.e., in Δ_p and L_1. The dependence of $\Delta C/C$ with time t_p is in the terms Δ_p and L_1. Of course it is the term $L_1(t_p)$ which prevents this dependence from being purely exponential (in the limit where the term Δ_p itself has a purely exponential dependence on t_p which is not always the case). Anyway we can use the theoretical expression (6.75) of the amplitude of the capacitance transient to fit the experimental curve (this can also be done with the exact theory). We do not detail here practical techniques used to achieve such a fit [6.5], but simply notice that one can extract information not only on σ_e but also on the energy level E_T since λ depends upon E_T through (6.78) [6.4,6].

A final remark can be made concerning situations where W_1 and $W_0 - \lambda$ are reversed with respect to the situation in Fig.6.11. In such cases the rectangular approximation can still be adapted. Details can be found in [6.7].

6.2.5 Exact Theory

In this section we derive exact expressions which generalize those obtained in the depletion approximation (following the method given by PONS [6.5]). We consider a Schottky barrier and an n-type semiconductor (Fig.6.6). We assume that the doping concentration is constant in the region of interest, i.e., onver a few Debye lengths near the end W of the depletion region. In this region we write Poisson's equation as

$$\frac{d^2V}{dx^2} = -\frac{e}{\varepsilon}\,[N_D(x) - n(x)] \quad . \tag{6.79}$$

In a narrow region near W, $N_D(x)$ can be approximated by $N_D(W)$ and $n(x)$ can be written

$$n(x) = N_D(\ell)\,\exp\left\{\frac{e}{kT}\,[V(x) - V(\ell)]\right\} \quad , \tag{6.80}$$

where $x = \ell$ is a point beyond W where the potential can be considered as constant. This point is close to W so that $N_D(\ell)$ is equal to $N_D(W)$. If we call $\nu(x)$ the ratio $n(x)/N_D(W)$ and $u(x)$ the quantity $\frac{e}{kT}\,[V(x) - V(\ell)]$, then Poisson's equation gives

$$\frac{d^2u}{dx^2} = \frac{1}{L_D^2}\,[\nu(x) - 1] \quad . \tag{6.81}$$

Multiplying both members by $(du/dx)dx$ and integrating between $x = \ell$ and x, we obtain

$$\left(\frac{du}{dx}\right)^2 = \frac{2}{L_D^2}\int_1^{\nu(x)}\left(1 - \frac{1}{\nu}\right)d\nu \quad . \tag{6.82}$$

As u is equal to $\ln\nu$, this expression leads to

$$\frac{dx}{L_D\sqrt{2}} = \frac{d\nu}{2\nu(\nu - 1 - \ln\nu)^{\frac{1}{2}}} \quad . \tag{6.83}$$

A first useful relation can then be obtained by integrating (6.83) between x and W

$$\frac{W - x}{L_D\sqrt{2}} = -\int_{\nu(W)}^{\nu(x)} \frac{d\nu}{2\nu(\nu - 1 - \ln\nu)^{\frac{1}{2}}} = \phi(\nu) \quad . \tag{6.84}$$

As we are improving on the depletion approximation we have to choose an exact definition for W. We take it to be

$$W = \frac{\varepsilon S}{C} = \frac{\varepsilon S d}{dQ}\,[V(x = 0) - V(\ell)] \tag{6.85}$$

as is the case in the depletion approximation. Here Q is given by (6.45) in which the upper bound of integration is extended to ℓ instead of W. This gives, for constant doping

$$Q = eN_DS\int_{x=0}^{\ell}(1 - \nu)\,dx \quad , \tag{6.86}$$

i.e., using (6.83)

$$Q = \frac{eN_DSL_D}{\sqrt{2}}\int_{\nu(0)}^{1}\frac{(1 - \nu)d\nu}{\nu(\nu - 1 - \ln\nu)^{\frac{1}{2}}} \quad . \tag{6.87}$$

Here Q is a function of the height of the potential barrier only through $\nu(0)$ equal to $\exp\{e[V(0) - V(\ell)]/kT\}$. Then W is given by

$$W = L_D\sqrt{2}\,\frac{[\nu(0) - 1 - \ln\nu(0)]^{\frac{1}{2}}}{1 - \nu(0)} \quad . \tag{6.88}$$

This expression relates W to $\nu(0)$, i.e., to the applied bias. It must be equal to (6.84) in the limit where x = 0. Equating these two expressions of W leads to an equation for $\nu(W)$. As soon as $\nu(0) \gtrsim 5$, the result is constant and the following is obtained:

$$\nu(W) \simeq 0.55 \quad . \tag{6.89}$$

This result remains valid for varying $N_D(x)$.

The final point concerns the generalization of (6.50). Starting from (6.44) with W replaced by ℓ, we can write the generalization of (6.50) as

$$0 = \int_0^\ell x\Delta n \, dx + \int_0^\ell x\Delta b \, dx \quad . \tag{6.90}$$

The term Δn can be written as $N_D(W)\Delta\nu$ and x, from (6.84), as $W - L_D\sqrt{2}\phi(\nu)$, so that

$$\int_0^\ell x\Delta n \, dx = N_D(W) \int_0^\ell (W - L_D\sqrt{2}\phi)\Delta\nu \, dx \quad . \tag{6.91}$$

Now, from (6.84) ΔW can be obtained as

$$\Delta W = L_D\sqrt{2} \frac{d\phi}{d\nu} \Delta\nu \quad , \tag{6.92}$$

and (6.83) can also be written

$$dx = -L_D\sqrt{2} \frac{d\phi}{d\nu} d\nu \tag{6.93}$$

so that we have

$$\Delta\nu dx = -\Delta W \, d\nu \quad . \tag{6.94}$$

This allows us to express (6.91) in the form

$$\int_0^\ell x\Delta n \, dx = -N_D(W)\Delta W \int_{\nu(0)}^1 [W - L_D\sqrt{2}\phi(\nu)] \, d\nu \quad . \tag{6.95}$$

Usually $\nu(0)$ is very small and can be equated to zero. The generalization of (6.50) obtained by combining (6.90,95) is thus

$$0 = -N_D(W)\Delta W\left[W - L_D\sqrt{2} \int_0^1 \phi(\nu)d\nu\right] + \int_0^\ell x\Delta b \, dx \quad . \tag{6.96}$$

However, as pointed out in [6.5], the corrective term containing the integral is so small that it can be ignored in most cases. The conclusion is that (6.50) remains valid in the general case.

Then the relative change in capacitance corresponding to that in Sect. 6.2.4 can be calculated from

$$\frac{\Delta C}{C} = -\frac{\Delta W}{W_0} \quad , \tag{6.97}$$

that is

$$\frac{\Delta C(t)}{C} = -\frac{N_T}{W_0^2 N_D(W_0)} \int_0^\ell x[f(x,t) - f_0(x)] \, dx \quad . \tag{6.98}$$

Exact values of $\Delta C/C$ can thus be numerically calculated using (6.61,66, and 69). This can serve to test the rectangular approximation described in Sect. 6.2.4. From this point of view it is interesting to notice that (6.71) can be solved more exactly than with the use of (6.73). We note that the left-hand side of (6.71) is just $\nu(W_1 - L_1)$ which is related to L_1 by (6.84), i.e., we obtain

$$L_1 = L_D\sqrt{2}\phi\left(\frac{n_1(W_1 - L_1)}{N_D(W_1)}\right) \quad , \tag{6.99}$$

which allows us to extract L_1 from (6.71).

6.2.6 Deep Level Transient Spectroscopy

We have seen in (6.16) that g_e varies exponentially with inverse temperature. At low enough temperature, g_e is exceedingly small, and in the depletion region of the reverse biased diode (Fig.6.6) the electrons remain frozen on the defect when $E_{Fe} > E_T$. When the temperature is raised, emission occurs at temperatures for which g_e has a nonnegligible value. The emission results in a change of the junction capacitance and in the injection of current in the circuit of the diode. The techniques which monitor the capacitance (or the current) changes as the temperature increases are called, respectively, TSCAP (for thermally stimulated capacitance) and TSC (for thermally stimulated current). They can be quantitatively interpreted only when the law which governs the variation of temperature with time (usually a linear variation) is known accurately. Such techniques are less sensitive than the one which measures directly the time constant of a transient capacitance or current at a given temperature [6.8]. The difficulty resides in the analysis of the transient $\Delta C(t)$; the associated time constant g_e^{-1} can only be easily obtained (from the plot of $\ln \Delta C$ versus t^{-1}) when the value for $\Delta C(\infty)$ can be determined. However, this requires that the emission rates due to other traps have negligible values in the same temperature range.

Recently, a method called DLTS (for deep level transient spectroscopy) has become very popular because it is sensitive and simple, i.e., it allows to some extent an easy separation of various emission rates occuring in the same temperature range. For this, periodical filling pulses (Sect.6.2.4) are applied on a reverse biased junction. The analysis of the resultant transient is performed using a filtering operation. As shown in (6.70), the transient capacitance resulting from the filling pulse has the form

$$\Delta C(t) = -\Delta C_0 \exp(-g_e t) \quad . \tag{6.100}$$

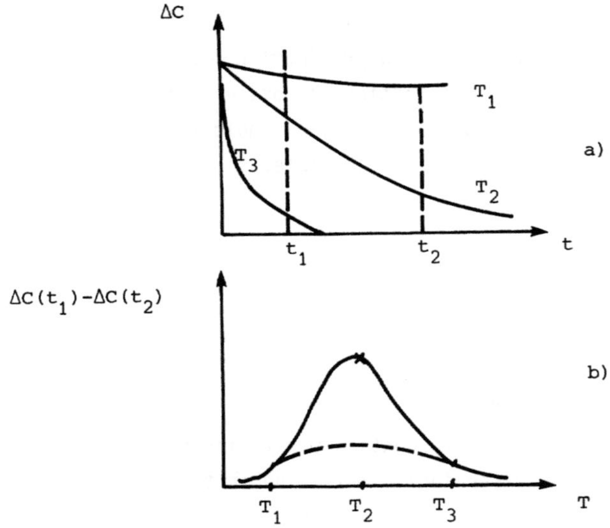

Fig.6.12a,b. Schematic description of the DLTS analysis: a) capacitance transient at different temperatures; b) resultant DLTS spectrum

The time constant of the transient capacitance can be obtained by taking the derivative of the signal at a given time t_0

$$\left(\frac{d\Delta C}{dt}\right)_{t_0} = \Delta C_0 g_e \exp(-g_e t_0) \quad . \tag{6.101}$$

When the temperature T varies, g_e varies accordingly and $|(d\Delta C/dt)_{t_0}|$ goes through a maximum for T such that $t_0 = g_e^{-1}$ since

$$\frac{d}{dg_e}\left(\frac{d\Delta C}{dt}\right)_{t_0} = \Delta C_0(1 - g_e t_0) \exp(-g_e t_0) \quad . \tag{6.102}$$

Thus, fixing t_0 is equivalent to the selection of a given value $g_e(T_0)$ of the emission rate. The temperature T_0 at which the maximum of the derivative occurs corresponds to $g_e(T_0) = 1/t_0$. In practice the derivative of the signal $\Delta C(t)$ is obtained by fixing the difference $\Delta C(t_1) - \Delta C(t_2)$ of the amplitudes at two times t_1 and t_2, measured with the help of a double boxcar (Fig.6.12). This is the method originally proposed by LANG [6.9].

Generally speaking, the DLTS analysis is a filtering operation. The measuring system gives a nonzero signal when g_e is situated in a given window, i.e., in a given temperature range. For a transient signal

$$S(g_e,t) = S_0 \exp(-g_e t) \tag{6.103}$$

the response is

$$R(g_e) = S_0 \int_0^{t_f} F(t) \exp(-g_e t)\, dt \quad , \tag{6.104}$$

where t_f is the period of the pulse repetition and F(t) represents the filter. The filter described above

$$F(t) = \delta(t - t_1) - \delta(t - t_2) \tag{6.105}$$

provides the following response

$$R(g_e) = S_0[\exp(-g_e t_1) - \exp(-g_e t_2)] \quad , \tag{6.106}$$

which is maximum for

$$g_e = \frac{\ln(t_2/t_1)}{t_2 - t_1} \quad . \tag{6.107}$$

Of course, any other filter can be used. The exponential function is difficult to use in practice. The filter such that $F(t) = 0$ for $0 < t < t_f/4$, $F(t) = 1$ for $t_f/4 < t < t_f/2$, $F(t) = 0$ for $t_f/2 < t < 3t_{f/4}$ and $F(t) = -1$ for $3t_f/4 < t < t_f$, which can be simply realized with a double lock-in detection, is interesting because the use of a lock-in, instead of a boxcar, enables one to work with signals in the μV range rather than in the mV range, increasing considerably the sensitivity of the technique.

With a standard capacitance meter whose sensitivity is $\lesssim 10^{-3}$, (6.57) gives the minimum defect concentrations which can be detected $\lesssim 2 \times 10^{-3}$ N_D (cm^{-3}). The DLTS technique, when applied to lightly doped material, is thus the most sensitive technique to detect electrically active defects (this sensitivity could be increased in principle by the use of a resonant circuit). However, in the case where the sample studied presents a resistance R in series with the junction capacitance C, the amplitude of the capacitance transient is decreased since the apparent capacitance C' measured (at pulsation ω) by the capacitance meter, the equivalent capacitance of the RC series circuit, is

$$C' = \frac{C}{1 + R^2C^2\omega^2} \quad . \tag{6.108}$$

Moreover the capacitance transient observed ΔC', related to ΔC through

$$\frac{\Delta C'}{C'} \Big/ \frac{\Delta C}{C} = \frac{1 - R^2C^2\omega^2}{1 + R^2C^2\omega^2} \quad , \tag{6.109}$$

can be of reverse sign, as compared to ΔC, when $RC\omega$ is larger than 1. This means that a majority carrier trap will appear like a minority carrier trap and vice versa [6.10].

The resolution Δg_e of the method is defined by the bandwidth of the filter, that is by the temperature width ΔT of a DLTS peak. It is related to Δg_e through

$$\frac{\Delta g_e}{g_e} = \frac{\Delta T}{T_0}\left(2 + \frac{E_c - E_T}{kT_0}\right) \quad . \tag{6.110}$$

This quantity depends only on the ratio t_1/t_2 in the case of filtering with (6.105).

When defects do not correspond to a single localized level but to a distribution of levels, i.e., to a density of states $D(E_T)$, the latter, when it is a slowly varying function of E_T, can be derived from a DLTS analysis. At a given temperature T, the response of the system is given by the generalization of (6.106)

$$R(T) = \int_{E_1}^{E_2} D(E_T)R(g_e)dE_T \tag{6.111}$$

where E_1 and E_2 are the bounds of $D(E_T)$. The response function $R(g_e,t)$ is given by (6.106). It depends on g_e but also on E_T since we have, from (6.16)

$$g_e = \beta \exp\left(\frac{E_T}{kT}\right) \quad . \tag{6.112}$$

We have seen that $R(g_e,T)$ has a peak at g_e, given by (6.107) i.e., E_T given by

$$E_{T_0} = kT \ln\left[\frac{\ln(t_2/t_1)}{\beta(t_2 - t_1)}\right] \quad . \tag{6.113}$$

If the halfwidth of this peak is small we can write

$$R(T) \sim S_0 D(E_{T_0}) \int_{E_1}^{E_2} [\exp(-g_e t_1) - \exp(-g_e t_2)]dE_T \quad , \tag{6.114}$$

where E_{T_0} is the energy of this peak, i.e., the energy for which g_e is given by (6.107). Inverting (6.112) we can rewrite

$$R(T) \sim S_0 D(E_{T_0})kT \int_{g_1}^{g_2} [\exp(-g_e t_1) - \exp(-g_e t_2)] \frac{dg_e}{g_e} \quad . \tag{6.115}$$

As the integrand is assumed to be a narrow peaked function, the bounds in the integral can be taken from $g_e = 0$ to $g_e = \infty$, then we obtain

$$R(T) = S_0 D(E_{T_0}) kT \ln \frac{t_2}{t_1} \quad . \tag{6.116}$$

The density of states is therefore obtained by dividing the amplitude of the DLTS spectrum by the quantity $kT_0 \ln(t_2/t_1)$. The determination of E_{T0} necessitates knowledge of β, i.e., of the capture cross section. However care must be taken that the above derivation is valid, i.e., that $D(E_T)$ is practically constant over the range in energy of the function $R(g_e,T)$. This condition depends upon t_1 and t_2 but we shall not make a detailed derivation of it here.

6.2.7 Admittance Spectroscopy

When a small ac voltage $\delta\gamma$ of frequency ω is applied to the reverse biased junction, the defect level sweeps through the electron Fermi level E_{Fe} and we expect alternative filling and emptying of the traps. The response of the traps to the perturbation satisfies the differential equation (6.35) which reduces to

$$\frac{db}{dt} = - g_e b + k_e s \tag{6.117}$$

when we consider again the case of an electron trap in n-type material. The occupancy depends on the ac potential at the point where the Fermi level intersects the trap level. The changes in population Δb and Δs from the equilibrium values b_0 and s_0 are given, to first order, by

$$\frac{d\Delta b}{dt} = - g_e \Delta b + k_e \Delta s + \Delta k_e s_0 \quad . \tag{6.118}$$

To obtain this equation we have taken into account the fact that $g_e b_0$ is equal to $k_e s_0$. The change Δk_e in the capture probability is associated with the change Δn of the free-carrier concentration. For a small amplitude of the perturbation we can treat each component separately at a given frequency ω. As $\Delta s = -\Delta b$ we can thus write (6.118) in the form

$$i\omega\Delta b = - (g_e + k_e)\Delta b + s_0 \Delta k_e \quad , \tag{6.119}$$

which leads to

$$\Delta b = \frac{s_0 \Delta k_e}{g_e + k_e + i\omega} \quad . \tag{6.120}$$

The free-carrier concentration in the conduction band remaining at thermal equilibrium, the perturbed concentration Δn is

$$\frac{\Delta n}{n} = \frac{\Delta k_e}{k_e} = \frac{e\delta v}{kT} \quad . \tag{6.121}$$

Substituting in (6.120) and introducing the occupancy f we get

$$\Delta b = \frac{(e\delta v/kT)N_T f(1 - f)}{1 + i\omega[(1 - f)/g_e]} \quad . \tag{6.122}$$

The function $f(1 - f)$ peaks at $f = 1/2$. It is thus possible to extract g_e from the determination of the frequency for which

$$\omega\left(\frac{1 - f}{g_e}\right) = 1 \quad , \tag{6.123}$$

using conductance or capacitance measurements to determine Δb. From the temperature variation of this frequency the temperature dependence of g_e can be derived, giving the energy level position $E_c - E_T$.

From this we see that the study of $C(\omega)$ or of the conductance $G(\omega)$ versus T provides the temperature variation of the emission rate, like the DLTS method. The main disadvantages of this method compared to the DLTS one are: i) a lower sensitivity since the traps detected are those situated within $\sim kT$ around the Fermi level and not those of the whole space charge region, and ii) the impossibility to detect minority carrier traps. Its main advantage is that it needs only simple electronic equipment (an ac generator of variable pulsation and a lock-in detector). More details about the theory of the response of traps to an applied ac voltage can be found in [6.11].

6.3 Nonradiative Recombination Processes

When a free carrier becomes trapped on the site of a defect, energy corresponding to the energy difference ΔH between a band and the defect level is involved. This energy can be i) dissipated radiatively, i.e., into photon emission, ii) transmitted to other free carriers: this is the Auger process or iii) transformed into phonons.

6.3.1 Auger Processes

Auger recombination is the inverse of impact ionization in the same way as radiative recombination is the inverse of optical excitation. There are two kinds of Auger recombination for band-to-band processes, with excitation of an electron (Fig.6.13a) or of a hole (Fig.6.13b). For the electron-electron recombination process of Fig.6.13a the rate of recombination is of course:

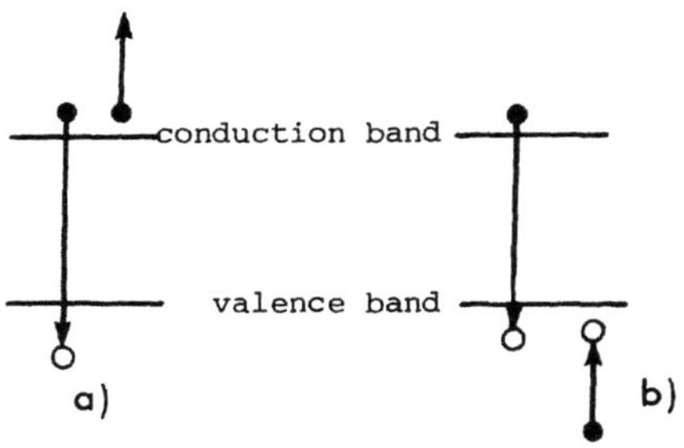

Fig.6.13. Representation of the Auger process, i.e., of recombinations with excitation of an electron (a) and of a hole (b)

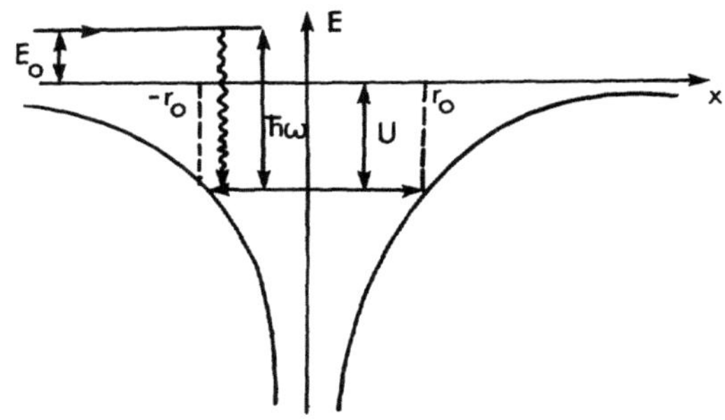

Fig.6.14. Motion of a carrier in the vicinity of a Coulomb center

$$R_{ee} = B_{ee}n^2p \quad , \tag{6.124}$$

while for hole-hole recombination process of Fig.6.13b, it is

$$R_{hh} = B_{hh}np^2 \quad . \tag{6.125}$$

In n-type material, the first term expresses Auger recombination with a minority carrier and the second one with a majority carrier. In extrinsic materials where the minority carrier concentration (p) is negligible compared to the majority carrier concentration, the term R_{ee} dominates and the constant B_{ee} is the coefficient which characterizes the Auger recombination. The recombination probability constant B_{ee} for Auger processes with donors is obtained from the ionization probability constant A_{ee}, due to impact ionization of the doping impurities, since, according to the principle of detailed balance, any ionization-recombination process is individually in equilibrium. For instance, in n-type material containing N_T donors (whose ionization energy is E_I), the equilibrium condition is written

$$A_{ee}nN_T^0 = B_{ee}n^2N_T^+ \quad , \tag{6.126}$$

where n is the free-carrier concentration, and N_T^+ and N_T^0 are the ionized and un-ionized donors, respectively. The ratio A_{ee}/B_{ee} is therefore the equilibrium ionization constant of the donors

$$\frac{A_{ee}}{B_{ee}} = gN_c \exp - \frac{E_I}{kT} \quad . \tag{6.127}$$

The thermal impact ionization probability constant A_{ee} is obtained from

$$A_{ee} = \int_{v_i}^{\infty} S(v)f(v)dv \quad , \tag{6.128}$$

where the lower limit for the carrier velocity v is taken to be the velocity v_i corresponding to the kinetic energy needed to ionize a donor

$$\frac{1}{2} m v_i^2 = E_I \quad . \tag{6.129}$$

The function f(v) is the Boltzman distribution. The impact ionization cross section S(v) is derived from the ionization cross section of hydrogen, corrected for the effective mass and for the dielectric constant since donors are hydrogen-like centers. It is given by [6.12]

$$S = 0.825 \frac{4\pi e^4}{E_I W} \ln\left(\frac{W}{0.012\, E_I}\right) \quad , \tag{6.130}$$

whereby the kinetic energy W of the free carriers is introduced. The calculation of A_{ee} has been performed by SOLAR and BURSTEIN [6.13] using the Born approximation [6.14], leading to

$$A_{ee} = 5.87 \times 10^{-4} \left(\frac{m}{m^*}\right)^{\frac{1}{2}} E_I^{-1} T^{-\frac{1}{2}} \left(\frac{1 + 0.522 \log_{10}\varepsilon}{\varepsilon}\right) \exp\left(-\frac{E_I}{kT}\right) \quad . \tag{6.131}$$

From the expression of A_{ee}, the recombination probability is readily obtained, using (6.127)

$$\underset{(\mathrm{cm}^3\,\mathrm{s}^{-1})}{B_{ee}} = 2.44 \times 10^{-19} \left(\frac{m}{m^*}\right)^2 E_I^{-1} T^{-2} \left(\frac{1 + 0.522 \log_{10}\varepsilon}{\varepsilon}\right) \tag{6.132}$$

when E_I is expressed in eV and T in K. Thus, for n-type silicon,

$$B_{ee} \sim 5.3 \times 10^{-18} T^{-2} \quad . \tag{6.133}$$

This result is, of course, different for nonhydrogenic states and can be considerably different when the defect is not electrically neutral after it captures an electron.

When the energy ΔH is not dissipated through an Auger process, it is released on the defect site and transformed into phonons. The question is then the following: how is this energy transformed into phonons and what are these phonons? Of course, the magnitude of the capture cross section and its dependence on temperature should reflect the mechanism of transformation and its study should provide information on the electron-phonon interaction at the defect site. Basically, there are two types of mechanisms: i) the cascade of phonons for shallow defects, which act as Coulomb attractive centers and ii) the multiphonon emission process, which seems to be the dominant process for deep defects. In the cascade process, the free carrier loses some energy by emitting one phonon each time it decays from an excited state to another less excited state of the defect. The multiphonon emission process corresponds to the simultaneous emission of different phonons. We now consider these two capture processes in some detail.

6.3.2 Cascade Capture

We detail here the original approach of LAX [6.15], derived in order to explain the very large cross sections (10 to 10^4 Å^2) which have been observed for Coulomb attraction centers. The principle of this capture can be best illustrated using a classical description for the motion of a carrier (Fig. 6.14), of kinetic energy E_0, in the vicinity of a Coulomb center, of attractive potential $V(x)$. At some time, it experiences a collision in which it emits a phonon of energy $\hbar\omega$. Thus, after the collision, the energy of the carrier becomes $E_0 - \hbar\omega$. When $E_0 - \hbar\omega = -U$ becomes negative, the carrier gets trapped with the binding energy U. Its new trajectory becomes then an oscillation between $+r_0$ and $-r_0$ where r_0 is the classical turning point at which $U + V(x)$ becomes negative. This simple representation is the one-dimensional equivalent of the model proposed by Lax.

The reason why a classical description can be used is, according to Lax, the following: the model assumes that capture occurs mainly through highly excited hydrogenic states which have very large cross sections. Because such states have a high quantum number, classical mechanics can be used. That such a point of view is valid or not is still an open question, which we do not discuss here since this does not change the physical nature of the explanation. Moreover the classical model leads to a much easier calculation [6.15].

Let us now see how the capture cross section can be calculated. Before doing this, we must define the physical problem we are looking at. Experimentally, the capture cross section is not directly observed but instead the capture-rate constant corresponding to the capture of free carriers by empty traps in their ground state. In a steady-state experiment, the constant capture rate is compensated by a constant thermal emission rate for instance. The capture cross section $\sigma(E_0)$ for a carrier of kinetic energy E_0 can thus be written

$$\sigma(E_0) = \sum_U \sigma(E_0,U)P(U) \quad , \tag{6.134}$$

where $\sigma(E_0,U)$ is the cross section for capture into a state of binding ener-

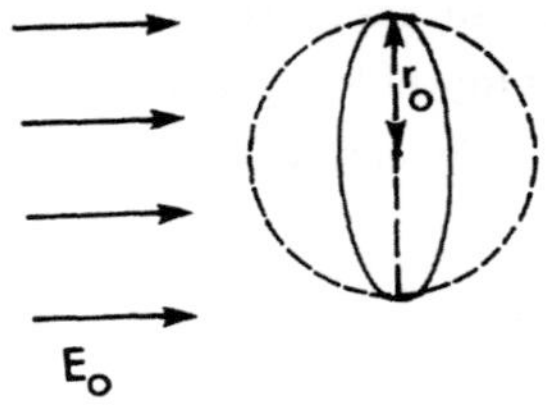

Fig.6.15. Sphere of radius r_0 in which capture is occurring

gy U, and P(U) is the "sticking probability", i.e., the probability for the carrier to reach the ground state before being reemitted.

The first problem is the calculation of $\sigma(E_0,U)$. We can get an idea of its expression by looking at Fig.6.15. Carriers of given kinetic energy E_0 emit phonons of energy $\hbar\omega$, i.e., have a fixed binding energy $U = \hbar\omega - E_0$, once trapped. According to Fig.6.14, trapping can occur only within a sphere of radius r_0, such that

$$V(r_0) = E_0 - \hbar\omega \quad . \tag{6.135}$$

Such a sphere is represented in Fig.6.15. Coming back to the definition of the cross section, the number of trapped carriers per unit time will be proportional to the flux J, times the section of a disc πr_0^2 where they can be captured, times the probability that they will be captured when they are in the sphere of radius r_0. This last probability is equal to the probability that a corresponding collision will occur, which itself is equal to the average distance across the sphere divided by the mean free path $\ell(E_0,U)$. The average distance across the sphere being equal to its volume $4\pi r_0^3/3$ divided by its perpendicular section πr_0^2, i.e., $4r_0/3$, the cross section $\sigma(E_0,U)$ is finally given by

$$\sigma(E_0,U) = \pi r_0^2 \, \frac{4}{3} \, \frac{r_0}{\ell(E_0,U)} \quad , \tag{6.136a}$$

that is

$$\sigma(E_0,U) = \frac{4}{3} \, \frac{\pi r_0^3}{\ell(E_0,U)} \quad . \tag{6.136b}$$

In this classical model, the energy U is continuous and the discrete sums have to be replaced by integrals. Thus (6.134) is replaced by

$$\sigma(E_0) = \int \sigma'(E_0,U)P(U)dU \tag{6.137}$$

where $\sigma'(E_0,U)$ is now the cross section for capture into states having energies comprised between U and U + dU. For Coulomb centers, LAX showed that

$$\sigma'(E_0,U) = \int_0^{r_0} 4\pi r^2 dr \, \frac{E(r)}{E_0} \, \frac{1}{\ell'[E(r),E_0 + U]} \quad , \tag{6.138}$$

where $E(r)$ is the kinetic energy at a distance r from the center and $\ell'[E(r),E_0 + U]$ is the differential mean free path. This expression exactly corresponds to (6.137). The differential mean free path for collisions between the carriers and acoustical or optical phonons was calculated in [6.15] but we shall not detail the expressions here.

The other important quantity to determine is the sticking probability $P(U)$. This probability can always be written

$$P(U) = \sum_{U'} K_{U,U'} P(U') \quad , \tag{6.139}$$

where $K_{U,U'}$ is the probability for the carrier to go from state U to U' by absorbing or emitting phonons. Expression (6.139) simply states that the probability for the carrier to go from state U to the ground state is the sum of the probabilities $K_{UU'}$ to go into other states U', times the corresponding probability $P(U')$ to go from U' to the ground state. U being continuous, (6.139) has to be replaced by an integral equation

$$P(U) = \int K(U,U')P(U')dU' \quad , \tag{6.140}$$

where the kernel $K(U,U')$ can be written in terms of the differential mean free path ℓ'

$$K(U,U') = \frac{\ell^{-1}(U,U' - U)}{\int dU' \ell^{-1}(U,U' - U)} \quad . \tag{6.141}$$

This expression states that $K(U,U')$ is the relative probability of collision with energy transfer $U' - U$ to phonons, correctly normalized.

The numerical calculation of LAX gives the right order of magnitude as shown on Fig.6.16. We can understand the results in an elementary way. First, the sticking probability $P(U)$ has the schematic shape of Fig.6.17 [6.15], which means that an electron of binding energy $U \lesssim kT$ is likely to be reemitted, as is quite natural. $P(U)$ is near to unity only for $U \gtrsim 4$ kT. Thus, the larger orbits will correspond to this case. The average kinetic energy E_0 of incident carriers also being of the order of kT, the larger orbits will correspond to states with $E_0 + U$ equal to $\sim$5 kT. The corresponding radius r_0 of the sphere will then be

$$\frac{1}{\varepsilon r_0} \simeq 5 \text{ kT} \tag{6.142}$$

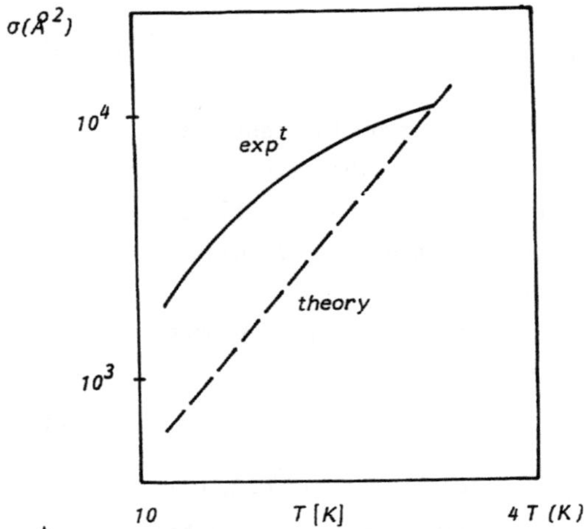

Fig.6.16. Experimental variation of the capture cross section of an hydrogenic impurity versus temperature compared with Lax's theory

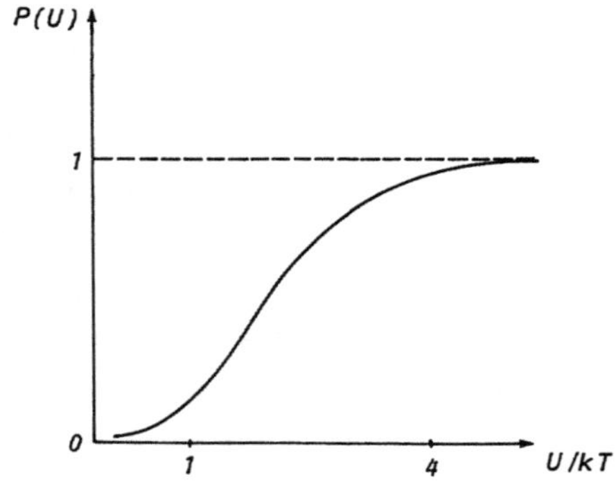

Fig.6.17. Sticking probability as a function of the binding energy in units of kT

leading, according to (6.136), to

$$\sigma \sim \frac{4}{3}\frac{\pi}{(5\varepsilon kT)^3}\frac{1}{\ell} \quad . \tag{6.143}$$

When only coupling to acoustic phonons is considered ℓ^{-1} has a temperature dependence in T^{-1} weaker than for r_0^3 which varies as T^{-3}. Then, σ increases as T^{-4} which is close to the experimental behavior. In the corresponding temperature range ($T \sim 5$ K), r_0 is of order 10^4 Å which explains why such enormous cross section for capture can be obtained.

LAX's original assumptions have been discussed and corrected by different authors. One approach [6.16] has been to use Boltzmann's equation, again in the classical description. All other works [6.17] have considered the discrete levels but retaining only the lowest s states whose contribution was considered to be predominant. Nevertheless, LAX's model gives the physical basis of the explanation of these giant cross sections and illustrates the cascade mechanism in a very simple way.

6.3.3 Carrier Capture by Multiphonon Emission

We consider now the case where the free carriers are captured at deep centers through multiphonon emission. The theory of this effect is, in many respects, similar to the theory of optical line shape developed in Sect. 4.3, except that here we have a nonradiative transition between band states and a defect-gap state. We shall see that this transition is induced by the

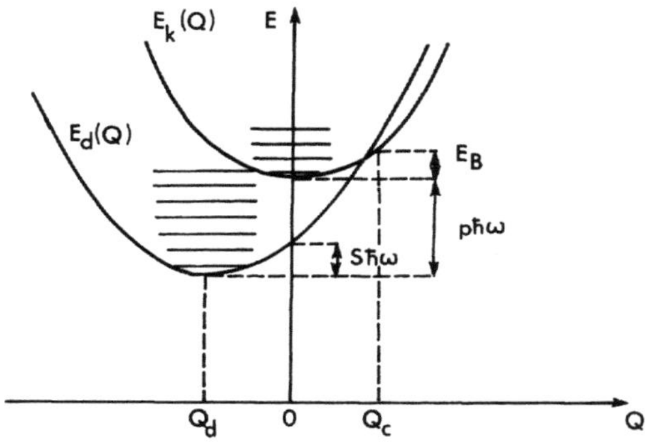

Fig.6.18. The configuration coordinate diagram

nonadiabatic terms in the Hamiltonian. We first discuss these nonadiabatic terms. We then derive the form of the capture rate in the low- and high-temperature limits. We also discuss briefly the Condon and non-Condon approximations.

Let us first investigate the physical origin of the nonradiative transitions. For this, we use again the simplified situation corresponding to the configuration coordinate diagram of Fig.6.18. It assumes coupling with one lattice coordinate Q; $E_d(Q)$, the defect electronic energy, is minimum for a value $Q = Q_d$, differing from $Q = 0$ in the presence of electron-lattice interaction. This is not so for the excited states $E_k(Q)$ which are completely delocalized and, thus, experience an infinitely small electron-lattice coupling with the local mode Q. Then, the minimum of $E_k(Q)$ occurs at $Q = 0$, i.e., for the perfect lattice configuration. We know from Sect.2.1.1 that when the two energy branches are not distant enough in energy, the total wave function is a combination of the Born-Oppenheimer products $\psi_{ed}(\underline{r},Q)\chi_{d,n}(Q)$ and $\psi_{ek}(\underline{r},Q)\chi_{k,m}(Q)$ (we have omitted phonon modes other than Q). We also know that the matrix element of the Hamiltonian between these Born-Oppenheimer products reduces to the matrix element of the nonadiabaticity operator H^{NA} defined by (2.9).

Usually H^{NA} is small enough to be treated by time-dependent perturbation theory, so that we can view the situation as follows: we assume that, initially the system is in state $E_k(Q)$ corresponding to the ionized defect with the electron in state k of the conduction band; then the probability P_k per unit time that the electron will be captured by the defect is given by

$$P_k = \frac{2\pi}{\hbar} \operatorname{Av}_m |\langle\psi_{ek}\chi_{km}|H^{NA}|\psi_{ed}\chi_{dn}\rangle|^2 \delta[(p + m - n)\hbar\omega] \quad , \qquad (6.144)$$

where Av_m denotes the thermal average over the vibrational quantum number in the initial state. There is here a difficulty related to the fact that we consider only one phonon mode of well defined frequency ω. In (6.144) m and n are integers while p is not, in general. The technique is then to use an average value $\bar{P}_k$ defined as the sum of the P_k over an energy range ΔE for the E_k, chosen in such a way that the conservation of energy in (6.144) can be realized. Obviously, we have $\Delta E = \hbar\omega$ so that we can write

$$\bar{P}_k = \frac{2\pi}{\hbar^2\omega} Av_m|\langle\psi_{ek}\chi_{km}|H^{NA}|\psi_{ed}\chi_{d,m+p}\rangle|^2 \quad . \tag{6.145}$$

This expression is quite similar to what we have derived for the broadening of optical absorption bands in the limit of zero frequency, since here the energies of the initial and final vibronic states have to be equal. To evaluate (6.145) the simplest approximation consists, again, in assuming the electronic matrix element $\langle\psi_{ek}|H^{NA}|\psi_{ed}\rangle$ to be independent of Q (this is often called the Condon approximation, but not by all authors [6.18]). In that case we obtain

$$\bar{P}_k = \frac{2\pi}{\hbar^2\omega} |\langle\psi_{ek}|H^{NA}|\psi_{ed}\rangle|^2 Av_m|\langle\chi_{km}|\chi_{d,m+p}\rangle|^2 \quad . \tag{6.146}$$

This quantity just corresponds to I_p defined by (4.68,70) and $\bar{P}_k$ can be rewritten in terms of the normalized probability W_p given by (4.68) as

$$\bar{P}_k = \frac{2\pi}{\hbar^2\omega} |\langle\psi_{ek}|H^{NA}|\psi_{ed}\rangle|^2 W_p \quad . \tag{6.147}$$

As discussed in [6.19], typical values for $\hbar\omega$ are of the order of 0.04 eV in GaAs while a deep level is such that $p\hbar\omega$ is of the order of 0.5 eV. This leads to values of p at least equal to 10 and sometimes amounting to 20. Under such conditions, the low-temperature approximation for W_p is valid over a wide range. The general mathematical condition for this is that [6.20]

$$\frac{S^2}{P}\bar{n}(\bar{n}+1) \ll 1 \tag{6.148}$$

with

$$\bar{n} = \left[\exp\left(\frac{\hbar\omega}{kT}\right) - 1\right]^{-1} \quad . \tag{6.149}$$

If we assume $\bar{n}$ to be smaller than 1 it is easily seen that the inequality (6.148) is satisfied for

$$\frac{\hbar\omega}{kT} \gg \ln\left|1 + \frac{S^2}{P}\right| \quad . \tag{6.150}$$

For $S^2/P \sim 1$, T must be much smaller than about 600 K. For $S \sim 1$, $S^2/P \sim 0.1$ to 0.05 and T has to be smaller than 3000 K, showing that the low-temperature limit is likely to be valid in many cases over most of the temperature range.

We have seen that in this low-temperature limit the quantity W_p takes the particularly simple form given by (4.44). Thus, $\bar{P}_k$ can be written in this limit as

$$\bar{P}_k = \frac{2\pi}{\hbar^2\omega} |\langle\psi_{ek}|H^{NA}|\psi_{ed}\rangle|^2(\bar{n}+1)^p \frac{S^p}{p!} \exp\left[-2S\left(\bar{n}+\frac{1}{2}\right)\right] \quad . \tag{6.151}$$

To obtain the capture cross section we have to divide $\bar{P}_k$ by the flux (or current density) J equal to $\langle v\rangle/\Omega$ where Ω is the total volume and $\langle v\rangle$ the average thermal velocity as shown in Sect.5.3.1. This leads for σ to a low-temperature form

$$\sigma = \sigma_0(\bar{n}+1)^p \frac{S^p}{p!} \exp\left[-2S\left(\bar{n}+\frac{1}{2}\right)\right] \tag{6.152}$$

with

$$\sigma_0 = \frac{2\pi}{\hbar^2\omega} \frac{\Omega}{\langle v\rangle} |\langle\psi_{ek}|H^{NA}|\psi_{ed}\rangle|^2 \quad . \tag{6.153}$$

Another interesting case is the high-temperature limit, valid if $p \gg 1$ (which is realized) and if [6.20]

$$\frac{4S}{p}\,[\bar{n}(\bar{n}+1)]^{\frac{1}{2}} \gg 1 \quad , \tag{6.154}$$

which is roughly equivalent to

$$\frac{\hbar\omega}{kT} \ll \ln\left(1+\frac{4S}{p}\right) \quad . \tag{6.155}$$

When $p \sim 10$-20 and $S \sim 1$ this inequality is only satisfied for T much larger than 300 K, which is not often the case. However, in situations where this can be realized (mainly for very large S), $\hbar\omega W_p$ becomes a Gaussian, given by (4.60), so that we have

$$\sigma \sim \sigma_0 \sqrt{\frac{\hbar\omega}{4\pi kTS}} \exp\left[-\frac{(p-S)^2\hbar\omega}{4kTS}\right] \quad . \tag{6.156}$$

It is not difficult to show that in the parabolic approximation for $E_k(Q)$ and $E_d(Q)$ the quantity $[(p-S)^2/4S]\hbar\omega$ is just equal to E_B (Fig.6.18), i.e., to the energy barrier for the electron in the conduction band between $Q = 0$ and $Q = Q_c$ where the two curves cross. This leads for σ to a law, of the form

$$\sigma \sim \sigma_0 \sqrt{\frac{\hbar\omega}{4\pi kTS}} \exp\left(-\frac{E_B}{kT}\right) \quad , \qquad (6.157)$$

as shown in [6.19] using different simple arguments. However, we emphasize the point that this law is not likely to be observed in practice and that the low-temperature form should be preferred in most cases.

We end this section with a short discussion of the validity of the so-called "Condon approximation". For this we consider the detailed form of the action of H^{NA} on $\psi_{ed}(\underline{r},Q)$ in (6.145). From Sect.2.1.1, we get

$$\langle\psi_{ek}|H^{NA}|\psi_{ed}\rangle = -\frac{\hbar^2}{2M}\left(2\langle\psi_{ek}|\frac{\partial}{\partial Q}|\psi_{ed}\rangle\frac{\partial}{\partial Q} + \langle\psi_{ek}|\frac{\partial^2}{\partial Q^2}|\psi_{ed}\rangle\right) \quad , \qquad (6.158)$$

where the matrix element corresponds to integration over the electron coordinates. It is hard to believe that such a matrix element can be considered as constant with respect to Q. To investigate this point we have to know the dependence of ψ_{ed} on Q. The simplest method (used originally by HUANG and RHYS [6.21]) is to start from Q = 0 and use first-order perturbation theory. Thus ψ_{ed} takes the form

$$\psi_{ed} = \psi_{ed}(Q=0) + \sum_k \frac{F_k Q}{(E_d - E_k)_{Q=0}}\frac{\partial}{\partial Q} \quad . \qquad (6.159)$$

To that order only the first term on the right of (6.158) is nonvanishing and we get

$$\langle\psi_{ek}|H^{NA}|\psi_{ed}\rangle = \frac{\hbar^2}{M}\frac{F_k}{(E_k - E_d)_{Q=0}}\frac{\partial}{\partial Q} \quad . \qquad (6.160)$$

This is what Huang and Rhys call the "Condon approximation" for the treatment of nonradiative transitions. It differs from the derivation given above by the fact that we have now to calculate quantities like $Av_m|\langle\chi_{km}|\partial/\partial Q|\chi_{d,m+p}\rangle|^2$. This can be done by the methods discussed in Sect.4.3, taking into account the fact that $\partial/\partial Q$ is equal to $\sqrt{M\omega/2\hbar}\,(a - a^+)$. Thus the matrix element $\langle\chi_{km}|\partial/\partial Q|\chi_{d,m+p}\rangle$ is a combination of $\sqrt{m+p+1}\langle\chi_{km}|\chi_{d,m+p+1}\rangle$ and $\sqrt{m+p}\langle\chi_{km}|\chi_{d,m+p-1}\rangle$. At low temperature $m\sim 0$, and for large p it is $\langle\chi_{km}|\chi_{d,m+p-1}\rangle$, which dominates since it behaves as $(-\sqrt{S})^{p-1}\sqrt{p/(p-1)!}$, while $\langle\chi_{km}|\chi_{d,m+p+1}\rangle$ behaves as $(-\sqrt{S})^{p+1}\sqrt{(p+1)/(p+1)!}$. The limit of zero temperature should then be dominated by the term $Av_m(m+p)|\langle\chi_{km}|\chi_{d,m+p-1}\rangle|^2$ which can be rewritten without error as $p\,Av_m\langle\chi_{km}|\chi_{d,m+p-1}\rangle|^2$. This comes back to the form (6.152) with p replaced by p - 1. The low-temperature limit should then be

$$\sigma \propto (\bar{n} + 1)^{p-1} S^{p-1} \frac{p}{(p-1)!} \left[\exp -2S(\bar{n} + \tfrac{1}{2})\right] \quad . \tag{6.161}$$

This is practically the result obtained by Huang and Rhys if we take the low-temperature limit of their expression. The high-temperature limit can also be obtained directly from the classical argument of Sect.4.3.5 where $\partial/\partial Q$ has to be replaced by $i/\hbar\, P$ (where P is the classical momentum). We thus obtain an integral similar to (4.58) but with p^2 in the numerator. Again, we obtain an activated cross section σ as in (6.157) but now the prefactor is proportional to $T^{\frac{1}{2}}$ instead of $T^{-\frac{1}{2}}$. Again, this is in good agreement with the high-temperature limit of Huang and Rhys.

Different authors have tried to go beyond Condon approximation of HUANG and RHYS [6.21]. It has been shown [6.18,19] that the most natural improvement is obtained by replacing in the denominator of (6.159) $(E_d - E_k)_{Q=0}$ by an expression varying linearly with Q. Among other authors [6.22], RIDLEY [6.19] finds that the effect of the non-Condon approximation is to enhance σ by a factor of order p^2, i.e., by a few orders of magnitude [6.23].

6.4 Experimental Determination of Ionization Energies, Entropies and Cross Sections

6.4.1 Capture Cross-Section

Consider the situation in Fig.6.19a in which defects present in the space-charge region of a reverse biased junction are ionized for $E_{Fe} < E_T$. The application of a voltage pulse which cancels the reverse bias leads, for instance, to the situation shown in Fig.6.19b. The case of a filling pulse has been treated in detail in Sect.6.2.4. We have seen there that, at time t_p of the end of the pulse, the occupancy function $f(t_p)$ is given by (6.66). However, under the conditions in Fig.6.19b we can write, for $0 < x < W_0 - \lambda$

$$\Delta_p = f(x,t_p) - f_0(x) = \frac{1 - \exp - c_e N_D t_p}{1 - \exp[-g_e(t_f - t_p)]\exp(-c_e N_D t_p)} \quad , \tag{6.162}$$

and also from (6.75)

$$\frac{\Delta C}{C} = - \frac{N_T}{2N_D(W_0)}\left(1 - \frac{\lambda}{W_0}\right)^2 \Delta_p \exp[-g_e(t - t_p)] \tag{6.163}$$

since, for the situation in Fig.6.19b, the occupancy function f_1 is unity for all values of x. The amplitude of the DLTS peak will be proportional to the quantity Δ_p (f_0 being the Fermi function corresponding to the steady-

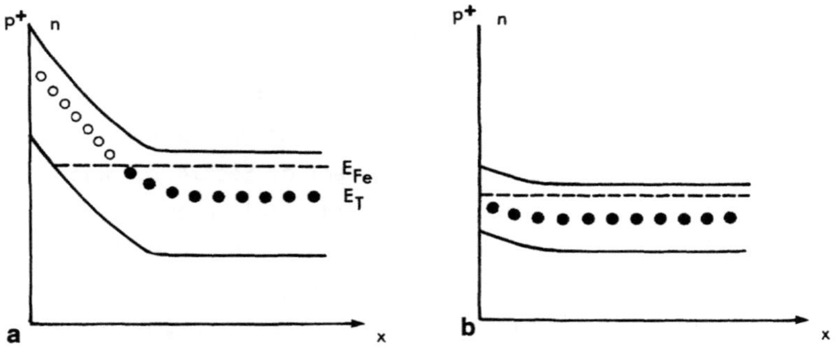

Fig.6.19a,b. Trap occupancy in a reverse biased p^+n junction (a), and in an unbiased junction (b)

state situation of Fig.6.19a). A measurement of the capture probability can thus be made by monitoring the amplitude $A(t_p)$ of the DLTS peak as a function of the time t_p during which the junction remains unbiased (the pulse width). This allows the determination of $|A(t_p) - A(\infty)|$, which, from (6.162, 163), is proportional to $\exp(-c_e N_D t_p)$, if $g_e(t_f - t_p) \ll 1$. Thus, the slope of the curve $\ln|A(t_p) - A(\infty)|$ versus t_p will give the value of k_e, from which the cross section σ_e can be calculated. A convenient method to determine k_e is to plot directly $\ln|d/dt_p\, A(t_p)|$ versus t_p. As $A(t_p)$ is proportional to $[1 - \exp(-c_e N_D t_p)]$ we have, k being a constant

$$\ln\left|\frac{d}{dt_p} A(t_p)\right| = K - c_e N_D t_p \quad . \tag{6.164}$$

However, in practice, flat band conditions are not realized. When some curvature in the bands remains, $k_e = c_e n_1(x)$ is not constant in the depleted zone. There is also the possibility that E_T crosses E_F in Fig.6.19b, i.e., that some of the traps are empty in that situation. Both cases correspond to the general situation described in Sect.6.2.4 where the capture cross section can be extracted from the experimental data by using a fitting procedure [6.5,7].

The variation of the capture cross section as a function of temperature has been measured for a variety of deep centers (most of them not identified) in various materials [6.24]. Although the temperature range in which the measurement can be performed is often small, $\sigma(T)$ seems to follow the law

$$\sigma = \sigma_\infty \exp\left(-\frac{\Delta E}{kT}\right) \tag{6.165}$$

except towards low temperature where σ becomes independent of T (Fig.6.20). As discussed in Sect.6.3.3, such behavior therefore strongly suggests that

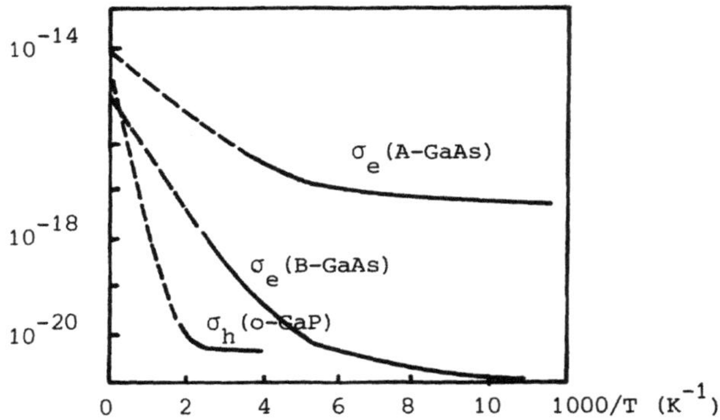

Fig.6.20. Experimental variations versus temperature of electron capture cross sections in GaAs (for the A and B centers) and of hole capture cross section in GaP (for the O center), after [6.12]

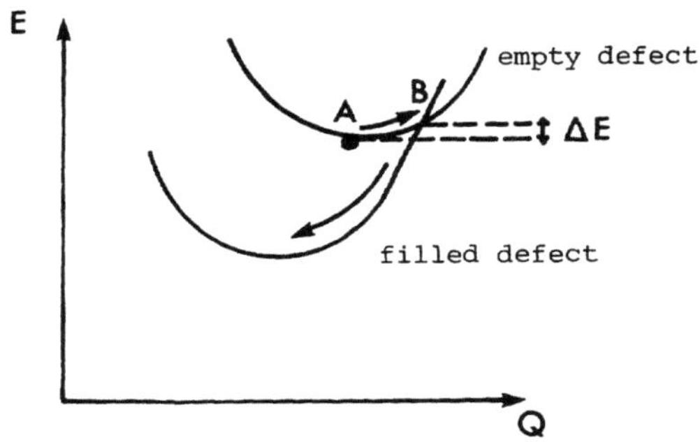

Fig.6.21. Schematic representation of carrier capture by a nonradiative recombination process

capture occurs through a multiphonon emission process. Then, the study of the cross section provides an interesting information concerning the configuration coordinate diagram: the activation energy ΔE (Fig.6.21), which can be viewed as the energy barrier that a carrier has to overcome to get trapped on the defect site, is the difference in energy between A and B. Therefore, when the vibrational energy is the same for both empty and filled states, knowledge of ΔE allows the determination of the positions of the two minima, i.e., of the quantity $S\hbar\omega$. However, as discussed in Sect.6.3.3, the high-temperature form (6.157) for σ must be used with caution. As shown in [6.19], the low-temperature expression (6.161) can also fit the experimental results and is presumably more appropriate in many cases. This may have important consequences as regards the deduced values of S.

In practice, the measurement of a capture cross section can easily be performed only in lightly doped material, unless the cross section is very small. The reason is the following. Typically a cross section is of the order of 10^{-13} cm^2, 10^{-15} cm^2 and 10^{-17} cm^{-2} for the capture of a carrier by a center which, respectively, has opposite charge to, is neutral and has the same charge as the captured carrier. As a result the capture rate varies between 10^{-6} n and 10^{-10} n (cm^{-3} s^{-1}) since $v \sim 10^7$ cm/s around 100 K, a typical temperature at which the measurements are performed. Thus, for a heavily doped material ($n \sim 10^{17}$ cm^{-3}), the pulse width necessary to saturate the defect population in the depletion region is $t_p \sim 10^{-11} - 10^{-7}$ s; for a lightly doped material ($n \sim 10^{13}$ cm^{-3}) it is $t_p \sim 10^{-7} - 10^{-3}$ s. It is therefore possible in heavily doped materials that the widths t_p of the pulses which must be applied are shorter than the RC time constant of the measuring circuit. One possible way to overcome the difficulty of a too high carrier concentration is to use the potential barrier of the junction to limit carrier concentration. The measurement of the cross-section is performed on the centers which lie at the edge of the space charge region [6.5, 7]. The analysis is then more complicated since the carrier concentration is not uniform but presents a distribution.

6.4.2 Experimental Determination of Ionization Energies and Entropies

We have mentioned in Sect.6.2.2 that an ionization enthalpy ΔH can be obtained by studying the emission rate g_e as a function of temperature, from the slope of the plot $\ln g_e$ versus T^{-1}. Once ΔH and the variation of the capture cross section σ with temperature have been determined, it is easy to see from (6.18) that knowledge of the emission rate enables calculation of the entropy of ionization ΔS.

Another way to measure ΔS consists in determining directly the free energy ΔG and its evolution with temperature and to derive ΔS as

$$\Delta S = - \frac{\partial \Delta G}{\partial T} \quad . \tag{6.166}$$

As mentioned in Sect.6.2.4, this free energy is deduced from the determination of the point at which the defect level crosses the Fermi level (6.78). This determination is made by monitoring the amplitude of the DLTS peak, $\Delta C/C$ versus the amplitude ΔV of the excitation pulse, the extrapolation of $\Delta C/C$ to zero providing the apparent threshold ΔV_0.

Due to errors made in the determination of the emission rate, the temperature, the capture cross section and the enthalpy of ionization, ΔS is usually obtained, at best, with an error of at least ± 1 k. This is thus a low accuracy since most centers have an entropy of ionization of few k.

6.4.3 DLTS Observation of a Negative U Center: The Vacancy in Silicon

We have seen in Sect.3.8 that the positive vacancy V^+, observed by EPR, in p-type silicon after photoexcitation, is found to be metastable and that its decay occurs with an activation energy of 0.057 eV. This observation has recently been completed [6.25] by a DLTS experiment in which a level is observed at $E_v + 0.13$ eV. Detailed considerations have shown that it corresponds to the simultaneous emission of two holes. The interpretation is based on the fact that in both cases there is emission of holes towards the valence band.

Let us first discuss the possible meaning of the activation energies observed for both processes. The general expression for the emission rate of a hole towards the valence band is given by (6.17), which we rewrite in our case as

$$g_h(n+1 \to n) = \frac{\sigma_h v_h}{\gamma} N_v \exp\left[\frac{E_v - E(n,n+1)}{kT}\right] \quad , \tag{6.167}$$

where σ_h and γ depend upon the pair $(n,n+1)$. We have also dropped the entropy part and used the notation defined in Sect.2.1.3 for the occupancy levels $E(n,n+1)$.

The simplest analysis is based on the assumption that $\sigma_h(T)$ is not thermally activated or at least that its temperature dependence can be neglected compared to $\exp\{[E_v - E(n,n+1)]/kT\}$. Then, in this simple limit, the measured activation energy directly gives the quantity $E(n,n+1) - E_v$, i.e., the position of the level with respect to the valence band. Thus, if the decay of V^+ observed by EPR corresponds to the reaction $V^+ \to V^0 + h$ we have

$$E(0,+) - E_v = 0.057 \text{ eV} \quad . \tag{6.168}$$

The case of the DLTS spectrum is more subtle. It is attributed to the reaction $V^{++} \to V^+ + h$. This would then give

$$E(+,++) - E_v = 0.13 \text{ eV} \quad , \tag{6.169}$$

which shows that $E(+,++)$ and $E(0,+)$ are in inverted order, indicating that the vacancy is a negative U center. However, since V^+ is unstable, the reaction $V^{++} \to V^+ + h$ is immediately followed by $V^+ \to V^0 + h$. Let us call $g_h(++)$ the rate of the first reaction and $g_h(+)$ the rate of the second one. It is thus not clear what is, a priori, the reaction which is seen in DLTS and, then, what corresponds to the measured activation energy. The answer to this question is given by Fig.6.22 on which are plotted the quantities g_h observed by EPR and by DLTS. As we discussed above, the decay of V^+ in EPR must correspond to $g_h(+)$. It is easy to see that the emission rate observed in DLTS is much lower than the extrapolated value of $g_h(+)$ in the same

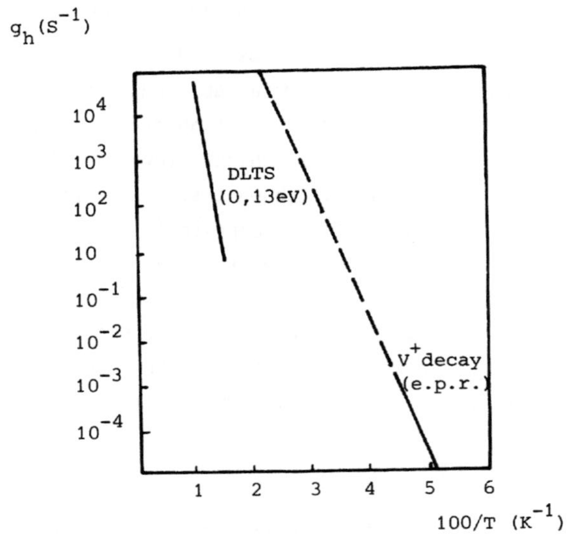

Fig.6.22. Comparison of the rate of decay of V^+ (following illumination), as observed by EPR, and of the emission rate of the center observed by DLTS after [6.26]

temperature range. The DLTS measurement corresponds to two phenomena: first $V^{++} \to V^+ + h$ [with the emission rate $g_h(++)$] and $V^+ \to V^0 + h$ [with the emission rate $g_h(+)$]. Since the rate observed by DLTS is much lower than $g_h(+)$, the measured value corresponds to $g_h(++)$. This means that the first hole emission is made in a time corresponding to $g_h(++)^{-1}$ and is followed immediately, in a time $g_h(+)^{-1}$, by the emission of a second hole. Thus, the DLTS peak corresponds to the emission of two holes but the rate is limited by the first emission. That is so can be seen from the solution of the three rate equations:

$$\frac{\partial c^{++}}{\partial t} = -g_h(++)c^{++} \quad ,$$

$$\frac{\partial c^{+}}{\partial t} = g_h(++)c^{++} - g_h(+)c^{+} \quad ,$$

$$\frac{\partial c^{0}}{\partial t} = g_h(+)c^{+} \quad ; \tag{6.170}$$

with the initial conditions

$$c^{++}(t = 0) \neq 0$$

$$c^{+}(t = 0) = 0$$

$$c^{0}(t = 0) = 0 \quad . \tag{6.171}$$

A complete mathematical solution of these equations is possible, but because $g_h(+) \gg g_h(++)$, it is easier to take the mathematical limit $g_h(+) \to \infty$. As $\partial c^+/\partial t$ must remain finite, this means that $g_h(+)c^+$ must also be finite, i.e., $c^+ \to 0$ and $\partial c^+/\partial t \to 0$ after a finite time. Then, $g_h(++)c^{++} = g_h(+)c^+$ and (6.170) reduces to:

$$\frac{\partial c^{++}}{\partial t} = -g_h(++)c^{++} \tag{6.172}$$

that is

$$\frac{\partial c^{++}}{\partial t} = -\frac{\partial c^0}{\partial t} \quad . \tag{6.173}$$

This is equivalent to saying that there is direct transfer of V^{++} into V^0 at a rate $g_h(++)$. Each time such transfer occurs, two holes are released in the valence band and the space charge is changed by $-2|e|$. Therefore, the DLTS experiment measures $g_h(++)$ and the peak amplitude must be twice the normal amplitude for the emission of one hole. This has effectively been demonstrated experimentally [6.25].

The above determination of $E(+,++)$ and $E(0,+)$ is based on the assumption that cross sections for captures are not thermally activated. BARAFF et al. have considered the opposite case, with a high temperature expression for σ [6.26], but this does not alter the conclusions very much.

6.5 Influence of the Electric Field on Emission Rates

6.5.1 The Frenkel-Poole Effect

When a defect lies in a region where there is an electric field F, the electron experiences the defect potential plus the perturbative potential $-eFz$, where z is the electron coordinate along the field. This situation is represented in Fig.6.23. The defect potential goes through a relative maximum at $z = z_m$. The characteristic of such a potential is that there can be bound states at energies lower than E_m and that there is a continuum of states for $E > E_m$ which can propagate towards $z \to \infty$. For a relatively weak electric field, the point z_m is far apart from the defect site and the energy E_T and wave function ψ_T of the ground state are the same as for $F = 0$. In such a case, there can be an emission process analogous to thermoelectronic emission from solids; there is a finite probability for the defect electron to be thermally excited to states higher than E_m, and then to give rise to a

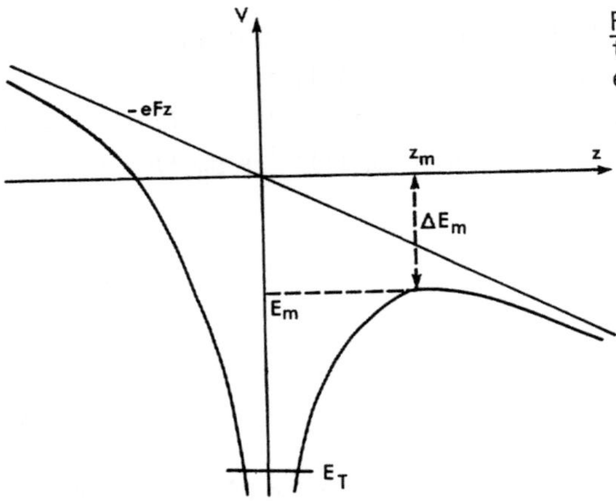

Fig.6.23. Schematic representation of a defect potential in an electric field

current density in the positive z direction. The probability for thermal excitation behaves as exp(-E/kT). This leads, as usual, to an emission rate of the form

$$g_e \propto \exp\left(-\frac{E_T - \Delta E_m}{kT}\right) \quad , \tag{6.174}$$

the prefactor coming from integration over states of energy higher than E_m, in a way similar to thermal emission. For a Coulomb potential $-e^2/\varepsilon r$, the quantities z_m and ΔE_m are given by [6.27]

$$z_m = \sqrt{\frac{e}{\varepsilon F}} \quad , \quad \Delta E_m = 2e\sqrt{\frac{eF}{\varepsilon}} \quad . \tag{6.175}$$

Typical values ($\varepsilon = 10^{-10}$ MKS, $F = 10^7\ \mathrm{Vm}^{-1}$) lead to $z_m \sim 100$ Å and $\Delta E_m \sim 0.3$ eV. Thus, the enhancement factor $\exp(\Delta E_m/kT)$ can take appreciable values since, at T = 300 K, it is already equal to exp(12).

Therefore, the linearity of ΔE_m with the square root of the electric field is characteristic of a charge leaving a center of opposite sign. Since this is the situation encountered for a donor or an acceptor, the study of ΔE_m versus F is a way to determine the donor or acceptor character of a defect [6.28].

There are corrections to this simple result (6.174) which concern the prefactor. Some of them are due to angular effects and have been evaluated using rough models [6.29,30]. However, they do not seriously affect the basis of the model and the fact that the field and temperature dependence of the enhancement factor are dominated by the term $\exp[2e(\sqrt{eF/\varepsilon})/kT]$. For shorter range defect potentials, the field dependence of ΔE_m lies between $F^{1/2}$ and F [6.31]. Finally, the effect for a square well is negligible, since $\Delta E_m = eFr_0$,

where r_0 is the radius of the potential well; indeed even for a high field $F = 10^7\ Vm^{-1}$, since r_0 remains small, ΔE_m is small ($\Delta E_m = 10^{-3}$ eV for $r_0 = 1$ Å). Thus the sensitivity of the emission rate to the electric field is a way to evaluate the range of a defect potential.

6.5.2 Tunneling Effect

The electric field F can also allow electron emission from a given defect by tunneling into the conduction band. This is represented schematically in Fig.6.24 where the defect potential is taken to be a square-well potential. At zero electric field, the defect potential is taken to have one bound state at the energy E_T, located at Δ_T below the conduction band. When a small electric field is applied, the bottom E_c of the conduction band is given by $E_c(0) - eFz$ and covers all the energy range. Thus, the level E_T now falls within a continuum to which it is coupled. As discussed in [Ref.1.1., Sect. 3.2.4], it becomes a resonance. In other words, an electron, which at time $t = 0$ is in the state ψ_T (the bound-state wave function), will tend to make transitions towards the conduction-band states lying at the same energy. The lifetime τ of the electron in this bound state is related to the width Γ of the resonance by the uncertainty relation $\tau\Gamma \sim \hbar$. We can also say that the electron tunnels through the barrier of height Δ_T.

One-dimensional treatments of tunneling through a triangular or square barrier are given in quantum-mechanics textbooks [6.32]. Here, we follow Oppenheimer's approach [6.33] in which the probability P per unit time that the electron makes a transition from the initial state ψ_T (of energy E_T) to final continuum states ψ_f (of energy E_f) is given by time-dependent perturbation theory, i.e.:

$$P = \frac{2\pi}{\hbar} \sum_f |\langle\psi_T|eFz|\psi_f\rangle|^2 \delta(E_T - E_F) \quad . \tag{6.176}$$

In one dimension, ψ_T decreases exponentially away from the defect, as

$$\psi_T \propto \exp\left(-\sqrt{\frac{2m\Delta_T}{\hbar^2}}\,|z|\right) \quad . \tag{6.177}$$

On the other hand, since F is small, final states ψ_f of energy $E_f = E_T$ can be treated in the WKB approximation. For $z < z_c$ they behave as

$$\psi_f \propto \exp\left[\int_{z_c}^{z} \sqrt{\frac{2m}{\hbar^2}\,eF(z_c - z)}\,dz\right] \tag{6.178}$$

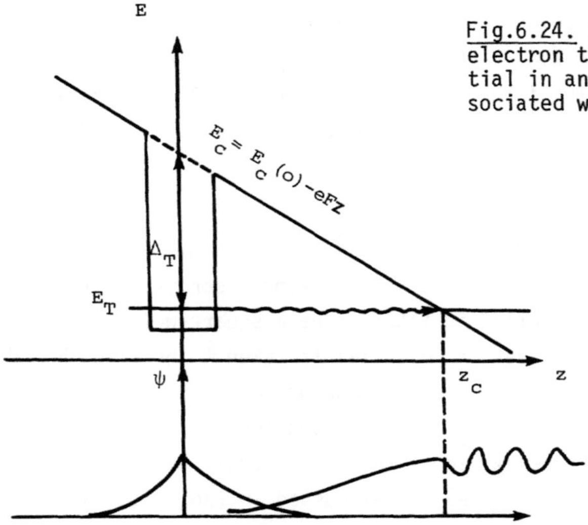

Fig.6.24. Schematic representation of electron tunneling from a defect potential in an electric field (a), and associated wave functions (b)

with $eFz_c = \Delta_T$. The product $\psi_T\psi_f^*$ thus behaves as

$$\psi_T\psi_f^* \propto \exp\left[-\sqrt{\frac{2m}{\hbar^2}\Delta_T}\left(|z| + \frac{2}{3}\frac{(z_c - z)^{3/2}}{z_c^{1/2}}\right)\right] \quad . \tag{6.179}$$

The exponential is maximum at z = 0 and the argument decreases very rapidly near this point. The matrix element $\langle\psi_T|eFz|\psi_f\rangle$ can be calculated by expanding this argument to second order in z^2, near z = 0, both for z positive and negative [6.34]. In this way, the matrix element $|\langle\psi_T|eFz|\psi_f\rangle|^2$ becomes proportional to $|\psi_T\psi_f^*|^2_{z=0}$. We can then write

$$P \propto \exp(-A) \tag{6.180}$$

with

$$A = \frac{4}{3}\sqrt{\frac{2m\Delta_T}{\hbar^2}}\, z_c \quad . \tag{6.181}$$

or, equivalently

$$A = \frac{4}{3}\sqrt{\frac{2m\Delta_T}{\hbar^2}}\,\frac{\Delta_T}{eF} \quad . \tag{6.182}$$

The prefactor depends on the shape of the potential well and the detailed integration around z = 0. Its field dependence follows some power law. The same calculation can be done in three dimensions and, again, we obtain (6.180) but with a different exponential prefactor.

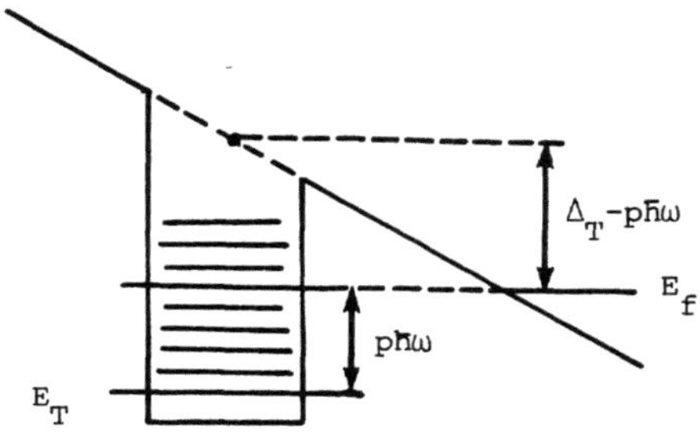

Fig.6.25. Schematic representation of phonon-assisted tunneling of an electron from a defect level in an electric field

6.5.3 Phonon-Assited Tunnel Emission

If we evaluate (6.182) with $\Delta_T \sim 0.5$ eV, $F \sim 10^3$ V cm^{-1} and m equal to the free-electron mass, then A in (6.180) is of order 10^5 and the tunnel emission is totally negligible. With effective masses m^* much smaller than m and high electric fields F (which can be of order 10^5 V cm^{-1} in reverse biased diodes), A can be as low as 50. With such a value, the tunneling probability is still too low to be an effective emission process. However, as shown in Fig.6.25, the tunneling rate can be enhanced for defects for which an appreciable electron-lattice coupling exists. The principle of phonon-assisted tunnel emission is simple: assume that the defect vibronic state is $\psi_T\chi_{Tp}$, where χ_{Tp} it the pth harmonic-oscillator wave function (for simplicity we again assume a coupling to only one lattice coordinate). The corresponding total energy is then $E_T + p\hbar\omega$. This corresponds to a reduced potential barrier $\Delta_T - p\hbar\omega$. We can expect that such a reduction in barrier height will significantly enhance the tunneling probability, when the matrix elements are not vanishingly small. To see if this is the case, we make again use of Oppenheimer's formula (6.176) which we now write

$$R = \frac{2\pi}{\hbar} Av_n \sum_{f,m} |\langle\psi_T\chi_{Tn}|eFz|\psi_f\chi_{fm}\rangle|^2 \delta[E_T - E_f + (n - m)\hbar\omega] \quad , \qquad (6.183)$$

where both initial and final states are Born-Oppenheimer products. The expression which is obtained is completely similar to those in optical absorption (Sect.4.3) and for multiphonon capture (Sect.6.3.3). Again, the simplest illustration is provided by the Condon approximation, in which the electronic matrix element is assumed to have some average value, independent of the lattice coordinate Q. This leads to

$$R = \sum_p Av_m |\langle\chi_{T,m+p}|\chi_{fm}\rangle|^2 \frac{2\pi}{\hbar} \sum_f |\langle\psi_T|eFz|\psi_f\rangle|^2 \delta(E_T - E_f + p\hbar\omega) \quad . \qquad (6.184)$$

In this expression, the term $\mathrm{Av}_m |\langle \chi_{T,m+p} | \chi_{f,m} \rangle|^2$, corresponding to the vibrational overlap part, takes the same form as in Sects.4.3 and 6.3.3, which we label W_p. The second part in (6.184) represents the tunneling probability but with the final energy state given by $E_f = E_T + p\hbar\omega$. It can be calculated along the same lines as in Sect.6.5.2 (for details see [6.34]) and, apart from a different prefactor, it is given by

$$P(\Delta_T - p\hbar\omega) \propto \exp\left[-\frac{4}{3}\sqrt{\frac{2m}{\hbar^2}}\ \frac{(\Delta_T - p\hbar\omega)^{3/2}}{eF}\right] \quad . \tag{6.185}$$

The final result for the phonon-assisted tunnel emission is thus

$$R \propto \sum_p W_p \exp\left[-\frac{4}{3}\sqrt{\frac{2m}{\hbar^2}}\,\frac{(\Delta_T - p\hbar\omega)^{3/2}}{eF}\right] \quad . \tag{6.186}$$

At zero temperature there are no excited phonons in the bound-state. Then, $p = 0$ and R has the same exponential argument as in the pure electronic case. At higher temperatures the average p increases, leading to an average decrease in barrier height, which can very much enhance the tunneling probability.

We have discussed here the quantum theory of this effect but it is worthwhile to mention the first theory, which was semiclassical [6.35]. The results in both cases are compared in [6.34] which shows that there can be significant differences in some cases.

7. Other Methods of Detection

In this chapter we discuss additional information that the use of a combination of optical, paramagnetic and electrical properties provide on defect characteristics and behavior. To begin with we consider photoexcited techniques, i.e., the effect of optical excitation on conductivity, paramagnetic resonance, deep level transient spectroscopy and optical absorption. In Sect.2, we consider optical detection of EPR. Finally, in Sect.3, we group the techniques which allow direct detection of phonons, i.e., which give a direct means to observe nonradiative recombination.

7.1 Photoexcitation

7.1.1 Principle

Any technique which detects a defect in a given charge state can be made spectroscopic by illuminating the material with monochromatic light of variable wavelength. When the energy of the light is such that it induces the transition of a carrier between the defect level and a band, the defect population is changed. This change is a measure of the position of the localized level associated with the defect within the forbidden gap. Moreover, as shown in Chaps.4 and 6, the time dependence of the variation of this occupancy when the excitation light is turned off or on gives the carrier-capture cross section and the optical cross section, respectively.

For instance, using such a method, the energy position of the singly ionized positive divacancy has been obtained with respect to the valence band [7.1], using EPR to detect the divacancy population. In the same way, IR absorption has been used to detect the population of intrinsic point defects in irradiated silicon [7.2]. Photocapacitance or photo-DLTS [7.3] have been used as well.

As shown in Chap.4, the probability per unit time for optical excitation of a carrier from a localized level g^0 can be written in terms of a quantum efficiency coefficient γ and of the optical cross section σ^0

$$g^0 = \gamma\sigma^0\phi(\lambda) \tag{7.1}$$

where ϕ is the photon flux.

Consider, for instance, a homogeneous semiconductor under steady-state optical excitation which induces transitions from a defect level to the conduction band. The carrier concentration in this band is $n = n_0 + \Delta n$, the concentration of filled defects $b = b_0 - \Delta n$, and of empty ones $s = s_0 + \Delta n$ (in the notations of Chap.6), n_0, b_0, and s_0 being the equilibrium concentrations. Under steady-state conditions, the optical generation rate is equal to the spontaneous recombination rate

$$g_e^0 b = c_e ns - g_e b \quad . \tag{7.2}$$

The quantities on the right-hand side have been defined in Sect.6.1 and are related through

$$g_e = c_e \frac{n_0 s_0}{b_0} \quad . \tag{7.3}$$

This allows us to rewrite (7.2) in the form

$$g_e^0 b = \frac{\Delta n}{\tau} \tag{7.4}$$

where τ is the lifetime of the excess carriers, which can be written in the detailed form

$$\frac{1}{\tau} = c_e\left(n_0 + \Delta n + s_0 + \frac{s_0 n_0}{b_0}\right) \quad . \tag{7.5}$$

In general this lifetime depends on Δn but, in the limit of small Δn (i.e., for small photon fluxes, which is usually the case), it reduces to the constant τ_0

$$\frac{1}{\tau_0} = c_e\left(n_0 + s_0 + \frac{s_0 n_0}{b_0}\right) \quad . \tag{7.6}$$

In the same limit, $\Delta n \ll b_0$, (7.4) gives the excess carrier density

$$\Delta n = \tau_0 g_e^0 b_0 \quad , \tag{7.7}$$

which can also be rewritten, using (7.1)

$$\Delta n = \gamma\sigma^0\tau_0\phi b_0 \quad . \tag{7.8}$$

7.1.2 Shockley-Read Recombination

The case treated above corresponds to excitation with light of a wavelength such that it does not induce band-band transitions. For photon energies larger than the band gap, electron-hole pairs are generated. They can recombine either directly or with the help of defect centers. We analyze this second situation, under steady-state conditions [7.4].

Let us call G^0 the band-band optical generation rate. In a way similar to (7.2) we can write

$$G^0 = c_e ns - g_e b \tag{7.9}$$

and

$$G^0 = c_h pb - g_h s \quad , \tag{7.10}$$

where the second equation describes the rates between the defect and the valence band. Again, c_e and g_e are related through (7.3), with the equivalent relation for holes. The neutrality condition imposes that

$$\Delta s = \Delta n - \Delta p \quad ,$$

that is (7.11)

$$\Delta s = -\Delta b \quad .$$

From this we can rewrite (7.9,10) respectively in the forms

$$G^0 = \frac{\Delta n}{\tau_n} = c_e\left[(\Delta n - \Delta p)\left(n_0 + \Delta n + \frac{n_0 s_0}{b_0}\right) + s_0 \Delta n\right] \tag{7.12}$$

$$G^0 = \frac{\Delta p}{\tau_p} = c_h\left[(\Delta p - \Delta n)\left(p_0 + \Delta p + p_0 \frac{b_0}{s_0}\right) + b_0 \Delta p\right] \tag{7.13}$$

which define the electron and hole lifetimes τ_n and τ_p. The excess carrier concentrations can be easily calculated as before, in the small excitation limit where G^0, Δn and Δp are small. Then we can drop terms other than first order in n and p in the right members of (7.12,13). These equations thus become linear in Δn and Δp, which can be expressed separately in terms of G^0 with proportionality coefficients τ_n and τ_p:

$$\tau_n = \frac{\frac{1}{c_e}\left[p_0\left(1 + \frac{b_0}{s_0}\right) + b_0\right] + \frac{1}{c_h} n_0\left(1 + \frac{s_0}{b_0}\right)}{(n_0 + p_0)(b_0 + s_0) + b_0 s_0} \tag{7.14}$$

$$\tau_p = \frac{\frac{1}{c_e} p_0\left(1 + \frac{b_0}{s_0}\right) + \frac{1}{c_h}\left[n_0\left(1 + \frac{s_0}{b_0}\right) + s_0\right]}{(n_0 + p_0)(b_0 + s_0) + b_0 s_0} \quad . \tag{7.15}$$

These expressions are strictly equivalent to those derived in [7.4]. They can be simplified further in the limit of a low density N_T of defects. For this, we notice that $b_0 + s_0$ is equal to N_T and $b_0 s_0$, of the order of N_T^2, is negligible. Introducing τ_{n0} and τ_{p0} given by

$$\tau_{n0} = \frac{1}{c_e N_T} \tag{7.16}$$

$$\tau_{p0} = \frac{1}{c_h N_T} \quad , \tag{7.17}$$

we get

$$\tau_n = \tau_p \sim \tau_{n0} \frac{p_0 + p_0 b_0/s_0}{n_0 + p_0} + \tau_{p0} \frac{n_0 + n_0 s_0/b_0}{n_0 + p_0} \tag{7.18}$$

which is a well-known expression [7.4], whose validity is however restricted to the limit of small numbers of defects.

7.1.3 Photoconductivity

Photoconductivity is a technique which measures changes in conductivity

$$\Delta\sigma = \Delta n e \mu_n \tag{7.19}$$

due to Δn carriers of mobility μ, which are photogenerated from the defect level E_T (Fig.7.1). The wavelength at which an increase of photoconductivity occurs gives the defect energy level.

In a real situation the material contains a doping level which, because it is shallow, is ionized at the same time as the level under investigation. Then the recombination of the photogenerated carriers occurs through two levels instead of one, as considered in the previous section. Sometimes, the situation is even more complicated, due to the presence of other additional trapping centers. The lifetime is no longer associated with the characteristics of the defect alone, and measurement of τ provides no information on this defect. Consider the simple case where the carriers recombine in majority through the doping level. This is a reasonable situation because i) the concentration of defects is small compared to the doping concentration N_D and ii) the trapping cross section of the dopant (which is an attractive Coulomb center) is very large, i.e., usually larger than the defect cross section. Then

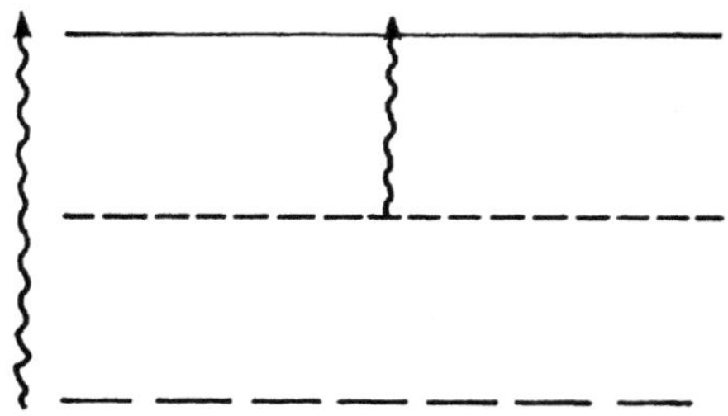

Fig.7.1. Photoinduced transition in a photoconductivity experiment

$$\tau = \sigma v N_D \quad . \tag{7.20}$$

The increase of carriers in the band

$$\Delta n = \tau g^0 \tag{7.21}$$

induces the following photoconductivity, proportional to the defect concentration:

$$\frac{\Delta\sigma}{\sigma} = \frac{\Delta n}{n_0} \quad . \tag{7.22}$$

7.1.4 Spin-Dependent Recombination

Consider the electron-hole pair of Fig.7.2. After recombination the total spin of the system (filled valence band) is zero. This is true both for cases of direct recombination or Shockley-Read recombination via a defect level [7.5,6]. Recombination thus only occurs between an electron-hole pair which is in a singlet state (of zero total spin). In practice this is not strictly true but the recombination rate W_S for the singlet state is much larger than W_T for the triplet state. In zero applied magnetic field, the spin directions of the injected carriers are completely random and the probabilities of getting a singlet or triplet state are 1/4 and 3/4 respectively. The corresponding recombination rate W is

$$W = \frac{1}{4} W_S + \frac{3}{4} W_T \quad . \tag{7.23}$$

These probabilities are altered if there is finite spin polarization, due to an applied magnetic field for instance. To express these probabilities it is interesting to notice that the operator π_S defined by

$$\pi_S = \frac{1}{4} - S_e S_h \tag{7.24}$$

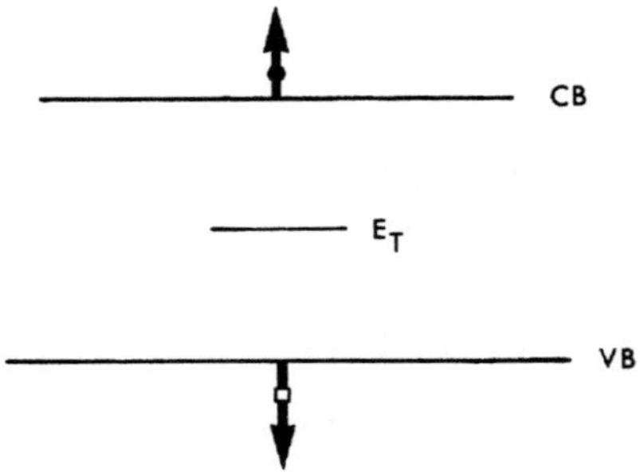

Fig.7.2. Schematic representation of a Shockley-Read recombination showing the spin of the recombining carriers

(where $\underline{S_e}$ and $\underline{S_h}$ are the electron and hole spin operators) is the projection operator onto the singlet state. On the other hand, the operator

$$\pi_T = \frac{3}{4} + \underline{S_e S_h} \tag{7.25}$$

is the projection operator onto the triplet state. Clearly the probabilities of finding singlet and triplet states are now given quite generally by the average values $<\pi_S>$ and $<\pi_T>$. This allows us to write the recombination rate in the form

$$W = W_S\left(\frac{1}{4} - <\underline{S_e}><\underline{S_h}>\right) + W_T\left(\frac{3}{4} + <\underline{S_e}><\underline{S_h}>\right) \quad , \tag{7.26}$$

which can be rewritten as

$$W = \frac{W_S + 3W_T}{4} + (W_T - W_S)<\underline{S_e}><\underline{S_h}> \quad . \tag{7.27}$$

The way the detection of spin-dependent recombination is performed is the following: a microwave field at the resonance induces transitions (at the rate W_{RF}) between Zeeman levels split in a magnetic field. The microwave power P_0 tends to equalize the populations of spins in both levels. As shown in Sect.3.2.2, the steady-state spin polarization

$$P = \frac{P_0}{1 + W_{RF}T_1} \tag{7.28}$$

(T_1 is the spin lattice relaxation time) can thus be modulated by the variation of P_0. The corresponding variation of the lifetime τ is detected by photoconductivity [7.6]. This technique provides an extremely sensitive detection of the magnetic resonance since it has the sensitivity of photoconductivity and not the usual limitations of the EPR technique. Spin polarization can also be induced by polarized light and spin-dependence effects are observable in luminescence [7.7]. Such lifetime studies thus provide information on the ability of a given defect to induce spin flip. Also, the dependence of the defect population on spin pumping indicates the sensitivity of charge-state transfer to spin polarization.

7.1.5 An Example: Photoconductivity of Boron in Diamond

The infrared absorption spectrum of boron-doped diamonds (natural type IIb or synthetic) reveal mainly peaks at 0.305, 0.347 and 0.363 eV due to transitions between the ground and excited states of the hole bound to boron [7.8]. As shown by conductivity measurements, the energy of 0.36 eV is the ionization energy of the hole into the valence band [7.9].

The photoconductivity spectrum exhibits three stages coinciding with these absorption lines (Fig.7.3) which must thus be attributed to transitions from the ground state to excited states of the boron center. The mechanism responsible for the occurence of photoconductivity in case of transitions to excited states is a two-stage process: the hole is first optically excited and then transferred into the valence band by thermal excitation. This process is usually referred to as photothermal ionization ([7.10] and [Ref.1.1, Sect.2.4.1]). The photoconductivity spectrum at higher energy exhibits an oscillatory behavior (Fig.7.3). The reason for this is that hole capture from the valence band to excited states occurs with emission of optical phonons [7.11]. Thus, holes whose energy in the valence band is an integer n of the optical phonon energy (0.165 eV) do not contribute to photoconductivity, and give rise to minima in the spectrum (at energies equal to the activation energy of the impurity plus $n \times 0.165$ eV). As shown in Fig.7.3, the spectrum of the synthetic boron-doped sample (provided by General Electric) shows up to 8 phonons replica of the excited states.

7.2 Optical Detection of Paramagnetic Resonance

We have seen in Chap.4 that optical spectroscopy is a powerful tool to study defects. However, it is really interesting only when it is possible to observe the sharp zero-phonon lines. Unfortunately, it often happens that in the case of internal transitions or of radiative recombination on deep levels, only a broad band is obtained due to the strong electronlattice interactions (Chap.4). Then it becomes impossible to get an accurate value of the position of the electronic energy levels.

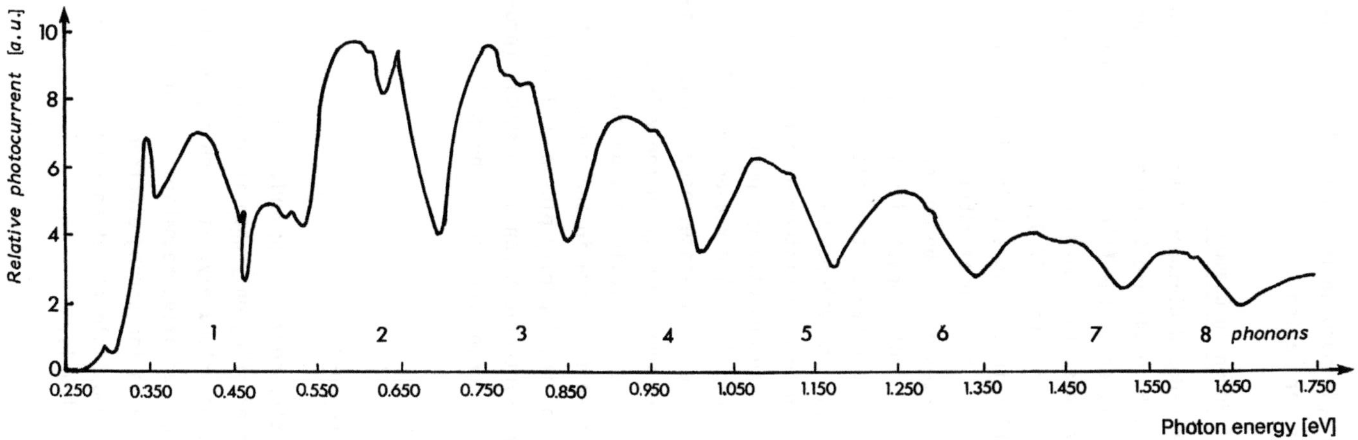

E_0	E_1	$E_1-0.165$	E_2	$E_2-2\times0.165$	E_3	$E_3-3\times0.165$	E_4	$E_4-4\times0.165$	E_5	$E_5-5\times0.165$	E_6	$E_6-6\times0.165$	E_7	$E_7-7\times0.165$	E_8	$E_8-8\times0.165$
	0.456	0.291	0.613	0.283	0.775	0.280							1.340	0.285	1.600	0.280
0.300	0.467	0.302	0.628	0.248	0.793	0.298	0.950	0.290	1.115	0.290	1.248	0.294				
0.348	0.512	0.347			0.852	0.357	1.007	0.347	1.73	0.348	1.340	0.350			1.665	0.345
	0.534	0.369	0.695	0.365									1.520	0.365		

Fig.7.3. Photoconductivity spectrum of synthetic boron-doped diamond at 4 K. The table gives the ground and excited states observed (labelled E_0) and the positions E_n of the nth phonon replica: ($E_i - n \times 0.165$ eV $= E_0$)

In such cases, the optical detection of paramagnetic resonance (usually called ODMR) appears to be very useful. This technique gives precise information since it enables the EPR of centers in their excited states to be observed. The ODMR technique has not yet been widely used for the study of deep levels [7.12]. Most of the results obtained concern donor-acceptor pairs responsible for a typical luminescence [7.13] or transitions between bands and shallow levels [7.14]. However, the method could be very useful to study internal transitions of deep levels in the same way as in ionic crystals (see for instance [Ref.7.15]). When combined with other optical techniques it could also give informations about the symmetry and the configuration coordinate diagram of the defect.

7.2.1 Principle of the Technique

Transitions are induced by an applied microwave power between the Zeeman sublevels of an excited state produced by optical excitation. These transitions are detected through a change of the luminescence intensity or through a change of the luminescence polarizability. The experimental set-up is thus simpler than in EPR experiments since there is no need for the detection of microwave absorption. The sample, inside a microwave cavity, is just placed in a magnetic field and excited with a laser. The ODMR spectrum consists in the variation of the luminescence intensity versus magnetic field. Therefore, only centers giving rise to luminescence can be detected and with a higher sensitivity than conventional EPR since a photon detector is far more sensitive than a detector in the microwave range. Correlation between a luminescence band and an ODMR spectrum can be made easily by studying the variation of the ODMR signal intensity versus wavelength.

7.2.2 ODMR of Deep Donor-Acceptor Pairs

Consider a sample containing donors D^+ which are ionized because they are compensated by deep (ionized) acceptors A^-. When band-gap light, which creates free electron and holes, is shined on the sample, luminescence occurs. The reason is that the free carriers are trapped on D^+ and A^- creating neutral donors D^0 and acceptors A^0 (Fig.7.4). Then, the electron of a donor and the hole of a nearby acceptor recombine together with emission of a photon whose energy [Ref.1.1, Sect.2.3] is

$$h\nu = E_G - E_D - E_A + E(r) \quad , \tag{7.29}$$

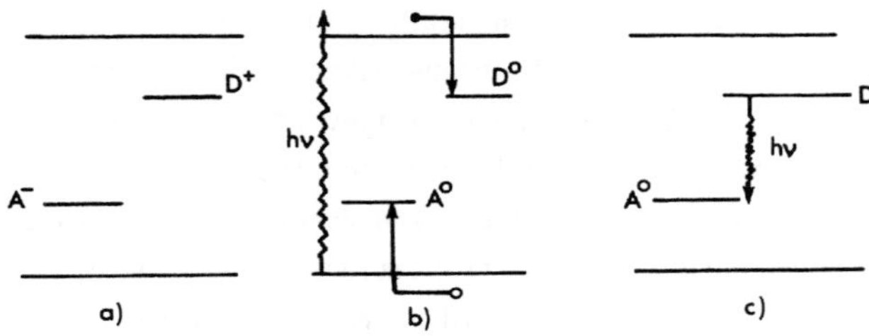

Fig.7.4a-c. Energy diagram of a donor-acceptor pair in the ground state a), in the excited state b), and showing radiative recombination c)

where E_G is the band gap, E_D and E_A the ionization energies of the isolated donor and acceptor, and E(r) the interaction energy between the donor and acceptor separated by the distance r. When the donor or the acceptor (or both) levels are deep, the electron-phonon interaction is large and a wide emission band is observed (Chap.4) in which it is impossible to distinguish the zerophonon line.

In the presence of a magnetic field, the spins of the electron (on D^0) and of the hole (on A^0) orient themselves to be parallel or antiparallel to H (Figs.7.5,6). A luminescence is observed originating from pairs for which the spins are antiparallel since only transitions which conserve the total momentum S = 0 can occur. As shown in Sect.7.1.4, this implies a weak spin orbit coupling, i.e., no spin flip during the recombination.

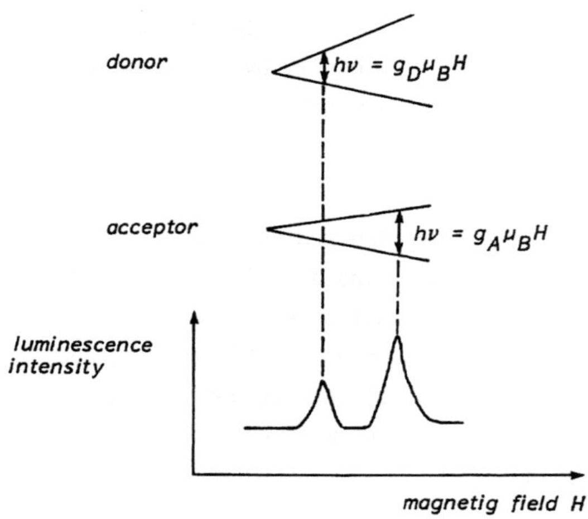

Fig.7.5. Schematic representation of the ODMR spectrum associated with a donor-acceptor pair

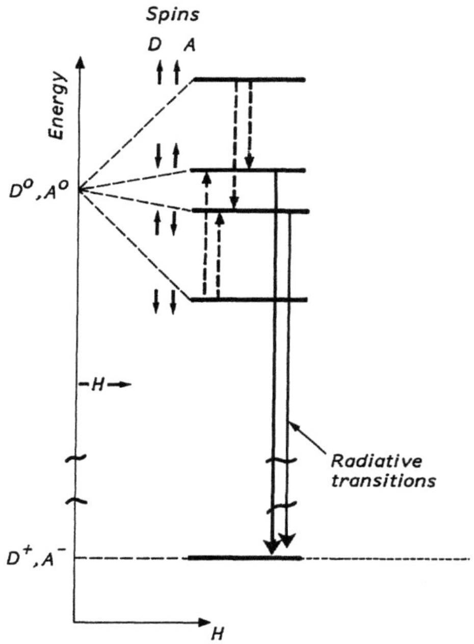

Fig.7.6. Energy levels of the ground (D^+, A^-) and excited states (D^0, A^0) of a donor-acceptor pair in a magnetic field, giving the RF and luminescence transitions. The different spin configurations under magnetic field H are shown

When a microwave power is applied at the frequency

$$h\nu = g_D \mu_B H \tag{7.30}$$

in order to induce a resonance between the Zeeman sublevels of the donor, the populations of electrons having spins up and down are modified. The result, as depicted in Fig.7.5, is an increase in luminescence intensity. Such increase occurs again when the magnetic field is changed and becomes such that

$$h\nu = g_A \mu_B H \quad . \tag{7.31}$$

7.3 Direct Detection of Phonons

We have seen throughout this book that one of the most effective ways to get information on defects is to study how photons interact with them. The information is obtained from the analysis of photons, but it can also be obtained from the analysis of phonons.

When a pulse of monochromatic light, of width Δt and intensity ϕ, is shined on a sample, the energy absorbed is

$$Q(\lambda) = \phi(1 - R)\,\frac{1 - \exp(-\alpha d)}{1 - R\,\exp(-\alpha d)}\,\Delta t \quad , \tag{7.32}$$

where α is the absorption coefficient at the wavelength λ, d the thickness of the sample and R its reflection coefficient (when $\alpha d \ll 1$, $Q \sim \phi \sim d\Delta t$).

If η is the internal quantum efficiency for radiative recombination,

$$\eta = \frac{\tau_t}{\tau_r} \tag{7.33}$$

with

$$\frac{1}{\tau_t} = \frac{1}{\tau_r} + \frac{1}{\tau_{nr}} \quad , \tag{7.34}$$

where τ_t is the total lifetime, τ_r and τ_{nr} the radiative and nonradiative lifetimes; then ηQ is reemitted into photons and $(1 - \eta)Q$ into phonons.

The quantity τ_r is obtained from luminescence spectroscopy (Chap.4). Phonon spectroscopy gives τ_{nr}. This last technique provides direct information on phonon emission, i.e., complementary information to luminescence spectroscopy. Thus it is particularly useful in the case of nonradiative recombination centers, which is a relatively common case for deep centers. There are two ways of measuring phonon emissivity: calorimetric absorption and acoustic detection, which we now describe.

7.3.1 Calorimetric Absorption

The measurement of the power absorbed into phonons must be made at low temperature (below ~4 K) to avoid thermal phonons and to reduce the specific heat of the sample (which has a T^3 dependence) by detecting the temperature rise of the sample. For this, the sample is suspended in vacuum, connected thermally to a carbon resistance, both surrounded by a radiation shield. Raising the temperature of the sample induces a change in resistance. Around 1 K, the sensitivity of a carbon resistance is of the order of $10^5\,\Omega\,K^{-1}$, i.e., $\sim 10^{-5}$ K can be detected. As in photoconductivity experiments, the light is chopped and the resistance change is measured using a lock-in detector. The minimum heat energy which has been thus detected, $\sim 10^{-12}$ J, has been enough to detect easily phonons emitted by nonradiative recombination [7.16] of nitrogen in GaP (Fig.7.7).

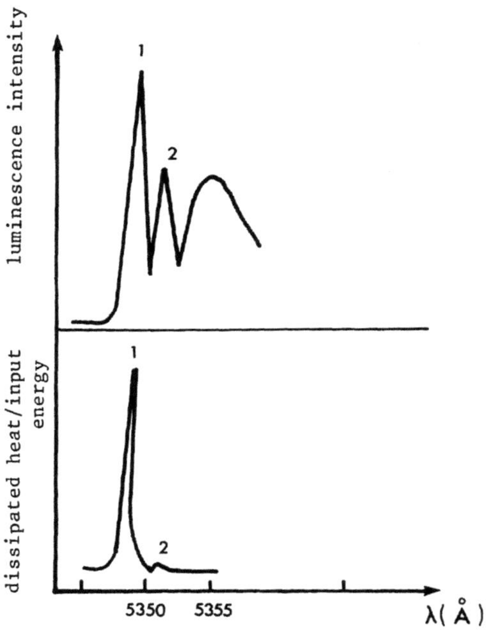

Fig.7.7. Calorimetric absorption and luminescence spectra in n-doped GaP showing radiative recombination of M-bound excitons after [7.16]. (The 1 and 2 lines are due to the decay of the J = 1 and 2 states)

7.3.2 Photoacoustic Spectroscopy

Photoacoustic spectroscopy is similar to calorimetric absorption spectroscopy but far less sensitive. The modulated light absorbed by the sample into phonons is detected using a microphone. Apparently, the acoustic signal is due to the periodic heat flow from the solid to the gas which surrounds it [7.17]. Because the magnitude of the acoustic signal is proportional to the amount of heat, there is again a direct relationship between the strength of the signal obtained and the amount of phonons emitted. Actually the photoacoustic spectrum seems to correspond rather closely to an optical spectrum [7.18,19]. Due to its sensitivity and to the fact that surface preparation strongly influences the quality of the spectrum, this technique which, like calorimetric absorption has not yet been widely used, seems less promising than the latter.

8. Defect Production by Irradiation

Defects or impurities are introduced in semiconductors, intentionally or unintentionally, during the growth process or following heat treatments. Quenching, plastic deformation and irradiation are other ways by which defects can be created. The problem with heating, quenching and plastic deformation [8.1] is that the defects produced, practically all unidentified up to now in most materials, are complexes resulting from the interaction of intrinsic defects (vacancies, interstitials, divacancies) with the various impurities present initially in the material. Moreover, the concentration of these defects is difficult, if not impossible, to control (see [Ref.1.1, Sect.6.4] for the discussion of defects resulting from quenching). On the contrary, the concentration, the distribution and (to some extent) the nature of defects produced by an irradiation can be controlled. The defect concentration is proportional to the dose of irradiation; the nature and the distribution of the defects is a function of the nature of the irradiating particle, their energy, and the impurities contained in the material which have the ability to trap the intrinsic defects originally produced by the irradiation. For this basic motivation, but also for practical reasons (knowledge of the behavior of electronic devices submitted to radiations in nuclear reactors, space, etc.) radiation effects in semiconductors have been widely studied [8.2].

In this chapter we consider the processes by which defects are produced in the case of irradiation with energetic electrons, neutrons and ions [8.3]. We describe first the way the number of defects produced by one irradiating particle is calculated. For this, we discuss the mechanisms of defect creation. We then consider the nature and the repartition of the defects produced. In the next chapter we shall describe how complex defects, i.e., defects resulting from the interaction of primary defects between themselves or with impurities, are formed, taking the case of silicon as an example. The reason silicon is chosen is that this is practically the only material in which the behavior of simple point defects is known, because they have been identified by electron paramagnetic resonance (Chap.3).

8.1 Interaction of Radiation with Solids

8.1.1 General Formalism

An incident energetic particle which enters a solid interacts with the electrons and the nuclei. It loses its energy by several processes which depend on the nature of the particle and on its energy. We first consider the elastic collisions the particle makes with the nuclei because this is the process which leads to atomic displacements, i.e., to the production of defects. The interaction of a particle with the electrons of the solid will be considered in Sect.8.3.3 where we consider its range, i.e., the depth up to which the particle penetrates.

The quantity of interest is the concentration ν of the defects produced per unit time by a flux

$$\phi = nv \quad , \tag{8.1}$$

where n is the density of particles and v their velocity, that is the number of collisions which give rise to atomic displacements, occuring per unit volume and unit time between the nuclei (in concentration N) and the particles

$$\nu = \sigma N \phi \quad . \tag{8.2}$$

The quantity σ is the cross section, characteristic of the interaction leading to atomic displacement. As already discussed in Sect.5.3, for the case of electrons scattered by defects, elastic collisions are not adequately described by a total cross section σ; knowledge of the differential cross section $d\sigma$ such that

$$\sigma = \int_\theta d\sigma(\theta) \tag{8.3}$$

which contains the angular (θ) distribution of the scattered particles, is necessary.

The distribution of collisions along the path x of an incident particle is not uniform. The probability p(x) that a particle, which has penetrated at depth x, makes a collision between x and x + dx being such that

$$-dp(x) = p(x)\sigma N\, dx \quad , \tag{8.4}$$

is therefore given by

$$p(x) = \exp(-\sigma N x) \quad . \tag{8.5}$$

The average penetration depth of the particle is then

$$\lambda = \frac{\int_0^\infty x p(x)\, dx}{\int_0^\infty p(x)\, dx} \quad . \tag{8.6}$$

8.1.2 Dynamics of a Collision

We consider here the overall dynamics of a collision; the collision itself will be described in the next section. Let E be the kinetic energy of an incident particle mass m. It strikes an atom (of mass M) of a solid. The kinetic energy T transmitted to this atom depends directly on the angular deflection θ of the incident particle (Fig.8.1).

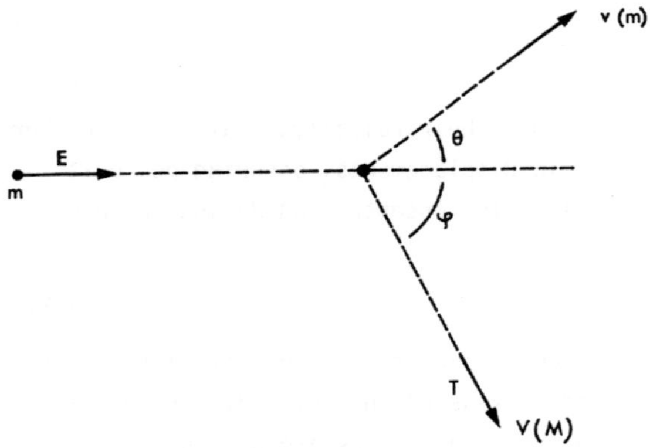

Fig.8.1. Dynamics of the collision between an incident particle of mass m and energy E with an atom of mass M

The expression which relates T and θ is obtained by writing the two conditions that a purely elastic collision must fulfill: conservation of kinetic energy and conservation of momentum. The result is [8.3]:

$$\frac{T}{E} = 2\,\frac{M}{m}\,\frac{1-\eta(\theta)}{(1+M/m)^2} \quad , \tag{8.7}$$

where $\eta(\theta)$ is the function given by

$$\cos\theta = \frac{1+(M/m)\eta}{\sqrt{1+2(M/m)\eta+(M/m)^2}} \quad . \tag{8.8}$$

The maximum energy T_m is transmitted for $\theta = 0$, i.e., for $\eta(\theta) = -1$. Thus

$$T_m = \frac{4Mm}{(m + M)^2} E \quad . \tag{8.9}$$

In the case where an atom of the solid collides with another one ($m = M$), $T_m = E$, $\eta(\theta)$ reduces to $2 \cos^2\theta - 1$ and (8.7) takes the simple form

$$T = E \sin^2\theta \qquad 0 \le \theta \le \pi \quad . \tag{8.10}$$

For a neutron, whose mass m is very small compared to M,

$$T = \frac{2m}{M} E(1 - \cos\theta) \qquad 0 \le \theta \le \pi \quad . \tag{8.11}$$

For an energetic electron ($m \ll M$) relativistic corrections are necessary. The conditions of conservation of the kinetic energy and of the total energy are then

$$m\gamma_i v_i = m\gamma_f v_f \cos\theta + MV \cos\varphi$$

$$0 = -m\gamma_f v_f \sin\theta + MV \sin\varphi$$

$$m\gamma_i c^2 = m\gamma_f c^2 + 1/2MV^2 \tag{8.12}$$

where $\gamma_{i,f}$ is the quantity

$$\gamma = \sqrt{1 - v^2/c^2} \quad . \tag{8.13}$$

Indices i and f relate to the initial and final states of the electron respectively.

The energy transmitted by the electron to the nucleus being of the order of eV, while its incident energy is of the order of MeV, the initial and final velocities of the electrons remain practically the same and we can use the approximation

$$v_i \sim v_f \qquad (\sim v) \quad , \tag{8.14}$$

that is

$$m\gamma_i c^2 \simeq m\gamma_f c^2 \quad (\sim m\gamma c^2) \quad . \tag{8.15}$$

Then the first two equations of (8.12) give

$$2\varphi + \theta = \pi \tag{8.16}$$

and the kinetic energy transmitted to the nucleus can be simply written in terms of the electron scattering angle

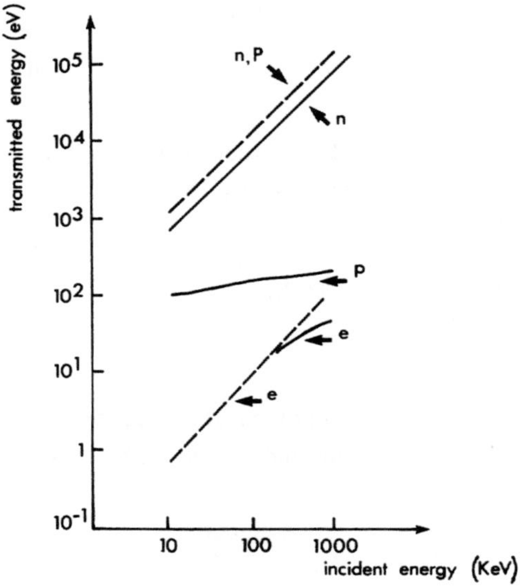

Fig.8.2. Maximum (dashed line) and average (full line) energy transmitted to a silicon atom as a function of the incident energy for electrons (e), protons (p), and neutrons (n)

$$T = T_m \sin^2 \frac{\theta}{2} = T_m \cos^2\varphi \tag{8.17}$$

with

$$T_m = 2\,\frac{m^2}{M}\,\gamma^2 v^2 \quad . \tag{8.18}$$

Expressed in terms of the incident energy

$$E = mc^2(\gamma - 1) \tag{8.19}$$

(8.18) becomes

$$T_m = 2\,\frac{m}{M}\,E\left(2 + \frac{E}{mc^2}\right) \tag{8.20}$$

with

$$mc^2 = 0.511\ \text{MeV} \quad . \tag{8.21}$$

In order to show the magnitude of the energy transmitted by electrons, protons and neutrons, Fig.8.2 gives the variation of T versus the energy of these particles.

8.1.3 Differential Scattering Cross Section

As already seen in the case of electron scattering by charged defects (Sect.5.3) the differential scattering cross section is a function of the interacting potential. In the case of irradiation of a solid by energetic particles we have to consider the interaction between the nuclei of the

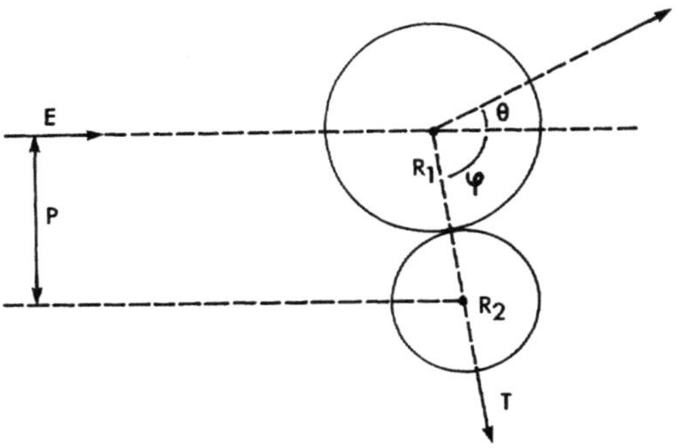

Fig.8.3. Collision between two spheres showing the impact parameter p

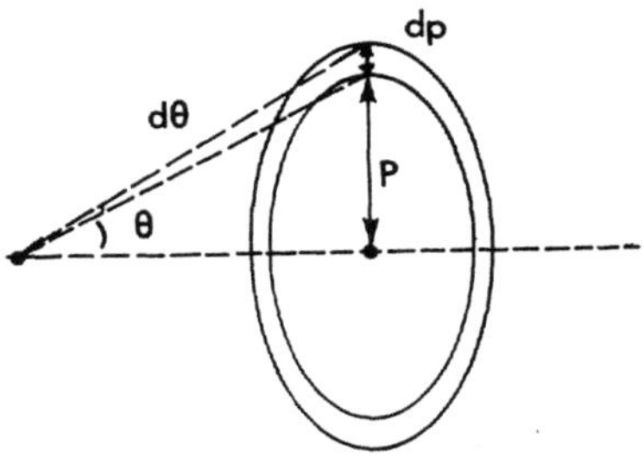

Fig.8.4. Scattering angle and impact parameter

solid and the incident particles. For charged particles the interaction can be represented by a Coulomb potential (Rutherford scattering); for neutral particles (neutrons) the interaction is rather equivalent to collisions between rigid spheres (hard-core potential). We now calculate in turn the corresponding differential cross sections.

a) *Hard Sphere Collision*

The collision between two rigid spheres of radii R_1 and R_2 is represented in Fig.8.3. It is clear on this figure that the impact parameter p determines the deflection angle θ since

$$\frac{p}{R_1 + R_2} = \cos\theta \quad . \tag{8.22}$$

Because

$$\varphi = \frac{\pi}{2} - \theta \quad , \tag{8.23}$$

the transmitted energy T can be written

$$T = T_m \cos^2\varphi \quad , \tag{8.24}$$

where T_m is given by (8.9). The differential scattering cross section $d\sigma$, i.e., the cross section for scattering in the solid angle $(\theta, \theta + d\theta)$ is directly related to the impact parameter (Fig.8.4)

$$d\sigma = 2\pi p \; dp \quad , \tag{8.25}$$

that is

$$d\sigma = 2\pi p \left| \frac{dp}{d\theta} \right| d\theta \quad . \tag{8.26}$$

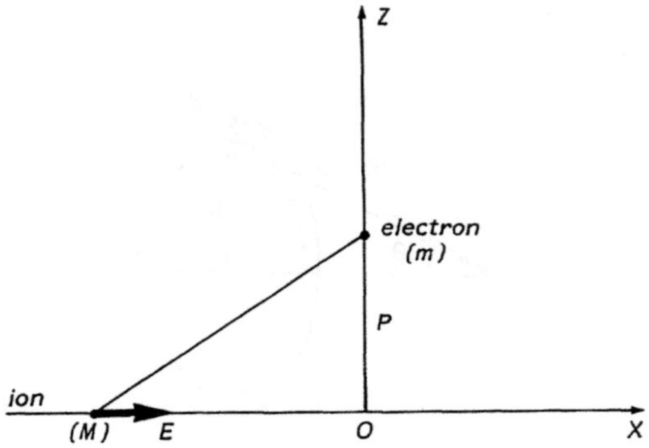

Fig.8.5. Collision between an energetic incident ion with an electron of the solid

From differentiation of (8.22,26) becomes

$$d\sigma = -2\pi(R_1 + R_2)^2 \cos\theta \sin\theta \, d\theta \tag{8.27}$$

which gives the following cross section

$$\sigma = \pi(R_1 + R_2)^2 \quad . \tag{8.28}$$

b) *Rutherford Scattering*

We calculate the Rutherford scattering cross section in a simple example in order to illustrate the approximation made (the so-called momentum approximation) to evaluate the slowing down of an incident particle by the electrons of the solid. We consider an energetic ion (of mass M and charge + Ze) moving in a solid, which interacts with an electron (of mass m and charge -e) bound to the nucleus of an atom. In such a situation we obtain the cross section associated with the energy an ion loses through electronic collisions along its path in the solid (this is the situation considered in Sect. 8.3.3). But, of course, the result will be identical for an incident energetic electron interacting with a nucleus of the solid; then the cross section will correspond to the energy transmitted to this nucleus. The electron is assumed to be stationary during the time the ion passes by. Then the forces applied on the electron before and after the ion goes through the position x = 0 cancel each other in the x direction and add in the z direction (Fig. 8.5). The forces $F_z(t)$ in the z direction

$$F_z(t) = \frac{Ze^2}{r(t)^2} \cdot \frac{P}{r(t)} \tag{8.29}$$

give to the electron the impulse

$$I = \int_{-\infty}^{+\infty} F_z(t)\, dt \tag{8.30}$$

and the energy T transmitted to the electron is

$$T = \frac{I^2}{2m} \quad . \tag{8.31}$$

Integration of (8.30)

$$I = \int_{-\infty}^{+\infty} \frac{Ze^2 p}{r^3}\, dt \quad , \tag{8.32}$$

that is, of the expression

$$I = \int_{-\infty}^{+\infty} \frac{Ze^2 p}{r^3} \frac{dt}{dr}\, dr \quad , \tag{8.33}$$

is carried out using the fact that

$$\frac{dr}{dt} = \frac{x}{r} \frac{dx}{dt} \tag{8.34}$$

with

$$\frac{dx}{dt} = \sqrt{\frac{2E}{M}} \quad . \tag{8.35}$$

The result is

$$I = \frac{Ze^2}{p} \sqrt{\frac{2M}{E}} \tag{8.36}$$

and finally

$$T = \frac{Z^2 e^4}{Ep^2} \frac{M}{m} \quad . \tag{8.37}$$

The resulting differential scattering cross section

$$d\sigma(E,T) = 2\pi p \frac{dp}{dT}\, dT \tag{8.38}$$

is given by the well-known Rutherford formula

$$d\sigma = \frac{\pi Z^2 e^4}{E} \frac{M}{m} \frac{dT}{T^2} \quad . \tag{8.39}$$

The energy momentum approximation consists in assuming that the energy transmitted T is small enough to leave the velocity of the incident particle unchanged.

8.2 Defect Production

8.2.1 Displacement of the Primary Knock-On Atom

The total cross section for displacement is

$$\sigma(E) = \int_{T_d}^{T_m} d\sigma(E,T) \quad . \tag{8.40}$$

The integration is made over T from the minimum energy T_d necessary to displace an atom from its site in the lattice to the maximum energy T_m which can be transmitted by the incident particle. The minimum energy T_d, known as the threshold energy for atomic displacement, implies that an atom of the lattice lies in a potential well of height T_d. The simplest approximation treats the probability P to displace an atom as (Fig.8.6)

$$P = 1 \quad \text{for} \quad T \geq T_d$$

and

$$P = 0 \quad \text{for} \quad T \leq T_d \quad . \tag{8.41}$$

The exact shape of the potential energy versus displacement is not important since only the knowledge of the height T_d of the barrier the atom has to overcome in order to leave its position is necessary. This justifies the use of a step function to represent the displacement probability.

Once T_d is known, the total cross section for displacement, from which the concentration of defects is determined, see (8.2), is calculated

$$\sigma(E) = \int_{T_d}^{T_m} d\sigma(E,T) \quad . \tag{8.42}$$

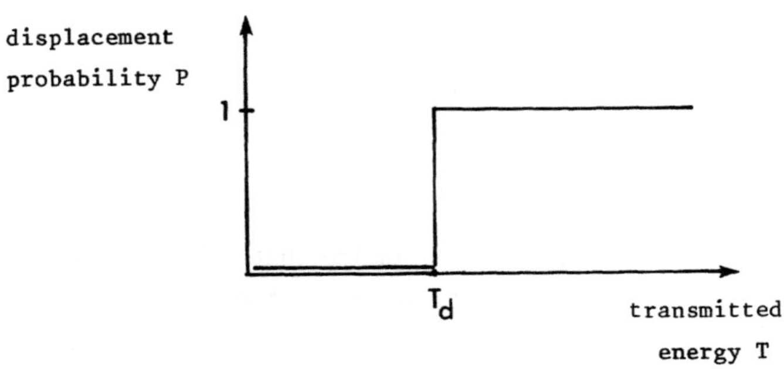

Fig.8.6. Displacement probability as a function of transmitted energy

8.2.2 Threshold Energy for Atomic Displacement

The minimum energy T_d required to displace an atom from its substitutional site is called the threshold energy. Usually T_d is assumed to be isotropic, i.e., independent of the direction in which the atom is displaced in the lattice. The justification for such assumption is that, experimentally, it is difficult to detect a significant orientation dependence of T_d (Sect.8.4). When an anisotropy of T_d is taken into account, the model used consists simply in considering that the displaced atom moves in a potential whose configuration depends on the orientation u of the impulse communicated to the atom compared to the cristallographic axes.

The first estimate of T_d was by SEITZ [8.4]. He argued that the displacement being a nonadiabatic process, T_d must be of the same order of magnitude as or larger than the energy required to remove an atom adiabatically from its site, an energy which he estimates to be ~25 eV for most solids. For KOHN [8.5] and BAUERLEIN [8.6] the formation of a vacancy-interstitial pair by collision (which is the result of an atomic displacement) is a three-step process. First, four bonds are broken; the energy E_1 required for this is four times the bond energy D. Then, the atom goes into an interstitial position; for this an energy E_2 is required in order to push away the atoms in the neighborhood of its path, i.e., to surmount the barrier these atoms oppose to its presence. Finally, there is a lattice relaxation around the empty site (vacancy) which involves an energy E_3. KOHN [8.5] has considered the case of an atom ejected in a $\langle\bar{1}\bar{1}\bar{1}\rangle$ direction which is, a priori, the "easiest" direction for atomic displacement. He estimated the work required to pass through the saddle point, which is the middle point of the triangle formed by three nearest neighbors, (Fig.8.7) to be

$$E_2 = \frac{\ell^3 c_{11}}{24} \tag{8.43}$$

(where ℓ is the lattice constant). The use of the elastic constant c_{11} to estimate E_2 implies that the potential energy is quadratic with the displacement, i.e., that the displaced atom does not come too close to its neighbors (otherwise the repulsive ion core would become important and modify the shape of the interaction potential). BAUERLEIN [8.6] felt that because there is sufficient room for interstitials in the diamond lattice, this energy E_2 can be neglected. As for the relaxation energy E_3, he says it is of the order of 0.1 eV and neglects it also (this is consistent in view of the results described in Sect.2.4.4). Then

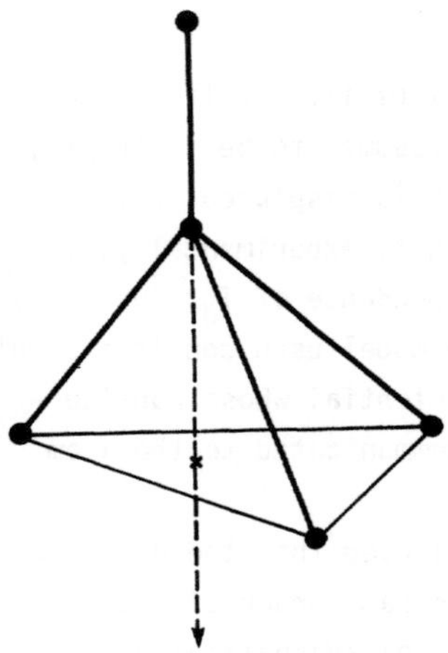

Fig.8.7. The $<\bar{1}1\bar{1}>$ direction in the covalent lattice

$$T_d \sim 4D \quad . \tag{8.44}$$

As shown in [Ref.1.1, Sect.6.1], the bond energy D can be evaluated from the cohesive energy E_c and from the $s^2p^2 \rightarrow sp^3$ promotion energy E_p. The result, using [Ref.1.1, Eq.(6.2)], is

$$T_d = 2(E_c + E_p) \quad . \tag{8.45}$$

Table 8.1, in which the estimated values of T_d are compared with experimental results, shows that the agreement between such estimation and experiment is reasonable in case of group 4 semiconductors. The situation is more complex in 3-5 compounds for which the experimental threshold energies are given in Table 8.2.

Table 8.1. Threshold energies estimated by KOHN (T_d^K) and BAUERLEIN (T_d^B), bond energy D and energy E_2 [from (8.43)] and experimental threshold energy (T_d) for group 4 semiconductors

Material	E_2 [eV]	D [eV]	T_d^K [eV]	T_d^B [eV]	T_d [eV]	Reference
Diamond	12.5	7.9	44.1	31.6	35 ± 5	[8.7]
Silicon	6.5	5.9	30.1	23.6	21	[8.8]
Germanium	5.8	6	29.8	24	27.5	[8.9]

Table 8.2. Experimental threshold energies for various compound semiconductors

Material	T_d [eV] experimental	References
GaAs	7-11	[8.10]
SbIn	6.4-9.9	[8.11]
InAs	6.7-8.3	[8.12]
GaSb	6.2-7.5	[8.13]

8.2.3 Primary Displacements

The number of displacements produced by the incident particle is given by the integration of the differential cross section (8.40)

$$\sigma(E) = \int_{T_d}^{T_m} \frac{d\sigma}{dT}\, dT \quad . \tag{8.46}$$

This integration must be made independently for each type of irradiating particle since $d\sigma/dT$ depends on the type of interaction, that is on the nature of the particle.

a) *Heavy Charged Particles (Ions)*

For heavy charged particles $d\sigma/dT$ is given by the Rutherford formula which we write, according to (8.39),

$$\frac{d\sigma}{dT} = \frac{a}{T^2} \tag{8.47}$$

with

$$a = \frac{\pi Z_1^2 Z_2^2 e^4}{E} \frac{M_1}{M_2} \quad , \tag{8.48}$$

where M_1, Z_1 relates to the incident particle and M_2, Z_2 to the atom of the solid. Integration of (8.47)

$$\sigma(E) = a \int_{T_d}^{T_m} \frac{dT}{T^2} \tag{8.49}$$

is straightforward

$$\sigma(E) = a\left(\frac{1}{T_d} - \frac{1}{T_m}\right) \quad . \tag{8.50}$$

Since M_1 and M_2 are of the same order of magnitude, the maximum transmitted energy T_m is, according to (8.9), of the order of the incident energy

E. This energy being usually in the range 0.1-1 MeV, T_m is therefore very large compared to T_d (~10 eV) and $\sigma(E)$ reduces to

$$\sigma(E) = \frac{a}{T_d} \quad . \tag{8.51}$$

b) *Neutrons*

For neutrons the interaction being described by the model of hard spheres collision, the differential scattering cross section $d\sigma$ is given by (8.27) and can be written

$$d\sigma = b\,\frac{dT}{T_m} \tag{8.52}$$

with, according to (8.24)

$$b = \pi(R_1 + R_2)^2 \quad . \tag{8.53}$$

The mass M_1 of the neutron being smaller than the mass M_2 of the atom, T_m reduces to

$$T_m = \frac{4E}{A} \tag{8.54}$$

where

$$A = \frac{M_2}{M_1} \tag{8.55}$$

is the atomic mass. Then

$$\sigma(E) = \int_{T_d}^{T_m} \frac{b}{T_m}\,dT \tag{8.56}$$

is

$$\sigma(E) = b\left(1 - \frac{T_d}{T_m}\right) \quad . \tag{8.57}$$

In practice, calculation of $\sigma(E)$ must take into account the fact that a neutron spectrum is seldom monoenergetic but presents a given distribution (for instance, an E^{-1} spectrum in case of a moderated reactor).

c) *Electrons*

Because their mass (m) is very small compared to the atomic mass (M), electrons must have a considerable energy (in the 0.1-10 MeV range) to transmit energy of the order of T_d. In this energy range, electrons are relativistic and the Rutherford formula (8.39) must be adapted [8.14], E being replaced

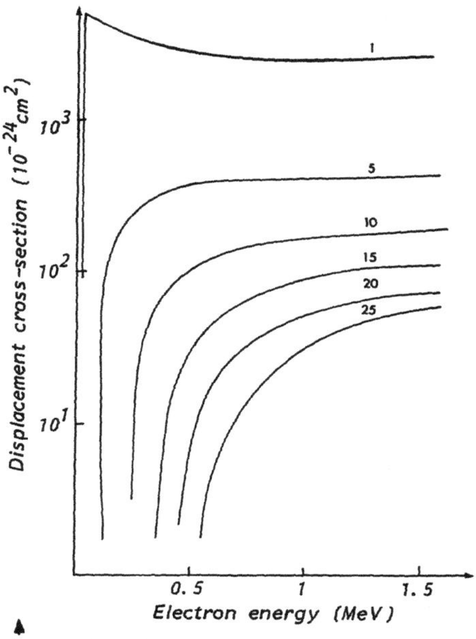

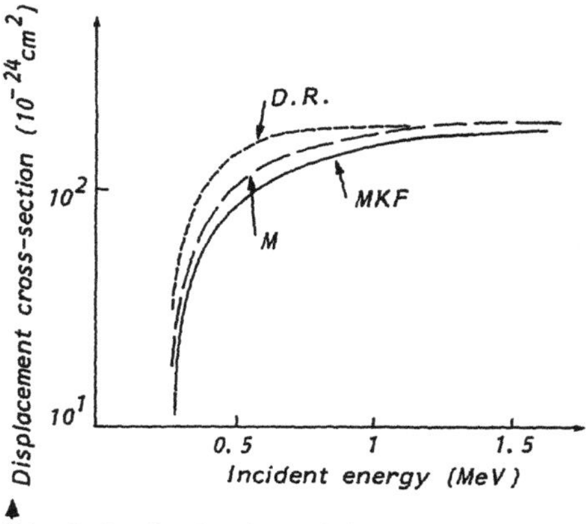

▲ Fig.8.9. Variation with the incident electron energy of the displacement cross section for a Ga atom, calculated for a threshold energy of 10 eV, using Darwin-Rutherford (DR), Mc Kinley-Feshbach (MKF) and Mott (M) formulas

▲ Fig.8.8. Displacement cross section as a function of electron energy of a Ga atom for various indicated values of the threshold energy

by (8.19). Then

$$\frac{d\sigma}{dT} = \frac{\pi Z^2 e^4}{mc^2(\gamma - 1)} \frac{M}{m} \frac{1}{T^2} \quad . \tag{8.58}$$

Introducing T_m given by (8.18), we finally have

$$\frac{d\sigma}{dT} = \pi\left[\frac{Ze^2\gamma}{mc^2(\gamma^2 - 1)}\right]^2 \frac{T_m}{T^2} \quad , \tag{8.59}$$

the integration of which gives

$$\sigma = \pi\left[\frac{Ze^2\gamma}{mc^2(\gamma^2 - 1)}\right]^2\left(\frac{T_m}{T_d} - 1\right) \quad . \tag{8.60}$$

Actually, the scattering of an electron by a nucleus necessitates a quantum treatment. The exact calculation, derived by MOTT [8.15], can be approximated by Mc KINLEY-FESHBACH's formula [8.16] for the case of atoms whose atomic number is not too large. The formula is

$$\sigma = \pi\left[\frac{Ze^2\gamma}{mc^2(\gamma^2 - 1)}\right]^2\left[\left(\frac{T_m}{T_d} - 1\right) - \frac{\gamma^2 - 1}{\gamma^2} \ln \frac{T_m}{T_d}\right.$$

$$+ \alpha\pi Z \sqrt{\frac{\gamma^2 - 1}{\gamma^2}} \left(2 \sqrt{\frac{T_m}{T_d}} - 2 - \ln \frac{T_m}{T_d} \right) \Bigg] \quad , \tag{8.61}$$

where α is the fine structure constant ($\alpha^{-1} = 137$). Figure 8.8 shows, for example, the variation of the cross section σ for the displacement of gallium atom versus electron energy for various values of T_d. In that case, the results given by the Mc Kinley-Feshbach formula differ by ~10% from the results provided by the exact formula of Mott (Fig.8.9).

8.2.4 Secondary Displacements

A primary knock-on atom, to which the energy T has been transmitted by the incident energetic particle, can in turn displace other atoms when T is large enough. The average number of displacements produced, i.e., of secondary defects, can be estimated simply using the cascade model proposed by KINCHIN and PEASE [8.17]. This model assumes hard-sphere collisions between atoms and a step-function displacement probability. Because of the hard-sphere scattering assumption, the kinetic energy of a knock-on atom is shared equally with the stationary lattice atom since they have the same mass.

As schematized in Fig.8.10, the primary knock-on atom (of energy T) produces a cascade which ends at the nth step when the transmitted energy, whose value is then $T/2^n$, becomes smaller than $2T_d$.

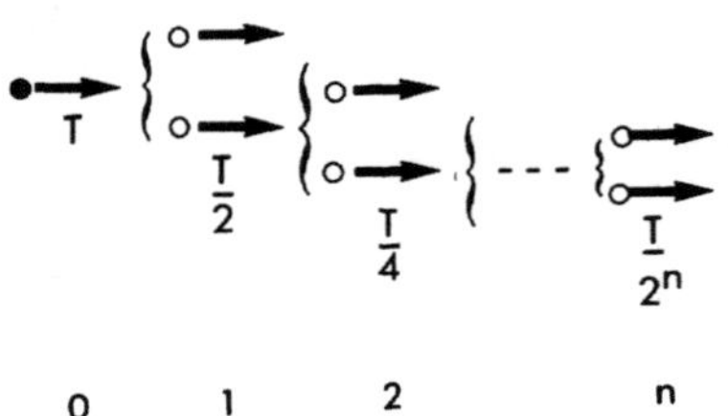

▲ Fig.8.10. Schematic representation of a cascade of displacements produced by a primary knock-on atom of energy T

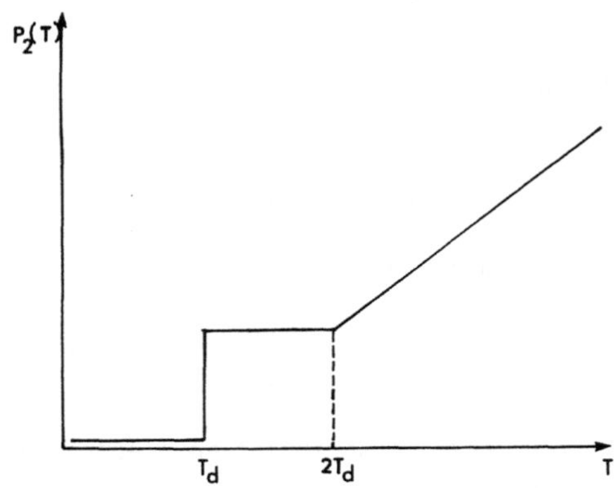

Fig.8.11. Probability of displacement as a function of the energy transmitted to a primary knock-on atom ►

As a result, the number of displacements in the cascade is

$$P_1(T) = \frac{T}{2T_d} \quad . \tag{8.62}$$

However, P(T) is accurate only for values of T very large compared to T_d, since for values of T such that $T_d \leq T \leq 2T_d$, it is obvious that only one displacement is possible. Therefore, a more accurate form of the displacement probability $P_2(T)$ is given in Fig.8.11. Thus, $P_1(T)$ can be used in case of energetic ions, when $T \gg T_d$. But in the case of electrons for which T remains on the order of Td, $P_2(T)$ should rather be used. Thus the number of secondary displacements is given by the cross section

$$\sigma(T) = \int_{2T_d}^{T_m} \frac{T}{2T_d} \frac{d\sigma}{dT} dT \quad . \tag{8.63}$$

More realistic calculations still use the notion of cascade but introduce the differential cross section for displacement $d\sigma$ adapted to the exact interaction potential.

8.3 Defect Nature and Spatial Distribution

The nature and spatial distribution of the defects created depend strongly on the nature and energy of the energetic particle.

8.3.1 Average Number of Defects per Particle

For electrons in the MeV range T_m remains of the order of T_d and, in practice, only single displacements which result in the formation of vacancy-interstitial pairs are produced. When $T_m \geq 2T_d$, the displacement of two neighboring atoms becomes possible, leading to the formation of divacancies. Consequently the defects obtained following electron irradiation are vacancies and interstitials in pairs, and at sufficiently high energy, divacancies and their interstitials. However, because vacancies and interstitials are mobile at low temperature these defects are difficult, if not impossible, to observe; only their association between themselves or with impurities are detected (Chap.9). These defects are uniformly distributed in widths of the order of several μm. Indeed, the energy the incident electrons lose to the lattice electrons is rather small, of the order of 0.1 eV $Å^{-1}$ (Sect.8.3.3).

For particles which transmit an energy very high compared to T_d, the total cross section for displacement can be written as the sum

$$\sigma = \sigma_p + \sigma_s \tag{8.64}$$

where σ_p, the cross section for the displacement of the primary atom, is given in Sect.8.2.3. The cross section σ_s for secondary displacements, i.e., for the displacement of atoms by the primary knock-on atom, is calculated using the displacement probability for the cascade process (Sect.8.2.4) using (8.63). In this expression, the differential cross section corresponds to the collision between atoms. When a hard-sphere collision model is used, $d\sigma$ is given by (8.52) and

$$\sigma_s(E) = \frac{b}{4T_dT_m}(T_m^2 - T_d^2) \tag{8.65}$$

which, in practice, reduces to

$$\sigma_s(E) = \frac{bT_m}{4T_d} \quad . \tag{8.66}$$

As a result the average number of displacements produced by a primary knock-on atom

$$\bar{\nu}(T) = \frac{\sigma_s(E)}{\sigma_p(E)} \tag{8.67}$$

is

$$\bar{\nu}(T) = \frac{T}{4T_d} \quad . \tag{8.68}$$

Thus, for a 0.1 MeV knock-on atom, produced for instance by a 0.1 MeV ion of the same mass, $\bar{\nu}(E) \sim 10^3$ when $T_d \sim 25$ eV (the case for silicon or germanium). In practice, a more realistic differential cross section must be used. This is not exactly the Rutherford cross section because the Coulomb potential does not take into account the electron screening of the nuclei. The potential is a Coulomb potential modulated by a function f of the quantity r/α (α is the Bohr radius)

$$V(r) = \frac{Ze^2}{r} f\left(\frac{r}{\alpha}\right) \quad . \tag{8.69}$$

Depending upon the energy, a Bohr potential $f = \exp(-r/\alpha)$, a Born-Mayer potential $f = (r/\alpha)\exp[-C_1(r/\alpha)]$, and a Thomas-Fermi potential $f = (r/\alpha)[(r/\alpha)^2 + C_2^2]^{-1/2}$ (C_1 and C_2 are adjustable constants) are used. It is easy to see that, for a Coulomb potential,

$$\bar{\nu}(T) = \frac{1}{2}\ln\left(\frac{T_m}{2T_d}\right) \quad . \tag{8.70}$$

8.3.2 Amorphous Layer Formation by Irradiation

The number of displacements produced by a single primary collision can thus be very high. Then a cascade of displacements consists of a region (which we call the heart of the cascade) containing a large concentration of vacancies, surrounded by a region containing interstitials. Of course, because the energy transmitted to one atom is very large compared to T_d, the probability of a displacement of two or more neighboring atoms is high and the heart of the cascade also contains divacancies, trivacancies, etc. Usually, the irradiation is performed at temperatures above which interstitials and (or) vacancies are mobile; therefore, after irradiation the interstitials have disappeared and the size of the heart of the cascade has decreased, due to vacancy out diffusion. The heart of the cascade is now surrounded by point defects formed by the association of the escaping vacancies with impurities (such as E or A centers). The inner part of the heart is composed of vacancy clusters since moving vacancies agglomerate.

When a material is uniformly irradiated by ions, isolated damaged regions are first created. Depending upon the size of the damaged regions and of the dose of irradiation, these regions start to overlap until a continuous disordered layer is formed. When the disorder is monitored, using any technique sensitive to the presence of defects [8.18], the disorder is found to increase as a function of the dose until it saturates at the critical dose. This critical dose ϕ_c depends on several parameters, in particular: i) the nature and energy of the incident ions since the size of the disordered region is a direct function of these two quantities (Sect.8.3.3); ii) the temperature, since vacancy annealing, when it occurs, reduces the original size of the cascade.

When the critical dose is reached, the disordered layer is often said to be amorphous. This conclusion should be taken cautiously because a disordered material is a material which contains a large concentration of vacancy clusters, i.e., a large amount of dangling bonds while, in an amorphous material, in which only short-range order is preserved, practically all bonds are satisfied. The criterion for the formation of a real amorphous layer should be the following: the transition from a crystalline to an amorphous phase must occur when the Gibbs free energy of the crystalline phase G_x plus the defects it contains, becomes larger than the free energy of the amorphous phase G_α. This condition can be written

$$G_x + n g_d \geq G_\alpha \tag{8.71}$$

when the defects, in concentration n, are all the same, g_d being the free energy per defect. Therefore, the crystalline to amorphous transition takes place when the concentration of defects is larger than the critical concentration

$$n_c = \frac{N(g_\alpha - g_x)}{g_d} \quad , \tag{8.72}$$

where N is the atomic concentration and g_x and g_α the free energies per atom in the crystalline and amorphous phases respectively. Such a condition is difficult to verify quantitatively. First, the critical concentration can be estimated from

$$n_c = \phi_c \frac{T_m}{2T_d} e \tag{8.73}$$

(e is the thickness of the disordered or amorphous layer) only when no annealing has occured. Second, all the defects are not identical; there are vacancies, divacancies, etc. Finally, the thickness of the layer can be calculated only when no annealing has occured. The critical dose of irradiation can be estimated from

$$\phi_c = 2Ne \frac{g_\alpha - g_x}{g_d} \frac{T_d}{T_m} \tag{8.74}$$

only at low temperature (so that vacancies remain immobile), neglecting the interstitial contribution, and with the assumption that the free energy of a cluster of i vacancies is equal to i times the free energy g_d of a vacancy.

8.3.3 Range of the Particle

Particles move through the solid and atomic displacements occur along their path (x). They are slowed down due to the energy they transfer to the electrons of the solid (which results in ionization). The rate of energy loss -dE/dx can be written in terms of the differential cross section $d\sigma(E,T)$ for the electron-particle scattering

$$-\frac{dE}{dx} = N_e \int_{T_{min}}^{T_m} T \, d\sigma(E,T) \quad , \tag{8.75}$$

where N_e is the density of electrons in the solid and T_{min} and T_m the minimum and maximum energy transferable to an electron (of mass m) by the incident particle (of mass M and energy E).

The range R(E) of the particle is obtained directly from this stopping power -dE/dx by

Fig.8.12. Schematic representation of electronic stopping power with the energy of a moving ion

$$R(E) = \int_0^E \frac{dE}{(-dE/dx)} \quad . \tag{8.76}$$

The minimum energy transmitted T_{min} is usually taken to be the ionization energy I of an electron from its atom because it is thought that energy transfers smaller than I do not contribute appreciably to the slowing down of the particle. Actually, atoms having many electrons of different I's, the ionization energy should be some appropriate average over all electrons. In case of incident charged particles, the Rutherford scattering law gives

$$-\frac{dE}{dx} = N_e \int_I^{4(m/M)E} \frac{T\pi Z^2e^4}{E} \frac{M}{m} \frac{dT}{T^2} \quad , \tag{8.77}$$

that is

$$-\frac{dE}{dx} = N_e \frac{\pi Z^2 e^4}{E} \frac{M}{m} \ln\left(\frac{4mE}{MI}\right) \quad . \tag{8.78}$$

This formula is approximate; it must be corrected for small values of E because when the incident energy of the particle is small the struck electrons have time to adjust to the field of the moving ion and are not ionized; moreover, the moving ion can be partly screened by lattice electrons. As a result the stopping power falls off for $E \to 0$ (Fig.8.12). For electrons, the stopping power is of the order of 10 MeV cm^{-1}, which implies that for samples of thickness $\sim 10^{-2}$ cm irradiated with ~1 MeV electrons, the energy loss is small and the defect distribution is uniform. When the thickness of the sample must be taken into account, the stopping power is usually estimated from empirical formula [8.19].

The notion of range of a particle is particularly important for ion irradiation since it determines the depth distribution of the implanted ions. Because of the fluctuations in energy loss and scattering angle at each collision, a range must be defined in terms of probability distribution of

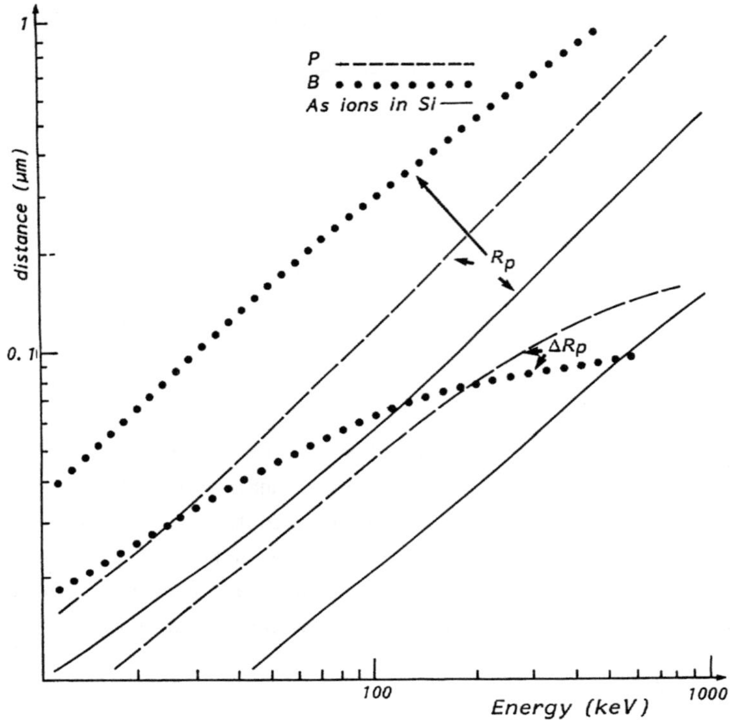

Fig.8.13. Variations of the projected range and range straggling of As, P and B ions in Silicon as a function of the incident ion energy

penetration depth. For an ion following a complicated path, the interesting quantities are: i) the projected range R_p, i.e., the projection of the total range on the incident direction of the ion. The perpendicular range is also useful for device applications; ii) the range straggling ΔR_p, i.e., the mean square fluctuation in range which characterizes the depth distribution of implanted ions in the direction of implantation; iii) the range straggling perpendicular to the direction of implantation. Ranges are obtained by integrating (8.75) when dE/dx, the sum of the electronic and atomic energy losses, is known as a function of E.

Numerical calculations have been performed, for instance, by LINDHARD, SCHARFF and SCHROTT (LSS) [8.20] using a realistic potential (Thomas-Fermi) to describe the collisions. Some results are illustrated in Fig.8.13. Details of the theory of ion ranges and their experimental measurement can be found in [8.21].

8.4 Experimental Determination of a Threshold Energy

The threshold energy T_d is obtained by comparing the experimental variation of the defect introduction rate (i.e., the defect concentration divided by the irradiation dose) versus the energy of irradiation E with the variations of the cross section $\sigma(E)$ for atomic displacement calculated for various values of T_d (Fig.8.15). Electrons in the 100 KeV - 2 MeV range are used because they are the only particles whose maximum transmitted energy is of the order of T_d (Fig.8.2) which, for practically all materials, ranges from ~1 eV to ~50 eV. Heavier particles, which transmit an energy large compared to T_d, do not create individual displacements, i.e., single vacancy-interstitial pairs, but cascades of defects in which the number of single displacements produced in a cascade is impossible to measure in practice.

There have been many attempts to determine the threshold energy in most semiconductors. The results obtained have a considerable scatter [8.22]. For instance, values of T_d ranging from 80-35 eV are found for diamond, from 30-12.7 eV for germanium and from 20.9-11 eV for silicon. The reason is the following [8.23]: when the defect introduction rate is measured at a temperature at which vacancies and (or) interstitials are mobile, the defects present are complex defects, formed by the association of vacancies and interstitials with impurities or with themselves. In order to be used as a measure of the concentration of the primary defects, it is necessary for the concentration of these complex defects to be proportional to the concentration of primary defects. Unfortunately, this is seldom true. Indeed, as we shall see in Chap.9, when one element of the vacancy-interstitial pair is mobile, the proportion of pairs which escape recombination and thus form complexes with impurities depends on several parameters: the temperature, the concentration and nature of the impurities, and the Fermi level position. It also depends on the energy of irradiation, since the distance between a vacancy and its interstitial (which strongly influences the recombination rate) increases with the transmitted energy.

Determination of the threshold energy must therefore be performed by monitoring directly the concentration of the primary defects, i.e., in practice, following irradiations at very low temperature. This is the reason why the experimental values of T_d given in Table 8.1 have been selected among all the various results. For germanium and diamond we have taken values which have been measured at temperatures at which vacancies and interstitials are thought to be stable. For silicon, we have taken the values obtained

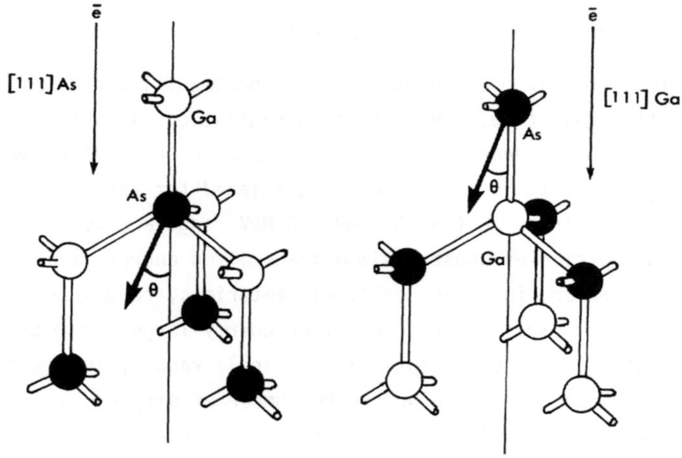

▲ Fig.8.14. Electron irradiation of GaAs in <111> As and <111> Ga directions

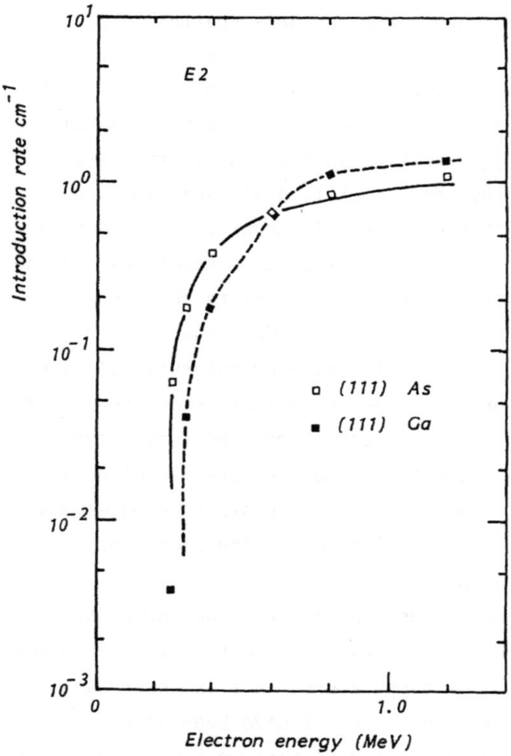

◄ Fig.8.15. The introduction rate of the so-called E_2 defect in GaAs for two opposite orientations of the electron beam. The full line is calculated for an isotropic threshold energy of 9 eV; the dashed line corresponds to the nonisotropic threshold energy described in [8.27]

by EPR observation of the vacancy itself; this result can therefore still be questioned since the interstitials are mobile even at 4 K (Chap.9).

Several attempts have been made to study the anisotropy of the threshold energies in group 4 and compound semiconductors. Only a very small orientation dependence of the defect introduction rate is observed in germanium [8.24] and silicon [8.25]. In silicon there are even a number of disagreements between the different authors. But, as discussed above, since the measurements are performed at high temperature, i.e., by measuring the concentration of complex defects, the results have to be taken with caution.

In compound semiconductors the orientation dependence of the defect introduction rate is a powerful means to determine the nature of the displaced atoms, i.e., to identify the sublattice to which the created defects belong. Consider for instance gallium-arsenide. A simple examination of its crystalline structure (zinc blende; Fig.8.14) suggests that when the irradiation is directed in the <111> Ga orientation, the displacement of an As atom by the incident electron should be difficult since a Ga atom is situated just behind it in the direction of its displacement. On the other hand, the displacement of a Ga atom should be easy since behind it, in the direction of the displacement, the lattice is open. The situation is of course reversed in the case of an irradiation in <111> As direction. Although this method seems rather naive it apparently works. For instance, in indium antimonide, the defects which anneal in a first stage (called stage I) have a larger defect-creation rate than other defects which anneal in a second stage (called stage II) when the irradiation is performed in a <111> Sb orientation. The reverse is observed when the irradiation is performed in a <111> In orientation. EISEN [8.26] therefore concluded that stage I defects are due to the displacement of Sb atoms while stage II defects are due to the displacements of In atoms.

This type of measurement must be performed with incident electron energies such that the maximum transmitted energy T_m is practically equal to T_d. Indeed, according to (8.17), only in this case is the scattering angle φ close to zero, i.e., the atoms are ejected in a direction close to the incident direction of the electrons. Measurements at high irradiation energy can lead to the wrong answer. Consider the results obtained for GaAs [8.27]: Figure 8.15 shows the variation of the defect introduction rate versus electron energy for one of the primary defects conventionally labeled E_2.

The curve corresponding to an irradiation in the <111> As direction fits well with the Mc Kinley-Feshbach formula (8.61) using a threshold energy of

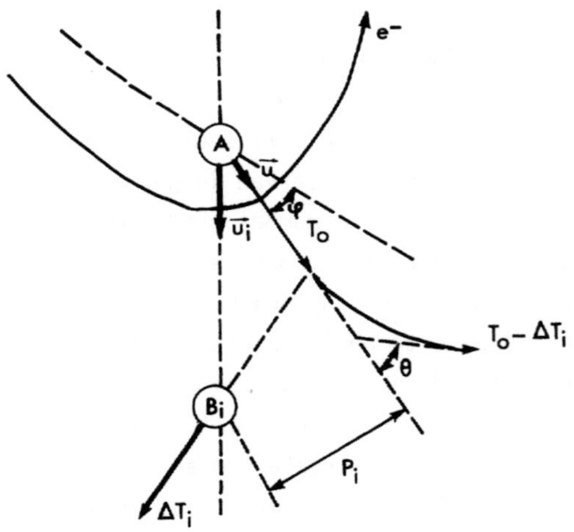

Fig.8.16. Schematic representation of the collision of a primary knock-on atom A with a neighbor B_i

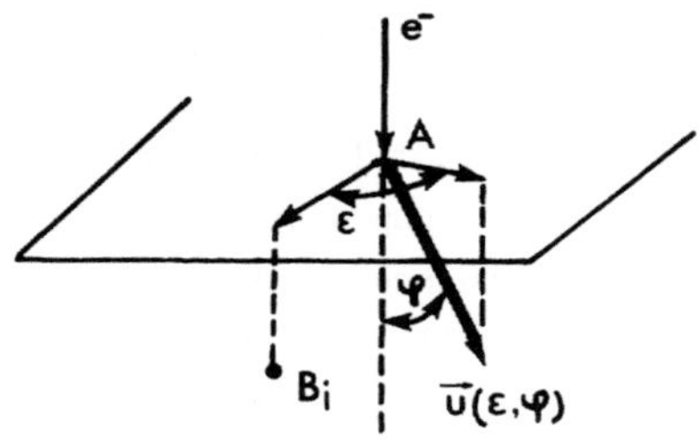

Fig.8.17. Angles ϕ and ε define the direction $\vec{u}$ of the primary knock-on as compared to the direction of the incident electrons and to the position of the neighbor B_i

9 eV. But the curve obtained for an irradiation in the <111> Ga direction cannot be fitted with a single threshold since, although the apparent threshold is smaller than for the <111> As direction, the defect-introduction rate at high energy is larger. The fact that the defect-introduction rate is larger for the <111> As direction than for the <111> Ga direction, for energies close to the threshold energy, demonstrates that the defects originate from the displacement of As atoms. The reverse conclusion drawn by LANG et al. [8.29] was due to the fact that they used an irradiation energy which was too large (1 MeV).

The model of orientation dependence of threshold energy which accounts for such behavior of the defect-introduction rate versus energy in the <111> As and <111> Ga directions is in essence different from the model of an anisotropic potential sink that the atom must experience to be expelled from its site. It assumes that T_d is isotropic (it is the energy to break four bonds); the anisotropy originates from the interaction the knock-on atom experiences with its neighbors during its displacement. In a collision with a neighboring atom B_i, the knock-on atom A loses a part ΔT_i of its kinetic energy T_0 (Fig.8.16). It is definitively displaced only when its net kinetic energy, $T_0 - \sum_i \Delta T_i$, the summation being taken on the i neighboring atoms, is larger than T_d. Because the energy ΔT_i depends on the direction $\vec{u}(\varepsilon,\varphi)$ in which atom A is expelled (Fig.8.17), the apparent threshold energy is anisotropic, since the condition of displacement is

$$T_0(\varepsilon,\varphi) - \sum_i \Delta T_i(\theta,\varepsilon,\varphi) \geq T_d \quad . \tag{8.79}$$

Consider Fig.8.16 which illustrates the collision of atom A with a neighbor B_i. Since the collision is elastic, according to (8.9,10) the lost energy ΔT_i is

$$\Delta T_i = T_0 \frac{4M_A M_{B_i}}{(M_A + M_{B_i})^2} \cos^2\theta \tag{8.80}$$

where T_0 is given by (8.17). The scattering angle θ being a function of the impact parameter

$$p_i(\theta) = a \sin(\vec{u},\vec{u}_i) \tag{8.81}$$

(a is the interatomic distance), ΔT_i is a well defined function of the primary recoil direction $\vec{u}(\varepsilon,\varphi)$ of atom A and of its initial kinetic energy T_0.

The displacement probability P to be used for the calculation of the total cross section

$$\sigma(E) = \int_0^\pi \sin\varphi \, d\sigma(\varphi) d\varphi \int_0^{2\pi} P \, d\varepsilon \tag{8.82}$$

is thus

$$P(\varphi,\varepsilon,T_0) = 1 \quad \text{when} \quad T_0(\varphi) - \sum_i \Delta T_i(\theta,\varepsilon,\varphi,T_0) \geq T_d \tag{8.83}$$

and

$$P(\varphi,\varepsilon,T_0) = 0 \quad \text{when} \quad T_0(\varphi) - \sum_i \Delta T_i(\theta,\varepsilon,\varphi,T_0) \leq T_d \quad . \tag{8.84}$$

In order to obtain θ as a function of P_i, the following relation [8.28] is used

$$\theta = \int_{r_m}^\infty \frac{P_i}{r^2}\left(1 - \frac{V(r)}{T_R} - \frac{b^2}{r^2}\right)^{-\frac{1}{2}} dr \quad , \tag{8.85}$$

where V(r) is the interacting potential between A and B_i, to be specified, and r_m the distance of closest approach of the two atoms. The energy T_R is defined by

$$T_R = \frac{T_0 M_A}{M_A + M_{B_i}} \quad . \tag{8.86}$$

As shown in Fig.8.18, the experimental results for several primary defects (labeled E_1, E_2 and E_3) in GaAs agree well with such a model [8.30]. The experimental anisotropy ratio, defined as the ratio of the defect intro-

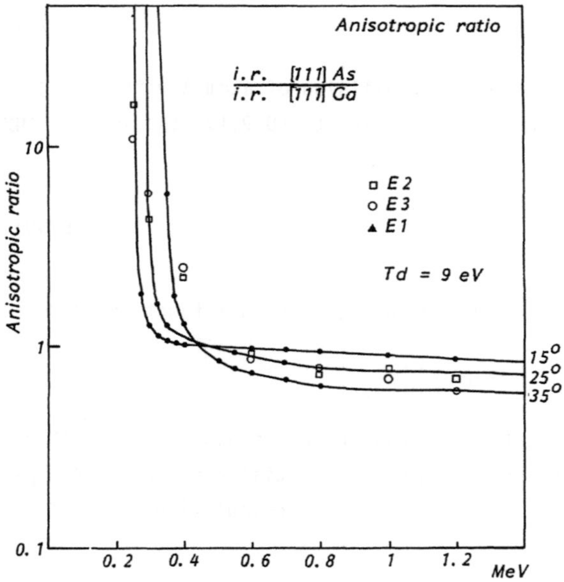

Fig.8.18. Ratio of the cross sections for displacement for irradiation in the <111> As and <111> Ga as a function of the energy of the incident electrons. The solid lines correspond to the calculation for various values of the hard core radius indicated on the figure

duction rates in the <111> As over the <111> Ga directions versus the incident electron energy, varies as predicted by this model when a value of 9 eV is assumed for the threshold energy T_d. This is shown in Fig.8.18 when the interacting potential $V(r)$ is a hard-core potential: $V(r) = \infty$ for $r < r_{hc}$ and $V(r) = 0$ for $r > r_{hc}$, where the hard-core radius r_{hc} is the sum of the radii of the two interacting atoms (Sect.8.1.3a). The fit with experimental values is obtained for $r_{hc} = 1 \pm 0.5$ A, a quite reasonable value. A good fit is also obtained for another type of potential, the empirical Born-Mayer potential

$$V(r) = V_0 \exp(-r/r_{bm}) \quad . \tag{8.87}$$

In that case, the radius r_{bm} is, as expected, of the order of the Thomas-Fermi screening radius which, in the case of Ga and As atomic interaction, is $\sim$0.1 Å; the constant V_0 is 3×10^4 eV.

The reason the two different interaction potentials give essentially the same result is that the displacement probability is sensitive to small impact parameters only, in which case hard-core interaction constitutes a reasonable approximation. Finally, we note that the fact that the anisotropy ratio is reversed at high energy (above 0.5 MeV in that case) does not depend on the type of interaction potential; it is just a consequence of the tetrahedral coordination of the crystal.

In conclusion, the measurement of the threshold energy is an interesting technique to obtain the nature of the defects created by irradiation. As we just discussed, in compound semiconductors the measurement of T_d allows one to determine the sublattice from which a defect originated. It is also possible to distinguish between defects related to vacancies or interstitials and defects related to divacancies. This is because the energy necessary to displace two neighboring atoms, i.e., to create a divacancy, is approximately twice the energy necessary to displace a single atom. The result is that divacancies (when directly created and not formed by the association of two vacancies) present a threshold energy which is twice the threshold energy for the formation of the vacancy-interstitial pair. This has been verified [8.31] by monitoring directly the vacancy and divacancy creation rates in silicon using EPR measurements. Finally, we note that the comparison between the experimental defect introduction rate and the calculated defect introduction rate (once the threshold energy has been determined) indicates if all the defects created are present, i.e., if these defects are vacancies and interstitials, or not.

8.5 Subthreshold Effects

There are instances where defects have been said [8.31] to be produced (in silicon, germanium, cadmium sulphide, indium antimonide and zinc sulphide) by electrons (or X-rays) which are unable to transmit to the lattice atoms an energy as large as T_d. Even illumination by light is said to create defects in cadmium sulphide [8.32]. The explanations proposed to account for defect formation in these conditions fall into two categories: i) The irradiation induces relocation of light impurities. Because they have a smaller mass than the lattice atoms, the displacement of such impurities can be induced with a smaller threshold energy; ii) The irradiation induces the ionization of valence electrons and results in a weakening of the bond energy and consequently in a decrease of the threshold energy. Such mechanism is analogous in essence to the ionization induced displacement mechanism which occurs in ionic materials. There, double ionization of an atom makes its charge equal to the charge of its neighbors, rendering its substitutional position unstable and the atom is ejected [8.33].

The existence of the first category of this mechanism has been clearly demonstrated: in germanium the existence of the subthreshold effect is directly related to the presence of hydrogen [8.34]. But this is not the case for the second category of mechanism. Experimental observations are made in such a way that the apparent defect production can also be attributed to spurious effects such as surface effects or impurity diffusion through an ionization enhanced mechanism [Ref.1.1, Sect.7.3]. Indeed, observations of defects introduced by subthreshold irradiation have been performed using conductivity measurements in thin samples or optical techniques which are both very sensitive to surface states. It has been shown that surface phenomena are responsible for the electrical changes observed under X ray irradiation [8.35]. It is also possible to interpret an apparent introduction of defects in terms of the enhancement of impurity diffusion [8.36], the impurities situated at the surface or at the electrical contacts diffusing in the bulk. When a spectroscopic technique such as EPR, is used to monitor the introduction of A centers [8.37], no subthreshold effects are found. The transient capacitance technique (Chap.6) is both well adapted to the detection of defects in a region below the surface, which is thus not perturbed by surface effects, and at the same time is very sensitive. When such a technique is used [8.38,39] no defects are found to be introduced by subthreshold irradiations.

9. Defect Annealing

The term "annealing is used here to denote the process of disappearance of defects. It thus occurs in materials containing defects in concentration larger than the thermodynamic equilibrium concentration. Thermodynamic driving forces reduce the concentration of the defects to the equilibrium concentration characteristic of the material and of the temperature [Ref.1.1, Sect.6.2].

An excess of defects disappears because: i) The defects become mobile. They migrate to sinks (surfaces, grain boundaries, etc.), they recombine with their counterpart (for instance an interstitial recombines with a vacancy) or they form new defects by association between themselves or with other types of defects (or impurities); ii) They dissociate when they are complexes.

Each of the processes: migration, recombination, complex formation and complex dissociation is characterized by an activation energy (Fig.9.1).

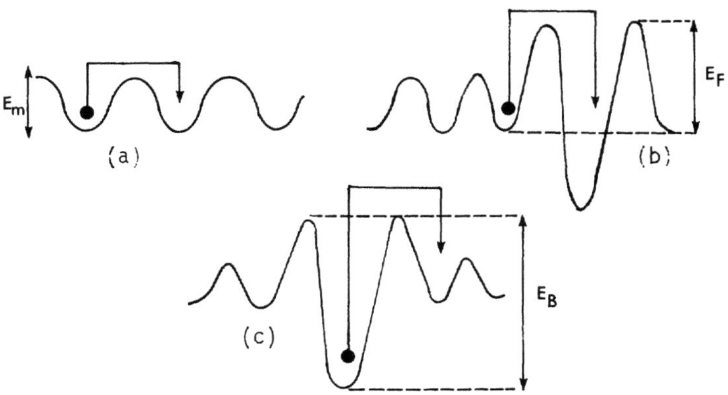

Fig.9.1a-c. Schematic representation of the mechanism of defect migration (a), complex formation (b), and complex dissociation (c) showing the corresponding activation energies involved in the process

At a given temperature T_a an annealing corresponds to a given decrease of the defect concentration versus time. As we shall discuss in Sect.9.1, these kinetics reflect the existence of the process involved in the annealing (at least the dominant one); they also reflect the initial distribution of the defects and eventually, the distribution of the various sinks which take part in the annealing. Therefore, knowledge of annealing kinetics can provide information on the defect distribution and, eventually, on their nature. In Sect.9.2, we describe how the parameters which characterize the annealing can be obtained. Finally, in Sect.9.3, we illustrate annealing processes using the case of the defects introduced by electron irradiation in silicon for which a large amount of information has been obtained.

9.1 Annealing Kinetics

9.1.1 Rate of Reaction

Consider N defects (per unit volume) which diffuse to sinks. Both the defects and the sinks are assumed to be distributed initially at random. The number of defects (per unit volume) which disappear in the sinks per unit time is proportional to the number of defects N(t) present at time t and can thus be written

$$\frac{dN}{dt} = -KN \quad . \tag{9.1}$$

It is proportional to N(t) because the sink concentration remains constant. Thus, the kinetics of the annealing is described by

$$N(t) = N_0 \, e^{-Kt} \tag{9.2}$$

where N_0 is the initial defect concentration. The rate constant K is proportional to the diffusion coefficient of the defects

$$K = \beta D \quad , \tag{9.3}$$

and the coefficient β depends on the geometrical shape of the sinks.

To see this and to illustrate the assumptions (9.3) implies, consider for instance the case where the sinks act as spherical traps of radius r_0 [9.1]. Because of the radial symmetry, the diffusion problem to be treated [Ref.1.1, Sect.7.4.1],

$$\frac{\partial c}{\partial t} = D\nabla^2 c \quad , \tag{9.4}$$

reduces to

$$\frac{\partial c}{\partial t} = \frac{D}{r} \frac{\partial^2}{\partial r^2} (rc) \quad , \tag{9.5}$$

that is

$$\frac{\partial (rc)}{\partial t} = \frac{\partial^2}{\partial r^2} (rc) \quad , \tag{9.6}$$

where $c(r,t)$ is the concentration of the diffusing defect at a distance r from the center of the sink and at time t.

In case the initial distribution $g(r)$ of the defects is a delta function centered at $r = r_0 + \xi$

$$g(r) = \delta(r - r_0 - \xi) \quad , \tag{9.7}$$

it is easy to verify that the solution of (9.6) is given by

$$rc(r,t) = \frac{1}{\sqrt{4\pi Dt}} \, r \exp\left[-\left(\frac{\xi + r_0 - r}{2\sqrt{Dt}}\right)^2\right] \quad . \tag{9.8}$$

For a more general distribution $g(r)$, the solution $c(r,t)$ is obtained by replacing $g(r)$ by a sum of δ functions.

In order to fulfill the boundary condition

$$c(r_0,t) = 0 \quad , \tag{9.9}$$

we add a distribution $-g(r)$, symmetric of $g(r)$ compared to r_0 (Fig.9.2). The solution is then given by

$$rc(r,t) = \frac{1}{\sqrt{4\pi Dt}}\left\{\int_0^\infty (\xi + r_0)g(\xi + r_0) \exp\left[-\left(\frac{\xi + r_0 - r}{2\sqrt{Dt}}\right)^2\right] d\xi \right.$$
$$\left. - \int_{-\infty}^0 (r_0 - \xi)g(r_0 - \xi) \exp\left[-\left(\frac{r_0 + \xi - r}{2\sqrt{Dt}}\right)^2\right] d\xi\right\} \quad , \tag{9.10}$$

an expression which, replacing ξ by $-\xi$ in the second integral, reduces to

$$rc(r,t) = \frac{1}{\sqrt{4\pi Dt}} \int_0^\infty (\xi + r_0)g(\xi + r_0)\left\{\exp\left[-\left(\frac{r - r_0 - \xi}{2\sqrt{Dt}}\right)^2\right]\right.$$
$$\left. - \exp\left[-\left(\frac{r - r_0 + \xi}{2\sqrt{Dt}}\right)^2\right]\right\} d\xi \quad . \tag{9.11}$$

The fractional recovery, i.e., the fraction of defects which have disappeared at time t, defined as

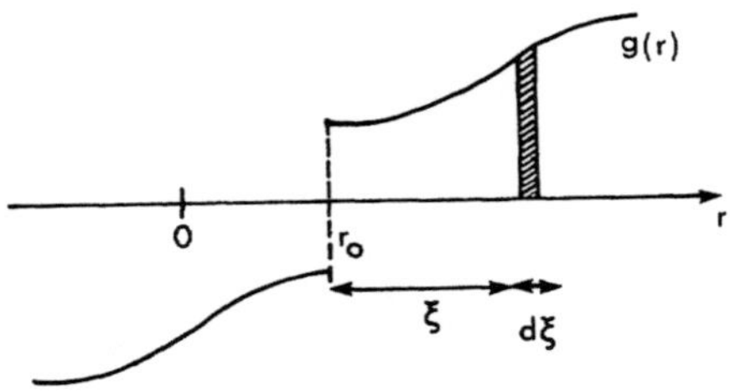

Fig.9.2. Schematic representation of the initial pair distribution function and of the symmetric function used to fulfill the boundary condition $c(r_0,t) = 0$

$$\phi(t) = \frac{N_0 - N(t)}{N_0} \quad , \tag{9.12}$$

is given by the flux of defects which have crossed the sphere of radius r_0

$$\frac{dN}{dt} = -4\pi r_0^2 N_0 D\left[\frac{\partial c(r,t)}{\partial t}\right]_{r=r_0} \tag{9.13}$$

at the time t

$$\phi(t) = 4\pi r_0^2 D\left[\frac{\partial c(r,t)}{\partial t}\right]_{r=r_0} \quad . \tag{9.14}$$

From (9.11), it can be shown that the expression of $\phi(t)$ is

$$\phi(t) = 4\pi r_0 \int_{r_0}^{\infty} r g(r) \operatorname{erfc}\left(\frac{r - r_0}{2\sqrt{Dt}}\right) dr \tag{9.15}$$

where erfc(x) is the complementary error function

$$\operatorname{erfc}(x) = 1 - \frac{2}{\sqrt{\pi}} \int_0^x e^{-u^2} du \quad . \tag{9.16}$$

We now consider the following simple examples:

i) g(r) is a delta function:

$$g(r) = N_0 \delta(r - r_i) \quad , \tag{9.17}$$

then

$$\phi(t) = \frac{r_0}{r_i} \operatorname{erfc} \frac{r_i - r_0}{2\sqrt{Dt}} \tag{9.18}$$

using:

$$N_0 = 1/4\pi r_i^2 \tag{9.19}$$

which expresses the fact that the defects are initially distributed on the sphere of radius r_i. The total probability of capture (at $t = \infty$) is then

$$\phi(\infty) = \frac{r_0}{r_i} \quad . \tag{9.20}$$

Thus, for instance, the probability for an interstitial to escape recombination with a vacancy, situated at a distance r_i and having a capture radius r_0, is $1 - r_0/r_i$.

ii) g(r) is a sum of δ functions

$$g(r) = \sum_i N_i \delta(r - r_i) \quad . \tag{9.21}$$

It is easy to see, from the preceding case, that $\phi(t)$ is given by

$$\phi(t) = \sum_i \frac{N_i}{N_0} \operatorname{erfc}\left(\frac{r_i - r_0}{2\sqrt{Dt}}\right) \tag{9.22}$$

where N_i/N_0 is the fraction of defects situated at $r = r_i$.

iii) g(r) is constant. When the defects are uniformly distributed, their average concentration is

$$g(r) = \frac{1}{\frac{4}{3}\pi r_m^3} \tag{9.23}$$

where r_m is the average distance between defects. Then

$$\phi(t) = \frac{3r_0}{r_m^3} \int_0^\infty r \operatorname{erfc}\left(\frac{r - r_0}{2\sqrt{Dt}}\right) dr \quad . \tag{9.24}$$

Once written as

$$\phi(t) = \frac{3r_0}{r_m^3} \int_0^\infty \left(r_0 + 2\sqrt{Dt}\,\frac{r - r_0}{2\sqrt{Dt}}\right) \operatorname{erfc}\left(\frac{r - r_0}{2\sqrt{Dt}}\right) 2\sqrt{Dt}\; d\,\frac{r - r_0}{2\sqrt{Dt}} \quad , \tag{9.25}$$

this integral can be integrated by parts, giving the following result

$$\phi(t) = \frac{6r_0}{r_m^3}\left(\frac{r_0}{\sqrt{\pi}} + \frac{\sqrt{Dt}}{2}\right)\sqrt{Dt} \quad . \tag{9.26}$$

Thus, only when $\sqrt{Dt} \gg r_0$, is $\phi(t)$ linear with time

$$\phi(t) = \frac{3r_0}{r_m^3} Dt \tag{9.27}$$

and the rate constant

$$K = \frac{d\phi}{dt} \quad , \tag{9.28}$$

written as (9.3) with

$$\beta = \frac{3r_0}{r_m^3} \quad . \tag{9.29}$$

9.1.2 Order of Reaction

A reaction of type (9.1) is said to be of first order because dN/dt is given by

$$\frac{dN}{dt} = -KN^{\alpha} \tag{9.30}$$

with $\alpha = 1$. The parameter α is called the order of the reaction. For a second-order reaction

$$\frac{dN}{dt} = -KN^2 \quad , \tag{9.31}$$

the kinetics are

$$N(r,t) = \frac{N_0(r)}{1 + N_0(r)Kt} \quad . \tag{9.32}$$

Second-order kinetics are encountered when two types of randomly distributed defects anneal by direct recombination. This is the case of vacancies (V) and interstitials (I), created in equal number by an irradiation, which recombine through a single jump. Indeed, if N_I and N_V are the concentrations of I and V, respectively, we have

$$\frac{dN_I}{dt} = \frac{dN_V}{dt} = -KN_I N_V \quad . \tag{9.33}$$

Since

$$N_I = N_V \tag{9.34}$$

this equation reduces to

$$\frac{dN_I}{dt} = -KN_I^2 \quad . \tag{9.35}$$

9.1.3 Description in Terms of Chemical Reactions

In cases where more than a defect and a sink are involved in the annealing kinetics, sums of reactions of various order, which are not independent, have to be considered. This leads to a system of coupled differential equations. In order to illustrate such a situation we consider the case of mobile defects I (interstitials for instance) which can annihilate on sinks V (vacancies) but also form complexes C with impurities S. Then, we have the following reactions

$$I + S \underset{k_2}{\overset{k_1}{\rightleftarrows}} C$$

$$I + V \xrightarrow{k_3} 0 \quad , \tag{9.36}$$

to which correspond the differential equations for the corresponding concentrations (N)

$$\frac{dN_c}{dt} = K_1 N_S N_I - K_2 N_c \tag{9.37}$$

and

$$\frac{dN_I}{dt} = -K_3 N_I - \frac{dN_c}{dt} \quad . \tag{9.38}$$

The solution of this set of nonlinear coupled differential equations gives the annealing kinetics when the rate constants K_1, K_2, K_3 are known.

9.1.4 Recombination of Correlated Pairs

When interstitials and vacancies are created by an irradiation, the interstitials I are situated in the neighborhood of the vacancies V from which they originate: V and I are distributed in pairs. The recombination occurs through the diffusion of one element of the pair and the kinetics of annealing reflect the distribution in distance of the pairs $g(r_p)$ (the number of pairs which are separated by a distance r_p). It can also be characterized by the so-called correlation factor γ, defined as the ratio of the distance R_{pq} between two pairs (p and q) over the distance r_p between two elements of one pair (p) (Fig.9.3):

$$\gamma = \frac{R_{pq}}{r_p} \quad . \tag{9.39}$$

It is easy to see that when the correlation factor is small ($\gamma \sim 1$), for instance because the concentration of pairs is large, the defects are randomly distributed and the kinetics of annealing follow (9.1); the activation energy associated with the kinetics is the activation energy of the diffusion coefficient. However, when γ is large, in case of close V-I pairs for instance, the kinetics are governed by a first-order reaction and the activation energy governing the annealing process is the barrier height separating the interstitial from its vacancy.

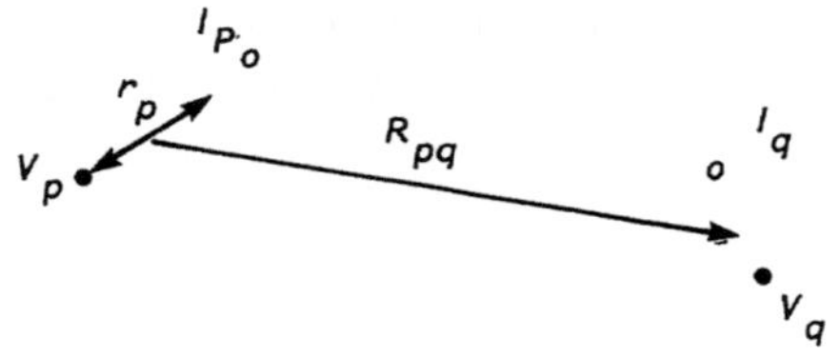

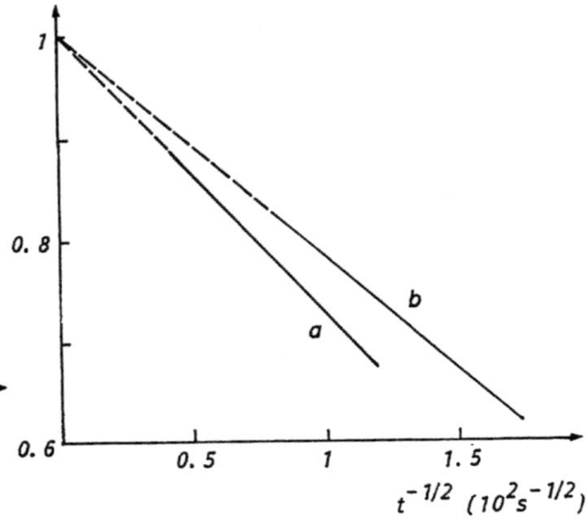

▲ Fig.9.3. Representation of a pair distribution

Fig.9.4. Asymptotic behavior of the annealed fraction versus $t^{-1/2}$ at 70.6 K (a), and 66.3 K (b), in 4×10^{15} As cm^{-3} doped germanium irradiated with 10^{15} (a) and 5×10^{14} (b), 1 MeV, electrons cm^{-2} (after [9.5]) ►

This case is treated in [9.2] because it is of importance for irradiation induced defects. Indeed, irradiation leaves V and I, which present a certain distribution g(r), depending on the mobility of V or I (or both) under irradiation. The way V and I recombine or escape the recombination determines the concentration of complex defects which are finally formed (by the association of the escaping V or I with an impurity for instance).

The problem is treated in Sect.9.1.1 and the solution given by (9.15) in which the distribution function $g(r_p)$ is a function which decreases rapidly as r_p increases to account for the correlation. The complete analysis requires exact knowledge of $g(r_p)$; the theoretical investigations of correlated reactions [9.3] have simply postulated a given functional form of $g(r_p)$. But, in practice, the question is: what is the kind of information concerning $g(r_p)$ which can be extracted from $\phi(t)$? It appears [9.4] that only the fraction of correlated pairs initially separated by the critical distance r_0 can be obtained from the time dependence of ϕ at an early stage. However, much later in the process the time dependence becomes a $t^{-1/2}$ law (equivalent to a third-order process) in which the correlation is still a well-defined concept. This occurs for times long enough so that

$$2\sqrt{Dt} \gg r_p - r_0 \quad . \tag{9.40}$$

Indeed, with this condition the complementary error function in (9.15) reduces to

$$\mathrm{erfc}\left(\frac{r - r_0}{2\sqrt{Dt}}\right) = 1 - \frac{r_p - r_0}{2\sqrt{Dt}} \tag{9.41}$$

and ϕ, the recovery fraction, becomes

$$1 - \phi(t) = 4\pi r_0 \frac{r_p - r_0}{2\sqrt{Dt}} \int_{r_0}^{\infty} g(r_p)dr_p \quad . \tag{9.42}$$

Thus the linearity of $\phi(t)$ versus $t^{-\frac{1}{2}}$ is a test of the presence of correlated pairs.

This type of behavior has been observed [9.5] in electron irradiated n-type germanium for the defects which disappear in the first annealing stage, around 65 K (Fig.9.4). This evidence for the existence of correlated pairs is one of the arguments used to conclude that this stage corresponds to the annihilation of vacancy-interstitial pairs. Since it is thought that this annihilation occurs through interstitial diffusion (Sect.9.3.3), the activation energy associated with the interstitial migration, i.e., the migration energy of the interstitial E_m^I, can be deduced from the slopes α_1 and α_2 of the plot $\phi(t^{-\frac{1}{2}})$ at two different temperatures (T_1 and T_2)

$$E = 2k \ln\left(\frac{\alpha_1}{\alpha_2}\right)\left(\frac{1}{T_1} - \frac{1}{T_2}\right)^{-1} \quad , \tag{9.43}$$

since α is proportional to $D^{-\frac{1}{2}}$. In the case of Fig.9.4 the result is E_m^I = 40 meV.

9.2 Determination of the Annealing Parameters

9.2.1 Rate Constant and Order of Reaction

As shown in Sect.9.1, the concentration N of a particular kind of defect, present at a time t, obeys a differential equation of the type

$$-\frac{dN}{dt} = Kf(N) \quad , \tag{9.44}$$

where f(N) can be any kind of function independent of temperature, which can be considered only as a monotonically increasing function of N. The rate constant has the form (Sect.9.1.1)

$$K = K_0 \exp\left(-\frac{E}{kT}\right) \tag{9.45}$$

where K_0 is a constant and E the associated activation energy. Depending on the precise process involved in the annealing, this activation energy is a migration energy, a dissociation energy, etc. The constant K_0 contains the vibrational frequency associated with the process [Ref.1.1, Sect.7.1.1]. The activation energy can be obtained experimentally in two ways (which of course, are equivalent): isothermal and isochronal annealing.

9.2.2 Isothermal Annealing

The variation of N versus time is recorded at a constant temperature T_1 to obtain the rate

$$\left(\frac{dN}{dt}\right)_1 = -K_0 \exp\left(-\frac{E}{kT_1}\right)f(N) \quad . \tag{9.46}$$

At a given time t_0, the temperature is abruptly changed to T_2 and then kept constant at this value so that the rate becomes:

$$\left(\frac{dN}{dt}\right)_2 = -K_0 \exp\left(-\frac{E}{kT_2}\right)f(N) \quad . \tag{9.47}$$

As f(N) is independent of temperature we can write

$$\left[\frac{(dN/dt)_1}{(dN/dt)_2}\right]_{t=t_0} = \exp\left(-\frac{E}{kT_1} + \frac{E}{kT_2}\right) \quad . \tag{9.48}$$

Thus, the ratio α_1/α_2 of the slopes α_1 and α_2 of N(t) at a given time t_0, for temperature T_1 and T_2 respectively (Fig.9.5), is related to E through

$$E = k\,\frac{T_1T_2}{T_1 - T_2}\,\ln\left(\frac{\alpha_1}{\alpha_2}\right) \quad . \tag{9.49}$$

Such procedure, being independent of f(N), is quite general: the activation energy can be obtained without making any assumption on the nature of annealing kinetics.

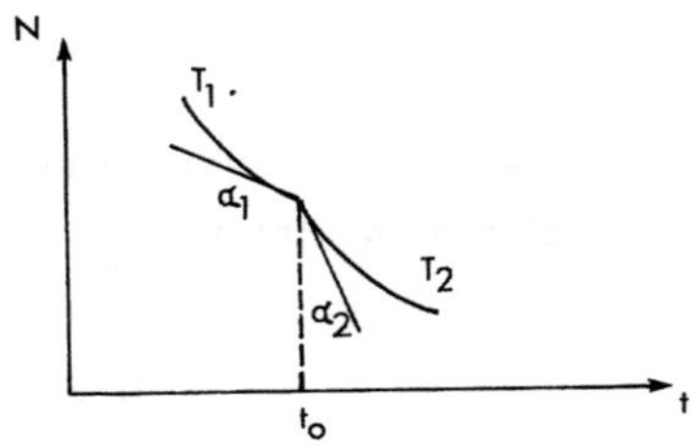

Fig.9.5. Schematic representation of isothermal annealing around a time at which the temperature is abruptly changed from T_1 to T_2

9.2.3 Isochronal Annealing

When the material is gradually warmed up, because the rate K is a thermally activated quantity, dN/dT given by (9.44) remains very small at low temperatures, then grows suddenly and, within a relatively small temperature range, practically all the defects disappear. The characteristic temperature at which this occurs is called the annealing temperature T_a. This annealing temperature T_a is an important quantity in practice because it is often the only parameter which allows one to correlate observations made on the same defect but with different techniques. Very often the annealing temperature is the only means by which a defect can be characterized.

When assumptions are made about the magnitude of K_0 and on the form of f(N), the measurement of T_a gives a fairly good determination of the activation energy. To see how this can be made, we assume that the temperature varies linearly with time, from T_0 at time t = 0

$$T = at + T_0 \quad . \tag{9.50}$$

Under these conditions

$$\frac{dN}{dt} = -K_0 \exp\left(-\frac{E}{kT}\right) f(N) \tag{9.51}$$

becomes

$$\frac{dN}{f(N)} = -K_0 \exp\left(-\frac{E}{kT}\right) dT \quad . \tag{9.52}$$

In order to integrate this equation it is necessary to know the precise form of f(N). Thus, isochronal annealing necessitates knowing the kinetics in order to determine E unambiguously. Consider for instance a first-order reaction

$$f(N) = N \quad . \tag{9.53}$$

Integration of (9.52) then gives

$$\ln\left(\frac{N}{N_0}\right) = -\frac{K_0}{a} \int_{T_0}^{T} \exp\left(-\frac{E}{kT}\right) dT \quad . \tag{9.54}$$

If T_0 and T are small enough to consider that $E/kT_0 \sim \infty$ and $E/kT \gg 1$, we can write (9.54) as

$$\ln\left(\frac{N}{N_0}\right) = \frac{E}{k} \frac{K_0}{a} \int_{\infty}^{E/kT} \frac{\exp(-x^2)}{x^2} dx \tag{9.55}$$

which, if we integrate by parts and retain only the first term, gives

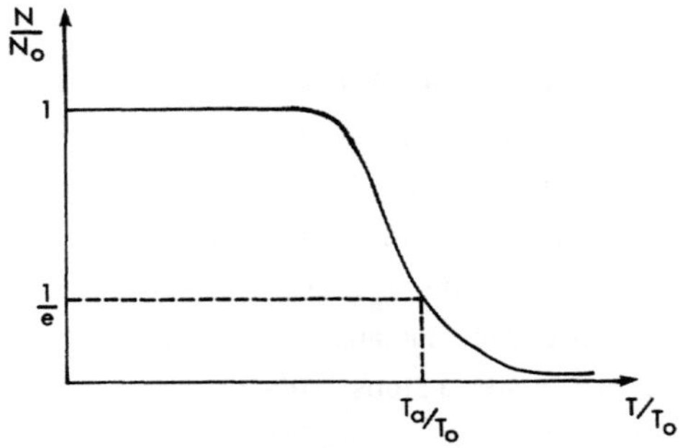

Fig.9.6. Schematic representation of the annealed fraction during isochronal annealing

$$\ln \frac{N}{N_0} = -\frac{K_0}{a}\frac{k}{E}T^2 \exp\left(-\frac{E}{kT}\right) \quad . \tag{9.56}$$

This expression gives the dependence of the defect concentration during isochronal annealing versus the parameters $T_0 = E/k$, K_0 and $\tau = T_0/a$

$$\ln \frac{N}{N_0} = -k_0\tau\left(\frac{T}{T_0}\right)^2 \exp\left(-\frac{T_0}{T}\right) \quad . \tag{9.57}$$

The abrupt change of N, from N_0 to 0, occurs around a temperature T_a (Fig. 9.6). The inflexion point occurring for $N/N_0 = e^{-1}$, the corresponding temperature is given by

$$1 = K_0\tau(T_a/T_0)^2 \exp(-T_0/T_a) \quad , \tag{9.58}$$

i.e.,

$$\frac{T_a}{T_0} \sim \frac{1}{\ln(K_0\tau)} \quad . \tag{9.59}$$

For higher order reactions the variation of N with temperature remains qualitatively similar.

In practice, the defect concentration must be measured at a reference temperature T_f. The linearity of T versus time is then realized in the following way: the sample is heated at the temperature $T_f + \Delta T$ for a given amount of time Δt (therefore $a = \Delta T/\Delta t$), then cooled down at T_f for the measurement of N. It is then heated at $T_f + 2\Delta T$ for the same amount of time Δt, and again cooled at T_f for measurement. The only difficulty of this type of measurement is to have the time of temperature rise short enough in order to minimize the amount of annealing which occurs during cooling and heating, as compared to that which takes place at the annealing temperature during the time Δt.

9.3 Annealing of Defects Induced by Electron Irradiation

9.3.1 Stability of the Vacancy-Interstitial Pair

Knowing the behavior of the vacancy interstitial pair is important because it determines the nature and concentration of the defects produced by an irradiation, which are stable at a given temperature. This behavior is apparently not simple in semiconductors: the proportion of pairs which recombine immediately after their creation, even at very low temperature, depends on the nature of the material, the temperature, but also on electronic parameters such as the Fermi level and the amount of ionization accompanying the irradiation. The proportion of pairs which escape recombination can be estimated from the comparison between the calculated defect introduction rate ν_c, and the observed one ν_0. Thus, in silicon, ν_0 is always small compared to ν_c: the vacancy interstitial pair is never observed even when the irradiation is performed at a temperature as low as 1.3 K [9.6]. In other materials, such as n-type Ge [9.7], p-type diamond [9.8], or n-type Ga As [9.9], the pairs seem to be stable (up to ~65 K in Ge, 280 K in diamond and 350 K in GaAs) since $\nu_c \sim \nu_0$. Evidence for electronic effects on pair stability is given by the variation of the defect introduction rate with the type and concentration of doping. For instance, $\nu_c \ll \nu_0$ in n-type Ge while $\nu_0 \sim \nu_c$ in p-type Ge.

The stability of a vacancy interstitial pair depends on the elastic and Coulomb interactions which exist between the elements of the pair. As illustrated in Fig.9.7, this stability can be schematized in the following way: if E_R is the barrier encountered by the interstitial to jump into the vacancy site and E_L the barrier encountered to be liberated, the probability to escape the recombination is:

$$P = \frac{P_L}{P_L + P_R} \quad , \tag{9.60}$$

where P_L and P_R are respectively the probabilities to overcome the barriers for liberation and recombination.

Then

$$P = \frac{1}{1 + \lambda \exp[-(E_L - E_R)/kT]} \quad , \tag{9.61}$$

where λ is the ratio of the statistical weights of the jumps that annihilate the pair to that of the jumps that lead to the liberation of the interstitial ($\lambda \sim 1$). This model was first introduced by WERTHEIM [9.10] to explain the

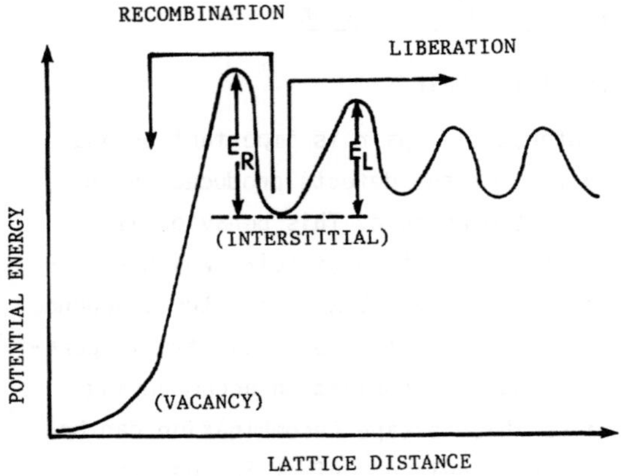

Fig.9.7. Schematic representation of the potential energy of an interstitial close to a vacancy, in the lattice, showing the energies involved in the liberation and recombination processes

fact that the dependence with the temperature of irradiation of the defect introduction rate ν in silicon follows the following law:

$$\gamma \,\alpha\, \exp\left(-\frac{\Delta E}{kT}\right) \quad . \tag{9.62}$$

It was extended later by MAC KAY and KLONTZ [9.11] in order to include the dependence of the defect creation rate with the type of doping, in the case of germanium.

The dependence of the probability P with Fermi level is due to the fact that the barrier E_R depends on the relative charge states of the vacancy and interstitial. In the simple case where the barrier to recombination becomes negligible when one element of the pair becomes ionized (because, for instance, this ionization results in a Coulomb attraction), the probability P becomes:

$$P = \frac{P}{1 + \gamma \exp[(E_F - E_T)/kT]} \tag{9.63}$$

where E_T is the ionization energy of the pair and γ a degeneracy factor.

The experimental evidence for this model is the following. In silicon the defects introduced by electron irradiation can be classified in two categories [9.12]: the "ITI" and "ITD" defects. The ITI (for irradiation temperature independent) defects have an introduction rate independent of the temperature of irradiation and they are not associated with impurities. They are intrinsic defects, such as divacancies, trivacancies, etc., directly created when the energy transmitted to the recoil atom is large enough to result in more than one displacement. They are not observed following low-energy (300 keV) electron irradiation and they form the majority of defects in case of high-energy (45 MeV) electron irradiation [9.13]. Their introduction rate

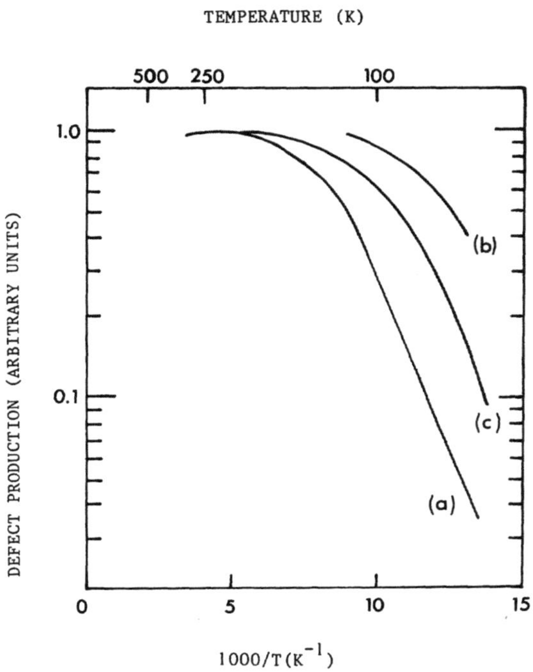

Fig.9.8. Variation of the defect introduction rate (measured using electrical techniques) as a function of the reciprocal irradiation temperature in C_Z|Ω cm (a), G_Z 40 Ω cm (b), and F_Z 10 Ω cm (c), n-type silicon after [9.12,14]

is temperature independent because they remain stable after their creation. The ITD (for irradiation temperature dependent) defects are impurity associated defects whose introduction rate exhibits an exponential dependence with the reciprocal irradiation temperature (Fig.9.8). STEIN and VOOK [9.14], further analyzing this variation of the defect production (for a given irradiation temperature) with the doping of the material, i.e., with the Fermi level, fitted the data of Fig.9.8 with a simple model in which P = 1 for the ionized state and P = 0 for the un-ionized state. The probability of defect formation is then the probability P' that the pair is ionized. The fit gives $E_C - E_T \sim 70$ meV. Of course, a model including no charge state dependence would give $E_R - E_L \sim 70$ meV, but would not explain the Fermi level dependence, i.e., would require that the values of the parameter λ vary with the doping concentration.

The fact that the same temperature law is observed in crucible-grown (C_Z) material (having a large oxygen concentration), and in floating-zone (F_Z) material (having a low oxygen content), in which the dominant defects produced are respectively the A center (vacancy + oxygen) and the E center (phosphorus + vacancy, for instance), suggests that this behavior is indeed a property of the defect production mechanism and not of the impurity involved in the resultant defect-impurity complex.

9.3.2 Mechanism for Complex Defect Formation

However, when the variation of the defect-introduction rate versus the temperature of irradiation is measured for individual defects and not for the sum of all the induced defects (given by electrical measurements), the picture becomes different. For some of the IR absorption bands associated with the vibrational mode of oxygen complexes, the activation energy ΔE is found to be $E = 50 \pm 5$ meV [9.15]; for less intense bands $\Delta E = 100 \pm 20$ meV [9.15]. In crucible-grown and zone-refined materials doped with lithium, the values of ΔE for Li-related defects are respectively 55 and 90 meV [9.16].

Such apparent variation of ΔE with the type of impurities contained in the material cannot be explained in terms of the charge state dependent stability of the vacancy-interstitial pair, using $E_C - E_T \simeq 70$ meV. Indeed i) in the case of absorption measurements the Fermi level always remains deep in the gap because such measurements require practically intrinsic materials (to prevent absorption by the free carriers); ii) in the case of the Li-doped materials the doping concentration is the same but the values of ΔE differ by a factor of two depending on whether the material is crucible grown or zone refined.

We also note that the behavior of the defect-introduction rate ν with the temperature of irradiation, described by (9.63), is observed only for temperatures larger than 70-80 K, i.e., for temperatures at which the vacancy is mobile (Fig.9.9). This suggests another explanation for the behavior of ν versus temperature: the activation energy ΔE is the potential barrier that that vacancy must overcome to get trapped on an impurity and form an impurity-vacancy complex (Fig.9.10). Such a barrier depends of course i) on the nature of the impurity involved in the complex (this explains why the barrier observed depends on the nature of the impurities contained in the material) and ii) on the Fermi level position, i.e., on the relative charge states of the vacancy and impurity. The dependence of complex formation with E_F will of course be given by an expression similar to (9.63).

Around 70 K the vacancies become mobile and can be trapped on impurities provided that the barrier ΔE for complex formation is of the order of few kT. First, complexes of vacancies with phosphorus and oxygen (for which $\Delta E \sim 50$ meV) are formed. With increasing temperature, complexes with other impurities, for which ΔE is larger, are formed. For temperatures lower than 70 K, the

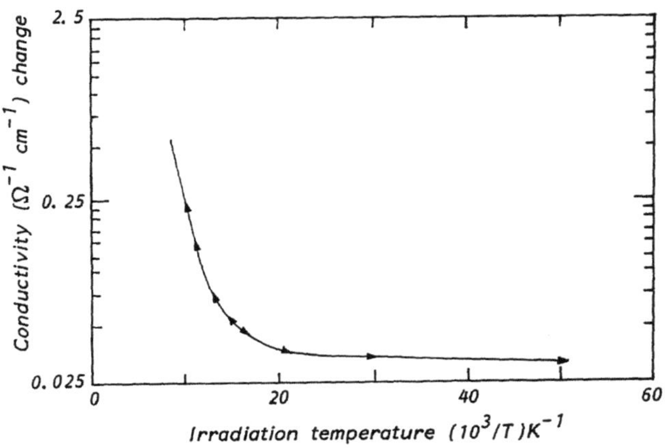

Fig.9.9. Variation of the conductivity (reflecting the defect introduction rate) in 10^{18} cm^{-3} n-type silicon irradiated with successive doses of 8×10^{16}, 1.5 meV, electrons cm^{-2} at each temperature (the measurements are performed at 20 K). It shows that the defect introduction rate is practically constant up to ~70 K, when the vacancy becomes mobile (Bourgoin, unpublished)

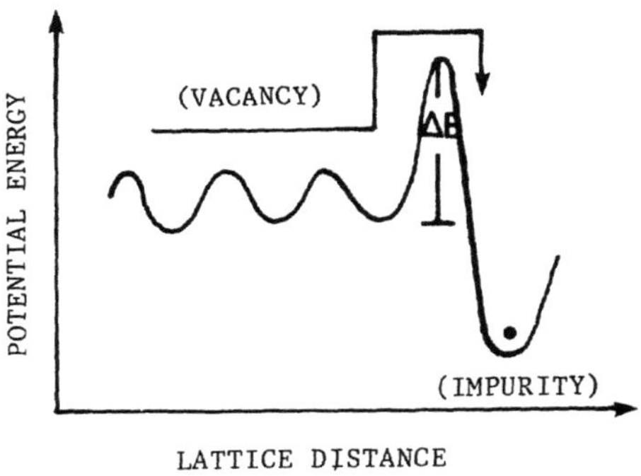

Fig.9.10. Schematic representation of the potential energy of a vacancy in the lattice in the vicinity of an impurity showing the activation energy involved in the formation of a vacancy-impurity complex

vacancies being immobile, the defect introduction rate is very small and constant; mobile interstitials recombine with vacancies and only the interstitials which escape recombination can form stable defects.

9.3.3 On the Mobility of the Interstitial

This brings us to the question: why is the interstitial mobile, even at very low temperature? More generally, why is the vacancy interstitial pair stable only in some materials and only in particular cases? If direct evidence of interstitial mobility at low temperature has been found only in silicon (Sect.9.3.4) we think that the reason for this behavior is given by the behavior of the vacancy-interstitial pair in germanium. Evidence that vacancy-interstitial pairs are present in n-type germanium is the following: the defect introduction rate measured [9.7] for electron irradiation at liquid helium temperature is practically equal to the calculated one (typically 1 cm^{-1} at 1 MeV). These pairs annihilate at 65 K since, at this temperature, more than 95% of the defects observed disappear [9.7]. WHAN's work [9.15], which strongly suggests that the vacancy becomes mobile around 100 K, implies that the annihilation of the pair is due to the mobility of the interstitial (thus induced around 65 K). On the other hand, in p-type germanium the defect introduction rate is very low compared to the calculated one, indicating that even at low temperature (4 K) the interstitial is already mobile. Such behavior can be understood if we make the assumption that a localized level (E_T) in the gap is associated with the interstitial. Let I_1 be the charge state of the interstitial when the Fermi level E_F is above E_T and $I_2 = I_1 + |e|$ its charge state when E_F is below E_T (Fig.9.11a). In doped n-type material, at low temperature, E_F is close to the conduction band and above E_T; in p-type material E_F is below E_T. Therefore, since in n-type material up to 95% of the defects disappear around 65 K this stage must be ascribed to the thermally induced mobility of I_1 (Fig.9.11b). When band-gap light is shined on the sample the 65 K stage is shifted to 27 K [9.5]; the reason is that the interstitials, in state I_1, trap the minority carriers (holes) generated by the light and pass into the I_2 state, in which state their mobility is thermally induced at 27 K (Fig.9.11c). But ionizing radiation (due to electron irradiation or light illumination) can also induce defect recovery at a temperature as low as 4.2 K [9.17]. In our view, this athermal migration is a manifestation of an ionization-enhanced migration [Ref.1.1, Sect.7.3]. It is a consequence of the fact that the interstitials alternately trap electrons and holes and therefore pass back and forth from the I_1 to the I_2 state (Fig.9.11d). Such athermal mobility is a function of the rates for trapping electrons and holes and therefore depends on their relative concentrations, i.e., on the level of injection of the minority carriers by the irradiation. Because in lightly doped n-type material the concentration of injected holes is not negligible compared to the concentration of electrons, the probability for the interstitial to trap a hole is large and its mobility large; the result is a small apparent defect introduction rate. In heavily doped n-type material, the electron concentration being very large compared to the hole concentration, the athermal mobility is considerably reduced and the apparent defect introduction rate

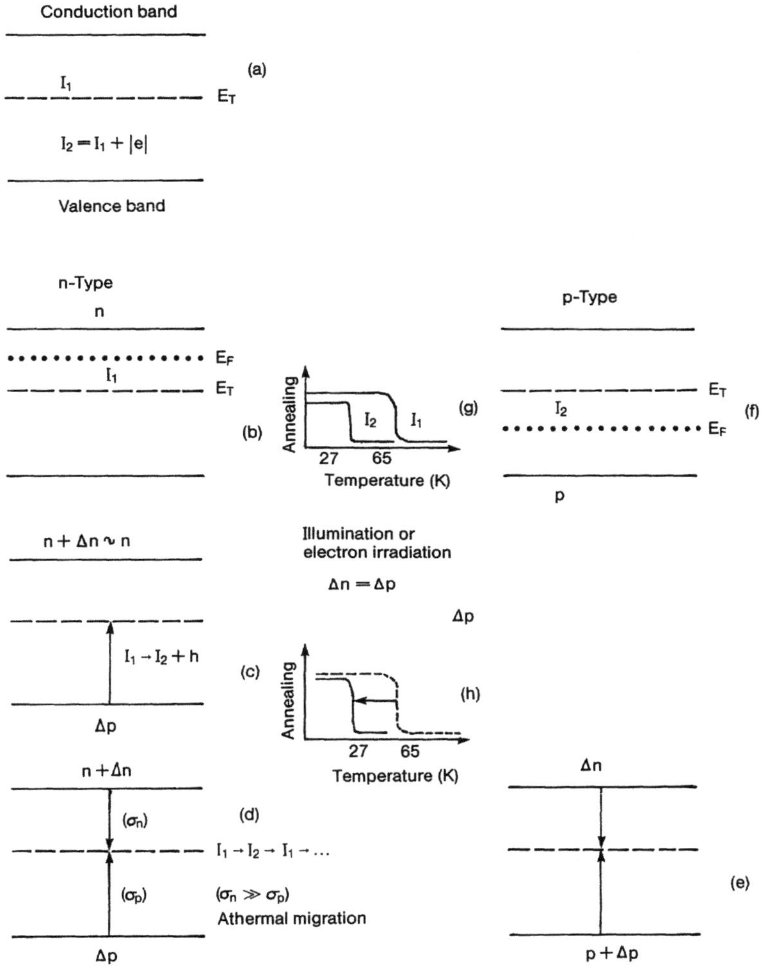

Fig.9.11a-i. The energy level associated with an interstitial (a) and the two associated charge states in the case of n-type (b) and p-type (f) materials and corresponding annealing behavior (g). The creation of electron hole pairs induces firstly a change of charge state (c) resulting in a change of annealing behavior (h) or secondly alternative charge states change (d) and (e) resulting in athermal migration

remains of the order of the theoretical one. In p-type material under irradiation, the hole concentration is larger than the electron concentration; the athermal mobility is then very large because the cross section for electron trapping is considerably larger than for hole trapping (Fig.9.11e).

There are strong indications that this qualitative picture is valid. In p-type material, irradiated at 4 K with neutrons or α particles, the 27 K stage is observed [9.18], i.e., the interstitial, in its I_2 state, was present after irradiation (Fig.9.11f). The reason is that neutrons or α particles create very few electron-hole pairs per displacement produced; the interstitials have a low probability of trapping electrons and their athermal mobility is negligible. The occurrence of the 35 K stage in heavily doped n-type material is another manifestation of the interstitial mobility. Indeed, in this kind of material, the defects which anneal in the 65 K stage can be converted by the irradiation itself or by a subsequent ionization into defects which anneal in a 35 K stage [9.19]. The reason is that the interstitials, mobile under irradiation or ionization, can be trapped on the doping impurities to form the 35 K defect. Proof that the doping impurity is involved in the 35 K defect is given by the fact that the 35 K introduction rate is proportional to the doping concentration [9.20]. Once the 35 K defects have been annealed, but not the 65 K defects, the use of ionizing radiation again induces mobility of the interstitials which are trapped by doping impurities and form new 35 K defects.

According to this picture the mobility of the interstitial is due to the fact that it traps electrons and holes created by the ionization induced by the irradiation. The reason why the interstitial is mobile in silicon in n-type as well as p-type material should thus be that the cross sections for electrons and holes trapping are of the same order of magnitude. No direct evidence for such ionization enhanced mechanism of the interstitial has been found. However, TROXELL and WATKINS [9.21] have shown that the lattice configuration of interstitial boron changes (from a bond-centered interstitial configuration to a split <100> interstitial configuration) when its charge states changes, resulting in its athermal migration occurring for cycles $B_i^+ \rightleftharpoons B_i^-$.

9.3.4 An Example: The Case of Silicon

As discussed in Sect.9.3.3, the interstitial is found to be mobile at all temperatures in silicon. The result is that the vacancy-interstitial pair cannot be observed even for irradiation at very low temperature. Since the interstitials are mobile they recombine with vacancies unless they become trapped on various sinks (dislocations, surfaces and impurities). The main trap for interstitials in p-type silicon is the doping impurity (boron, aluminum): direct proof is provided by EPR experiments which show that electron irradiation at 4 K results in the creation of Al or B interstitials in a concentration very large compared to the concentration which should result from their direct displacement by the irradiation [9.22]. Nothing is known about the self interstitial. In p-type material, all the interstitials exchange with the doping impurity, giving rise to interstitial Al, or interstitial B, since their introduction rate is equal to the introduction rate of the vacancies [9.23]. The characteristics of these interstitials, as well as of interstitial carbon, which is another identified interstitial produced by electron irradiation, are given in Table 9.1. When these interstitials become mobile they get trapped on the corresponding substitutional impurity, giving rise to complexes:

$$C_i + C_S \text{ [9.24]} , \quad Al_S + Al_i \text{ [9.22]} , \quad Ga_s + Ga_i \text{ [9.22]} \quad \text{(Fig.9.12)} .$$

Practically no other impurity-plus-interstitial complex has been firmly identified. But it is possible that clusters of interstitials exist.

Table 9.1. Interstitial defects

Defect	Charge state	Energy-level [eV]	Migration energy [eV]	Annealing temperature [K]	Technique of observation	Remarks
Al	2+	E_c-0.5 [9.26]	0.6 [9.25]	470 [9.25]	EPR [9.22]	
B		E_c-0.15 [9.27] E_c-0.45 [9.27]			EPR [9.25] DLTS [9.27]	Behaves as a negative U center [9.27] Sensitive to IED
C	+	E_v+0.3 [9.28]	0.8 [9.28]	350 [9.28]	EPR [9.29]	

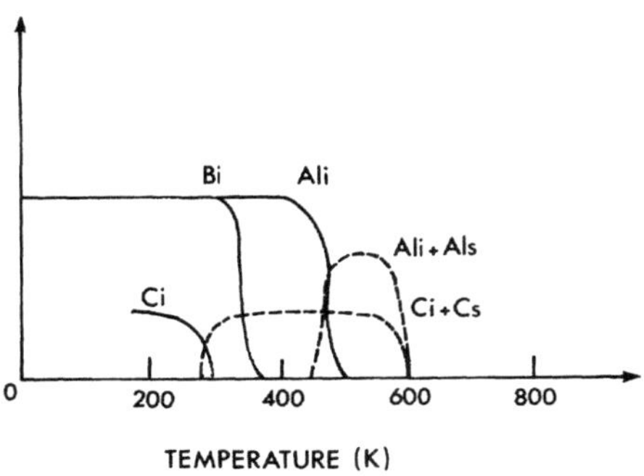

Fig.9.12. Annealing behavior of interstitial-related defects in silicon

a)

b)

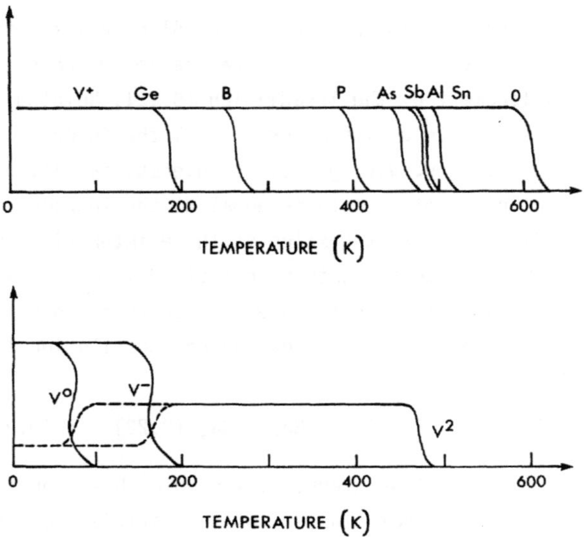

Fig.9.13a,b. Annealing behavior of (a) vacancy impurity complexes, and (b) vacancy clusters

The vacancies which have escaped recombination with interstitials become mobile around 100 K. The exact temperature and the activation energy associated with this mobility depend on the charge state. Once mobile, these vacancies interact between themselves, forming divacancies, or with impurities. In case the material contains oxygen (Cz-grown material) the dominant trap is oxygen: the vacancy-oxygen complex is known as the A center. In case the concentration of oxygen is low (in floating-zone material) the dominant trap is the doping impurity: the vacancy doping impurity complex is the E center. The vacancy also interacts with carbon which is a common impurity in Si. At high temperature (above ~400 K) these complex defects disappear (Figs.9.13a,b) giving rise to other new ones. Because these new complexes are vacancy clusters (divacancy, then quadrivacancy, pentavacancy, etc.) this disappearance is due to their dissociation: vacancies are liberated and form clusters. The main characteristics (energy level, capture cross section, migration or dissociation energy, charge state) of the defects which have been firmly identified are given in Table 9.2 and the techniques which have enabled us to observe them are also indicated.

Table 9.2. Vacancy related defects

Defect	Charge state	Energy level [eV]	Capture cross section [cm^{-2}]	Migration or dissociation energy [eV]	Annealing temperature [K]	Technique of observation	Remarks
	2-	E_c-0.09 [9.30]					
	-		$\sigma_n=10^{-19}-10^{-20}$ [9.30]	0.18 [9.22]	70-20 [9.22]	EPR [9.22]	Sensitive to IED, behaves as a negative U center [9.21]
Vacancy	0			0.33 [9.22]	170 [9.22]	DLTS [9.30]	
	+	E_v+0.12 [9.31]	$\sigma_p=5\times10^{-17}$ [9.30]				
	2+					photoconductivity [9.31]	
	+	E_c-0.23 [9.30]	$\sigma_n=10^{-16}$ [9.30]			EPR [9.32]	
Divacancy	0	E_c-0.39 [9.30]	$\sigma_n=10^{-15}$ [9.30]	1.3 [9.32]	570 [9.30, 32]	IR [9.33]	
	-	E_v+0.21 [9.30,31]	$\sigma_p=2\times10^{-16}$ [9.30]			DLTS [9.34,35]	
A center V + O	- 0	E_c-0.17 [9.30,36]	$\sigma_n=10^{-14}$ [9.30]	1.1-1.7 [9.30,37,38]	620 [9.12]	EPR [9.39,40] IR [9.41,42] DLTS [9.30]	
E center							
V+Ga				0.66 [9.44]		EPR [9.46]	
V+B				0.42 [9.44]	260 [9.43]	EPR [9.43]	
V+Al		E_v+0.27 [9.34]	$\sigma_p=6\times10^{17}$ [9.34]	0.77 [9.44]	470 [9.30]	EPR [9.45], DLTS [9.30]	

V+P	E_c-0.43 [9.30]	σ_n=5.10^{-16} [9.48]	0.95-1.25 [9.49]	400 [9.49]	EPR [9.47], IR [9.51]	Migration energy dependent on Fermi level [9.49]
V+As	E_c-0.42 [9.52,55]	σ_n=10^{-17} [9.52]	1.1 [9.54]	450 [9.54]	EPR [9.50]	
V+Sb	E_c-0.43 [9.55]		1.3 [9.53]	470 [9.53]	EPR [9.53]	
V+Sn	E_v+0.35 [9.43]			500 [9.43]	EPR [9.43]	Charge state dependent configuration [9.50]
V+Ge				200 [9.56]	EPR [9.56]	
V+O+C				670 [9.57]	EPR [9.57] IR [9.58]	Complex of unknown configuration [9.57]
V_2+O	E_c-0.5 [9.59]			620 [9.59]	EPR [9.59]	
V_3+O	E_v+0.4 [9.59]			670 [9.59]	EPR [9.59]	
V_2+O_2	E_c-0.4 [9.59] or E_v+0.4 [9.59]			670 [9.59]	EPR [9.59]	
V_3+O_3	E_c-0.4 [9.59] or E_v+0.4 [9.59]			870 [9.59]	EPR [9.59]	

References

Chapter 1

1.1 M. Lannoo, J.C. Bourgoin: *Point Defects in Semiconductors I, Theoretical Aspects*, Springer Series in Solid-State Sciences, Vol.22 (Springer, Berlin, Heidelberg, New York 1981)

Chapter 2

2.1 H.A. Jahn, E. Teller: Proc. Roy. Soc. (London) A*161*, 220 (1937)
2.2 A.M. Stoneham: *Theory of Defects in Solids* (Clarendon Press, Oxford (1975)
2.3 M.D. Sturge: In *Solid State Physics*, ed. by F. Seitz and D. Turnbull (Academic, New York, London 1967)
2.4 R. Englman: *The Jahn-Teller Effect in Molecules and Solids* (Wiley, New York 1972)
2.5 O. Madelung: *Introduction to Solid-State Theory*, Springer Ser. Solid-State Sci, Vol.2 (Springer, Berlin, Heidelberg, New York 1978) Sect.9.1.3
2.6 M. Born, K. Huang: Dynamical Theory of Crystal Lattices (Clarendon Press, Oxford 1968)
2.7 G.A. Baraff, E.O. Kane, M. Schlüter: Phys. Rev. Lett. *43*, 956 (1979); Phys. Rev. *21*, 3563 (1980)
2.8 G.A. Baraff, E.O. Kane, M. Schlüter: Phys. Rev. *21*, 5662 (1980)
2.9 P. Hohenberg, W. Kohn: Phys. Rev. B*136*, 864 (1964)
2.10 W. Kohn, L.J. Sham: Phys. Rev. A*140*, 1133 (1965)
2.11 J.C. Slater: *The Self Consistent Field for Molecules and Solids* (Mc Graw-Hill, New York 1974)
2.12 J.F. Janak: Phys. Rev. B*18*, 7165 (1978)
2.13 U. Opik, M.H.L. Pryce: Proc. Roy. Soc. A*238*, 425 (1957)
2.14 For a detailed analysis see: A.M. Stoneham, M. Lannoo: J. Phys. Chem. Sol. *30*, 1769 (1969)
M. Lannoo: Ann. Phys. Paris *3*, 391 (1968) (Thesis)
2.15 H.C. Longuet-Higgins, U. Öpik, M.H.L. Pryce, R.A. Sack: Proc. Roy. Soc. A*244*, 1 (1958)
2.16 M.C.M. O'Brien: Proc. Roy. Soc. A*281*, 323 (1964)
2.17 L. Landau, E.M. Lifshitz: *Quantum Mechanics* (Pergamon, Oxford 1959)
2.18 F.S. Ham: Phys. Rev. A*138*, 1727 (1965)
2.19 M. Caner, R. Englman: J. Chem. Phys. *44*, 4054 (1966)
2.20 G.D. Watkins: J. Phys. Soc. Japan *18*, Suppl. II, p.22 (1963)
2.21 G.D. Watkins: *Effet des Rayonnements sur les Semiconducteurs* (Dunod, Paris 1964) p.97
2.22 J. Friedel, M. Lannoo, G. Leman: Phys. Rev. *164*, 1056 (1967)
2.23 M. Lannoo: Ann. Phys. Paris *3*, 391 (1968), Appendix B (Thesis)
2.24 R.P. Messmer: *Lattice Defects in Semiconductors* 1974 (Inst. Phys. Conf. Ser. No. 23 (1975) p.44

2.25 G.D. Watkins, R.P. Messmer: Phys. Rev. Lett. *32*, 1244 (1974)
2.26 C.A. Coulson, M.J. Kearsley: Proc. Roy. Soc. A*241*, 433 (1957)
2.27 C.A. Coulson, F.P. Larkins: J. Phys. Chem. Solids *32*, 2245 (1971)
2.28 G.T. Surratt, W.A. Goddard III: Solid State Commun. *22*, 413 (1977)
2.29 M. Lannoo, G.A. Baraff, M. Schlüter: Phys. Rev. *24*, 955 (1981)
2.30 M. Lannoo: In *Defects and Radiation Effects in Semiconductors* 1978 (Inst. Phys. Conf. Ser. No. 46 (1979) p.1; see also [Ref.1,1 Sect.4.2.2]
2.31 F.P. Larkins, A.M. Stoneham: J. Phys. C*4*, 143 (1971)
2.32 M. Lannoo, G. Allan: Phys. Rev. B*25*, 4089 (1982)
2.33 G.D. Watkins, J.R. Troxell: Phys. Rev. Lett.*44* , 593 (1980)
2.34 P.W. Anderson: Phys. Rev. Lett. *34*, 953 (1975)

Chapter 3

3.1 C.P. Schlichter: *Principles of Magnetic Resonance*, 2nd ed., Springer Ser. Solid Sci., Vol.1 (Springer, Berlin, Heidelberg, New York 1980)
3.2 A. Abragam, B. Bleaney: *Resonance Paramagnetique Electronique des Ions de Transition* (Presses universitaires de France, Paris 1971)
3.3 Applications of EPR to semiconductors were treated, for instance by G.W. Ludwig, H.H. Woodbury: In *Solid State Physics*, ed. by F. Seitz and D. Turnbull, Vol.13 (Academic, New York 1962) or G. Lancaster: *Electron Spin Resonance in Semiconductors* (Plenum, New York 1967)
3.4 This subject has recently been treated by G.D. Watkins: In *Point Defects in Solids*, Vol.2, ed. by J.H. Crawford and L.M. Slifkin (Plenum, New York 1975) Chap.4
3.5 This is also the case for other simple defects produced by electron irradiation such as the divacancy [G.D. Watkins, J.W. Corbett: Phys. Rev. *138*, A543 (1965)], the vacancy-oxygen center [G.D. Watkins, J.W. Corbett: Phys. Rev. *121*, 1001 (1961)], the so-called E-center [E.L. Elkin, G.D. Watkins: Phys. Rev. *174*, 881 (1968)], etc. EPR identification of some of these defects will also be detailed here
3.6 For details and justification on this question see textbooks on quantum mechanics: For instance, L.I. Schiff: *Quantum Mechanics* (Mc Graw-Hill, New York 1965); L.D. Landau, E.M. Lifshitz: *Quantum Mechanics* (Pergamon, London 1968)
3.7 For details see T.H. Wilmshurst: *Electron Spin Resonance Spectrometers* (Plenum, New York 1968)
3.8 J.H. Van Vleck: *Electric and Magnetic Susceptibilities* (Univ. Press, Oxford 1932)
3.9 M.H.L. Pryce: Proc. Phys. Soc. (London) A*63*, 25 (1950); see also A.M. Stoneham: *Theory of Defects in Solids* (Clarendon Press, Oxford 1975) Chap.13
3.10 G.D. Watkins, J.W. Corbett: Phys. Rev. *121*, 1001 (1961)
3.11 J. Christiansen (ed.): *Hyperfine Interactions of Radioactive Nuclei*, Topics Current Physics, Vol.31 (Springer, Berlin, Heidelberg, New York 1983)
3.12 G.D. Watkins: In *Radiation Damage in Semiconductors*, ed. by P. Baruch (Dunod, Paris 1965) p.97
3.13 W.V. Smith, P.P. Sorokin, I.L. Gelles, G.J. Lasher: Phys. Rev. *115*, 1546 (1959)
3.14 K.L. Brower: Phys. Rev. B*1*, 1908 (1970)
3.15 G. Feher: Phys. Rev. *114*, 1219 (1952)
3.16 See, for instance: G.D. Watkins: Phys. Rev. *155*, 802 (1967)
3.17 K.L. Brower: Phys. Rev. B*4*, 1968 (1971)
3.18 A. Carrington, A.D. McLachlan: *Introduction to Magnetic Resonance* (Harper and Row, New York 1967) p.116

3.19 K.L. Brower: In *Radiation Effects in Semiconductors*, ed. by J.W. Corbett and G.D. Watkins (Gordon and Breach, New York 1971) p.189
3.20 J.W. Corbett: In *Solid State Physics*, Suppl.7, ed. by F. Seitz and D. Turnbull (Academic, New York 1966) p.79
3.21 G.D. Watkins, J.W. Corbett: Discussions Faraday Soc. *31*, 86 (1961); Phys. Rev. *134*, A 1359 (1964)
3.22 G.D. Watkins: J. Phys. Soc. Jpn. *18*, Suppl.2, 22 (1963)
3.23 M. Lannoo, J.C. Bourgoin: In *Defects and Radiation Effects in Semiconductors 1982* (Inst. Phys. Conf. Ser., to be published)
3.24 G.D. Watkins: In *Radiation Effects in Semiconductors*, ed. by F.L. Vook (Plenum, New York 1968) p.97
3.25 M. Lannoo, G.A. Baraff, M. Schlüter: Phys. Rev. *24*, 955 (1981)
3.26 G.D. Watkins: In *Lattice Defects in Semiconductors* (Inst. Phys. London 1975) Conf. Ser. 23, p.1
3.27 E.L. Elkin, G.D. Watkins: Phys. Rev. *174*, 881 (1968)
3.28 R.A. Swalin: J. Phys. Chem. Solids *18*, 290 (1961)
3.29 F.P. Larkins, A.M. Stoneham: J. Phys. C (Solid State Physics) *4*, 143 (1971)
3.30 M. Lannoo, G. Allan: Phys. Rev. B *25*, 4089 (1982)
3.31 G.A. Baraff, E.O. Kane, M. Schlüter: Phys. Rev. *21*, 5662 (1980)
3.32 G.D. Watkins, J.R. Troxell: Phys. Rev. Lett. *44*, 593 (1980)

Chapter 4

4.1 K. Seeger: *Semiconductor Physics*, Springer Ser. Solid-State Sci., Vol.40 (Springer, Berlin, Heidelberg, New York 1982)
4.2 J.I. Pankove (ed.): *Electroluminescence*, Topics Appl. Phys., Vol.17 (Springer, Berlin, Heidelberg, New York 1977)
4.3 P. Bräunlich (ed.): *Thermally Stimulated Relaxation in Solids*, Topics Appl. Phys., Vol.37 (Springer, Berlin, Heidelberg, New York 1979)
4.4 J.I. Pankove: *Optical Processes in Semiconductors* (Dover, New York 1971)
4.5 R.H. Bube: *Electronic Properties of Cristalline Solids* (Academic, New York 1974).Chap.11
4.6 G. Harbeke: In *Optical Properties of Solids*, ed. by F. Abelès (North Holland, Amsterdam 1972) Chap.2
4.7 M. Balkanski (ed.): *Handbook on Semiconductors*, Vol.2 (North Holland, Amsterdam 1980)
4.8 L.I. Schiff: *Quantum Mechanics* (McGraw-Hill, New York 1955)
4.9 A.M. Stoneham: *Theory of Defects in Solids* (Clarendon Press, Oxford 1975) p.272
4.10 M. Lax: J. Chem. Phys. *20*, 1752 (1952)
4.11 K. Huang, A. Rhys: Proc. Roy. Soc. (London) A*204*, 406 (1950)
4.12 J.J. Markham: Rev. Mod. Phys. *31*, 956 (1959)
4.13 M.H.L. Pryce: *Phonons in Perfect Lattices and in Lattices with Point Imperfections* (Oliver and Boyd, Edinburgh 1966)
4.14 M. Lannoo, A.M. Stoneham: J. Phys. Chem. Sol. *29*, 1987 (1968)
4.15 D.M. Eagles: J. Phys. Chem. Sol. *16*, 76 (1960)
H.B. Bebb: Phys. Rev. *185*, 116 (1969)
A.A. Kopylov, A.M. Pikhtin: Sov. Phys. Sol. State *16*, 1200 (1975)
H.G. Grimmeiss, L.A. Lebedo: J. Appl. Phys. *46*, 2155 (1975)
B. Monemar, L. Samuelson: Phys. Rev. B*18*, 809 (1978)
4.16 G.M. Martin, A. Mitonneau, A. Mircea: Electron. Lett. *13*, 191 (1977)
4.17 D.V. Lang, R.A. Logan: J. Elect. Mat. *4*, 1053 (1975)
4.18 A. Mircea, A. Mitonneau: Appl. Phys. *8*, 15 (1975)
4.19 A. Humbert, L. Hollern, D. Bois: J. Appl. Phys. *47*, 4137 (1976)
4.20 A.M. White, P. Porteaux, W.F. Sherman, A.A. Stadtmüller: J. Phys. C*10*, L473 (1977)

A. Zylberztejn, R.H. Wallis, J.M. Besson: Appl. Phys. Lett. *32*, 764 (1978)
4.21 A. Mircea, A. Mitonneau: J. Phys. *40*, L31 (1979)
4.22 A. Mircea, A. Mitonneau, J. Hallais, M. Jaros: Phys. Rev. B*16*, 3665 (1977)
4.23 For a review on these studies see H.G. Grimmeiss: Ann. Rev. Mater. Sci. *7*, 377 (1977)
4.24 A. Chantre, G. Vincent, D. Bois: Phys. Rev. *15*, 23, 5335 (1981)
4.25 G. Lucovsky: Sol. State Commun. *3*, 299 (1965)
4.26 H. Bethe, P. Morrisson: *Elementary Nuclear Physics*, 2nd ed. (Wiley, New York 1956)
4.27 J. Walker, L.A. Vermeulen, C.D. Clark: Proc. Roy. Soc. London A*341*, 253 (1974)
4.28 C.D. Clark, J. Walker: Diamond Research 1972 (Diamond Information Bureau, London) p.2
4.29 S.A. Solin: Phys. Rev. Lett. *38*A, 101 (1972)
4.30 C.D. Clark, J. Walker: Proc. Roy. Soc. London A*334*, 241 (1973)
4.31 A.M. Stoneham: Solid State Commun. *21*, 339 (1977)
4.32 M. Lannoo, G.A. Baraff, M. Schlüter: Phys. Rev. B*24*, 943,955 (1981)
4.33 G. Davies: Rep. Prog. Phys. *44*, 788 (1981)
4.34 F.P. Larkins, A.M. Stoneham: J. Phys. C*4*, 143,154 (1971)

Chapter 5

5.1 For a detailed treatment of semiconductor statistics, see J.S. Blakemore: *Semiconductor Statistics* (Pergamon, London 1962)
5.2 E.A. Guggenheim: Proc. Phys. Soc. (London) A*66*, 121 (1953); P.T. Landsberg: Proc. Phys. Soc. (London) A*66*, 662 (1953)
5.3 F.J. Morin, J.P. Maita: Phys. Rev. *98*, 28 (1954)
5.4 P.P. Debye, E.M. Conwell: Phys. Rev. *93*, 693 (1954)
5.5 R.A. Smith: *Semiconductors* (University Press, Cambridge 1968)
5.6 R. Coates, E.W.J. Mitchell: Adv. Phys. *24*, 593 (1975)
5.7 N.F. Mott: In *Electronic and Structural Properties of Amorphous Semiconductors*, ed. by P.G. Le Comber, J. Mort (Academic, London 1973) Chap.1
5.8 See, for instance, J.M. Ziman: *Models of Disorder* (Cambridge Univ. Press, Cambridge 1972) Chap.9
5.9 A. Miller, E. Abrahams: Phys. Rev. *120*, 745 (1960)
5.10 B.I. Shklovskii, A.L. Efros, I.Y. Yanehel: JETP Lett. *14*, 233 (1971)
5.11 J.A. Johnson, K. Lark-Horovitz: Phys. Rev. *72*, 531 (1974); *71*, 374 (1947); also
I. Esterman, A. Foner, J.A. Reindall: Phys. Rev. *72*, 530 (1947)
5.12 T.A. Longo, R.K. Ray, K. Lark-Horovitz: J. Phys. Chem. Solids *8*, 259 (1959)
5.13 H. Fritzsche, M. Cuevas: Phys. Rev. *119*, 1238 (1960)
5.14 B. Massarani, M. Caillot, J.C. Bourgoin: Phys. Rev. B*15*, 2224 (1977)
5.15 J.C. Bourgoin, G. Frossati, A. Ravex, D. Thoulouze, M. Vandorpe, B. Vaksmann: Phys. Stat. Sol. (a) *92*, 585 (1979)
5.16 B. Massarani, J.C. Bourgoin, R.M. Chrenko: Phys. Rev. B*17*, 1758 (1978)
5.17 See, for instance, P. Kireev: *La Physique des Semiconducteurs* (MIR, Moscow 1975)
5.18 D.L. Dexter, F. Seitz: Phys. Rev. *86*, 964 (1952)
5.19 See, for instance, R.H. Bube: *Electronic Properties of Crystalline Solids* (Academic, New York 1974) Chap.8
5.20 E. Conwell: Phys. Rev. *77*, 388 (1950)
5.21 E. Conwell, V.F. Weisskopf: Phys. Rev. *77*, 388 (1950)

5.22 H. Brooks: *Adv. in Electronics and Electron Physics*, Vol.7, ed. by L. Marton (Academic, New York 1955) p.85
5.23 A.D. Boardman: Proc. Phys. Soc. *85*, 141 (1965)
5.24 See, for instance, H.S.W. Massey, E.H.S. Burhop: *Electronic and Ionic Impact Phenomena*, Vol.1 (Clarendon Press, Oxford 1969)
5.25 See, for instance, L. Schiff: *Quantum Mechanics*, 2nd Ed. (MacGraw Hill, New York 1949)
5.26 C. Erginsoy: Phys. Rev. *79*, 1013 (1950)
5.27 P.P. Debye, E.M. Conwell: Phys. Rev. *93*, 693 (1954)

Chapter 6

6.1 C.T. Sah: Proc. IEEE *55*, 654 (1967)
6.2 E.H. Rhoderick: *Metal-Semiconductor Contacts* (Clarendon Press, Oxford 1978)
6.3 R.A. Smith: *Semiconductors* (University Press, Cambridge 1968)
6.4 D. Pons: Thesis 3rd Cycle, Paris VI University (1979), unpublished
6.5 D. Pons: To be published
6.6 D. Pons: Appl. Phys. Lett. *37*, 413 (1980)
6.7 D. Stievenard: Thesis 3rd Cycle, Lille (1982)
6.8 C.T. Sah, L. Forbes, L.L. Rosier, A.F. Tasch, A.B. Tole: Appl. Phys. Lett. *15*, 145 (1969)
6.9 D.V. Lang: J. Appl. Phys. *7*, 3023 (1974)
D.V. Lang: In *Thermally Stimulated Relaxation in Solids*, ed. by P. Bräunlich, Topics Appl. Phys., Vol.37 (Springer, Berlin, Heidelberg, New York 1979) Chap.3
6.10 A. Broniatowski, A. Blosse, B.C. Srivastava, J.C. Bourgoin: J. Appl. Phys. (to be published)
6.11 For more details see, for instance, W.G. Oldham, S.S. Naik: Solid State Electronics *15*, 1085 (1972)
6.12 N.F. Mott, H.S. Massey: *Theory of Atomic Collisions* (Clarendon Press, Oxford 1949)
6.13 N. Solar, E. Burstein: Phys. Rev. *98*, 1757 (1955)
6.14 M. Born: Z. Physik *38*, 803 (1926)
6.15 M. Lax: Phys. Rev. *119*, 1502 (1960)
6.16 D.R. Hamann, A.L. McWorther: Phys. Rev. A*134*, 250 (1964)
6.17 G. Ascarelli, S. Rodriguez: J. Phys. Chem. Solids *22*, 57 (1961); Phys. Rev. *124*, 1321 (1961); Phys. Rev. *127*, 167 (1962)
6.18 K. Huang: Scientia Sinica *24*, 27 (1981)
6.19 B.K. Ridley: J. Phys. C: Sol. State Phys. *11*, 2323 (1978)
6.20 J.J. Markham: Rev. Mod. Phys. *31*, 956 (1959)
6.21 K. Huang, A. Rhys: Proc. Roy. Soc. (London) *204*A, 406 (1950)
6.22 V.A. Kovarskii: Sov. Phys. Sol. State *4*, 1200 (1962)
6.23 We are indebted to Dr. Burt for enlightening discussions about the comparison of the different works from [6.18 to 6.22]
6.24 D.V. Lang, C.H. Henry: Phys. Rev. Lett. *35*, 1525 (1975)
C.H. Henry, D.V. Lang: Phys. Rev. *15*, 989 (1977)
6.25 G.D. Watkins, J.R. Troxell: Phys. Rev. Lett. *44*, 593 (1980)
6.26 G.A. Baraff, E.O. Kane, M. Schlüter: Phys. Rev. *21*, 3563 (1980)
6.27 J. Frenkel: Phys. Rev. *54*, 647 (1938)
6.28 L.C. Kimerling, J.L. Benton: Appl. Phys. Lett. *39*, 410 (1981)
6.29 A.K. Jonsher: Thin Solid Films *1*, 213 (1967)
6.30 J.L. Hartke: J. Appl. Phys. *39*, 4871 (1968)
6.31 A. Chantre, G. Vincent, D. Bois: Phys. Rev. *15*, 23, 5335 (1981)
6.32 L.D. Landau, E.M. Lifshitz: *Quantum Mechanics* (Pergamon, London, Paris 1958)

6.33 J.R. Oppenheimer: Phys. Rev. *31*, 66 (1928)
6.34 S. Makram-Ebeid, M. Lannoo: Phys. Rev. Lett. *48*, 1281 (1982); Phys. Rev. *25*, 6406 (1982)
6.35 D. Pons, S. Makram-Ebeid: J. Physique Appl. *40*, 1161 (1979)

Chapter 7

7.1 G.D. Watkins, J.W. Corbett: Phys. Rev. *138*, A543 (1965)
7.2 F. Carton Merlet, B. Pajot, P. Vajda: In *Defects and Radiation Effects in Semiconductors* (Inst. Phys. London 1979) Conf. Ser. 46, p.311
7.3 C.H. Henry, H. Kukimoto, G.L. Miller, F.R. Merritt: Phys. Rev. B*7*, 2499 (1973)
7.4 W. Shockley, W.T. Read: Phys. Rev. *87*, 35 (1952)
7.5 I. Solomon: In *Amorphous Semiconductors*, ed. by M.H. Brodsky, Topics Appl. Phys., Vol.36 (Springer Berlin, Heidelberg, New York 1979) Chap.7
7.6 D. Lepine: Phys. Rev. B*6*, 436 (1972)
7.7 B.C. Cavenett: In *Luminescence Spectroscopy*, ed. by M.D. Lumb (Academic, London 1978) Chap.5
7.8 S.D. Smith, W. Taylor: Proc. Phys. Soc. *79*, 1142 (1962)
7.9 B. Massarani, J.C. Bourgoin, R.M. Chrenko: Phys. Rev. B*17*, 1758 (1978)
7.10 SH.M. Kogan, T.M. Lifshitz: Phys. Stat. Solidi (a) *39*, 11 (1977)
7.11 A.T. Collins, E.C. Lightowlers, P.J. Dean: Phys. Rev. *183*, 725 (1969)
7.12 For a review see B.C. Cavenett: In *Defects and Radiation Effects in Semiconductors* (Inst. Phys., London 1981) Conf. Ser. 59, p.69
7.13 K.M. Lee, Le Di Dang, G.D. Watkins: Sol. State Commun. *35*, 527 (1980)
7.14 R.F. Brumoin, B.C. Cavenett, J.J. Davies, J.E. Nicholls: Sol. State Commun. *18*, 1283 (1976)
7.15 P. Edel, C. Hennies, Y. Merle d'Aubigne, R. Romestain, Y. Twarowski: Phys. Rev. Lett. *28*, 1268 (1972)
7.16 D. Bimberg, A. Bubenzer: Appl. Phys. Lett. *38*, 803 (1981)
7.17 A. Rosencwarg: Physics Today *28*, 23 (1975)
7.18 L. Eaves, H. Vargas, P.J. Williams: Appl. Phys. Lett. *38*, 768 (1981)
7.19 V.S. Letokhov, V.P. Zhaorov: *Optoacoustic Laser Molecular Spectroscopy*, Springer Ser. Opt. Sci., Vol.37 (Springer Berlin, Heidelberg, New York 1983) (to be published)

Chapter 8

8.1 The interaction of point defects with dislocations, not considered here, is treated in B. Henderson: In *Defects in Crystalline Solids*, ed. by E. Arnold (The University Press, Belfast 1972) Chap.7
8.2 For reviews on radiation effects in semiconductors, see
J.W. Corbett: In *Solid State Physics*, Suppl.7, ed. by F. Seitz and D. Turnbull (Academic, New York 1966)
D.S. Billington, J.H. Crawford: *JR in Radiation Damage in Solids* (University Press, Princeton 1961)
8.3 The interaction of energetic particles with solids is treated in C.H.R. Lehman: *Interaction of Radiation with Solids and Elementary Defect Production* (North Holland, Amsterdam 1977).
A. Sosin, W. Bauer: In *Studies in Radiation Effects in Solids*, Vol.3, ed. by G.J. Dienes (Gordon and Breach, New York 1969)
F. Seitz, J.S. Kochler: In *Solid State Physics*, Vol.2, ed. by F. Seitz and D. Turnbull (Academic, New York 1956) p.307
G.J. Dienes, G.H. Vineyard: *Radiation Effects in Solids* (Interscience New York 1957)

8.4 F. Seitz: Disc. Faraday Soc. *5*, 271 (1949)
8.5 W. Kohn: Phys. Rev. *94*, A 1409 (1954)
8.6 R. Bäuerlein: In *Radiation Damage in Solids*, ed. by D. Billington (Academic, New York 1962) p.358
8.7 J.C. Bourgoin, B. Massarani: Phys. Rev. B*14*, 3690 (1976)
8.8 J.W. Corbett, G.D. Watkins: Phys. Rev. *138*, A555 (1965)
8.9 R.A. Calcott, J.W. Mac Kay: Phys. Rev. *161*, 698 (1967)
8.10 D. Pons, J.C. Bourgoin: J. Appl. Phys. *51*, 2038 (1980)
8.11 F.H. Eisen: Phys. Rev. *135* A, 1394 (1964)
8.12 R. Bäuerlein: Z. Naturforsch. *14*a, 1069 (1959)
8.13 K. Thommen: Phys. Rev. *174*, 938 (1968); *179*, 920 (1969)
8.14 C.G. Darwin: Phil. Mag. *25*, 301 (1913)
8.15 N.F. Mott: Proc. Roy. Soc. A*124*, 426 (1929); A*135*, 429 (1932)
8.16 W.A. Mc Kinley, Jr., H. Feshbach: Phys. Rev. *74*, 1759 (1948)
8.17 G.W. Kinchin, R.S. Pease: Repts. Prog. Phys. *18*, 1 (1955)
8.18 J.C. Bourgoin, J.F. Morhange, R. Beserman: Rad. Effects *22*, 205 (1974)
8.19 C. Katz, A.S. Penfold: Rev. Mod. Phys. *24*, 281 (1952)
8.20 J. Lindhard, M. Scharff, H.E. Schrott: Kgl. Danske Vid. Selsk. Matt. Fys. Medd. *33*, no 14 (1963)
8.21 G. Dearnaley, J.H. Freeman, R.S. Nelson, J. Stephen: *Ion Implantation Series Defects in Crystalline Solids*, Vol.8 (North Holland, Amsterdam 1973)
8.22 A compilation of these data can be found in: J.W. Corbett, J.C. Bourgoin: In *Point Defects in Solids*, Vol.2, ed. by J.H. Crawford and L.M. Slifkin (Plenum, New York 1975) Chap.1
8.23 J.C. Bourgoin: Rad. Effects Lett. *15*, 50 (1972)
8.24 Y. Chen, J.W. Mac Kay: Phil. Mag. *19*; 357 (1969)
W.L. Brown, W.M. Augustyniak: J. Appl. Phys. *30*, 1300 (1952)
8.25 I.V. Kryukova, V.S. Vavilov: Sov. Phys. Solid State *6*, 266 (1964)
I.M. Addad, P.C. Banbary, J.A. Grimshaw: Phil. Mag. *12*, 1203 (1965)
G.G. George, E.M. Gunnersen: In *Radiation Damage in Semiconductors*, ed. by P. Baruch (Dunod Paris 1965) p.385
J.W. Corbett, G.D. Watkins: Phys. Rev. *138*, A555 (1965)
8.26 F.H. Eisen: Phys. Rev. *135*, A1394 (1964)
8.27 D. Pons, J.C. Bourgoin: Phys. Rev. Lett. *47*, 1293 (1981)
8.28 Demonstration of this formula can be found in C. Lehman: *Interaction of Radiation with Solids* (North Holland, New York 1977) Chap.1
8.29 D.V. Lang, R.A. Logan, L.C. Kimerling: Phys. Rev. B*15*, 4874 (1977)
8.30 D. Pons: Unpublished
8.31 J.W. Corbett, G.D. Watkins: Phys. Rev. *138*A, 555 (1965)
8.32 V.S. Vavilov, A.E. Kiv, O.R. Niyazova: Phys. Stat. Solidi (a) *32*, 11 (1975)
8.33 N.E. Korsunkaya, I.V. Markovich, M.K. Sheinkman: Proc. 9th Intern. Conf. Physics of Semiconductors, Vol.2, ed. by S.M. Ryvkin (Mauka, Leningrad 1968) p.1095
8.34 J.H.O. Varley: Nature *174*, 886 (1954); J. Nucl. Energy *1*, 130 (1954); J. Phys. Chem. Solids *23*, 985 (1962)
8.35 Y. Chen, J.W. Mac Kay: Phys. Rev. *167*, 745 (1968)
8.36 E.W. Kreutz: Phys. Stat. Solidi (a) *34*, 489 (1976)
8.37 V.I. Panov, L.S. Smernov: Sov. Phys. Semicond. *5*, 1439 (1972)
8.38 J.C. Bourgoin, J.W. Corbett: Rad. Effects *36*, 157 (1978)
8.39 P.M. Mooney, J.C. Bourgoin: Unpublished

Chapter 9

9.1 A complete treatment can be found in H.M. Simpson, A. Sosin: Rad. Effects *3*, 1 (1970)
9.2 T.R. Warte: Phys. Rev. *107*, 463 (1957); also J. Chem. Phys. *28*, 103 (1958)
9.3 D. Peak, J.W. Corbett: Phys. Rev. B*5*, 1226 (1972)
9.4 D. Peak, H.L. Fresch, J.W. Corbett: Rad. Effects *11*, 149 (1971)
9.5 J. Zizine: In *Radiation Effects in Semiconductors*, ed. by F.L. Vook (Plenum, New York 1968) p.186
9.6 R.E. Mac Keiphen, J.S. Koehler: In *Radiation Effects in Semiconductors*, ed. by J.W. Corbett and G.D. Watkins (Gordon and Breach, New York 1971) p.141
9.7 R.A. Callcott, J.W. Mac Kay: Phys. Rev. *161*, 698 (1967)
9.8 B. Massarani, J.C. Bourgoin: Phys. Rev. B*14*, 3682 (1976)
9.9 D. Pons, P. Mooney, J.C. Bourgoin: J. Appl. Phys. *51*, 2038 (1980)
9.10 G.K. Wertheim: Phys. Rev. *115*, 568 (1959)
9.11 J.W. Mac Kay, E.E. Klontz: In *Radiation Damage in Semiconductors*, ed. by P. Baruch (Dunod, Paris 1964) p.11
9.12 B.L. Gregory, C.E. Barnes: In *Radiation Effects in Semiconductors*, ed. by F.L. Vook (Plenum, New York 1968) p.124
9.13 F.L. Vook: Phys. Rev. *163*, 790 (1967)
9.14 F.L. Vook, H.J. Stein: In *Radiation Effects in Semiconductors*, ed. by F.L. Vook (Plenum, New York 1968) p.99
9.15 R.E. Whan, F.L. Vook: Phys. Rev. *153*, 814 (1967)
9.16 G.J. Brucker: Phys. Rev. *183*, 712 (1969)
9.17 I. Arimura: Trans. Nucl. Sciences NS *21*, 21 (1974)
9.18 M.L. Swanson: Can. J. Phys. *44*, 2181 (1966); G.W. Gobeli: Phys. Rev. *112*, 732 (1958)
9.19 H.M. De Angelys, R.E. Penczer: Phys. Rev. *39*, 5842 (1968)
9.20 J. Bourgoin, F. Mollot: Phys. Stat. Solidi (b) *43*, 343 (1971)
9.21 J.R. Troxell, G.D. Watkins: Phys. Rev. B*22*, 921 (1980)
9.22 G.D. Watkins: In *Radiation Damage in Semiconductors*, ed. by P. Baruch (Dunod Paris 1964) p.97
9.23 G.D. Watkins: *Lattice Defects in Semiconductors* (Inst. Phys., London 1974) Conf. Ser. 23, p.1
9.24 K.L. Brower: Phys. Rev. B*4*, 1968 (1971)
9.25 G.D. Watkins: Phys. Rev. B*12*, 5824 (1975)
9.26 G.D. Watkins, J.W. Corbett, R.M. Walker: J. Appl. Phys. *30*, 1198 (1959)
9.27 J.R. Troxell, G.D. Watkins: Phys. Rev. B*22*, 921 (1980)
9.28 L.J. Cheng, M. Süle, J.D. Gerson, J.W. Corbett: Phys. Rev. B*15*, 3836 (1977)
9.29 G.D. Watkins, K.L. Brower: Phys. Rev. Lett. *36*, 1329 (1976)
9.30 L.C. Kimerling: In *Radiation Effects in Semiconductors* (Inst. Phys. London 1977) Conf. Ser. *31*, p.221
9.31 P. Vajda, L.J. Cheng: J. Appl. Phys. *42*, 2453 (1971)
9.32 G.D. Watkins, J.W. Corbett: Phys. Rev. *138*, A 543 (1965)
9.33 L.J. Cheng, J.C. Corelli, J.W. Corbett, G.D. Watkins: Phys. Rev. *152*, 761 (1966)
9.34 P. Mooney: Phys. Rev. B*15*, 3836 (1977)
9.35 A.D. Evwaraye: In *Defects and Radiation Effects in Semiconductors* (Inst. Phys., London 1979) Conf. Ser. *46*, p.267
9.36 K.L. Brower: Phys. Rev. B*1*, 1908 (1970)
9.37 C.D. Clark, A. Fernandez, D.A. Thompson: Phil. Mag. *20*, 951 (1969)
9.38 J.W. Corbett, R.S. McDonald, G.D. Watkins: J. Phys. Chem. Solids *25*, 873 (1964)
9.39 G. Bemski, G. Feher, E. Gere: Bull. Am. Phys. Soc. *4*, 135 (1968)
9.40 K.L. Brower: Phys. Rev. B*5*, 4274 (1972)

9.41 J.W. Corbett, G.D. Watkins, R.M. Chrenko, R.S. McDonald: Phys. Rev. *121*, 1015 (1961)
9.42 A.R. Bean, R.C. Newman: Solid State Commun. *9*, 271 (1971)
9.43 G.D. Watkins: Phys. Rev. B*12*, 4383 (1975)
9.44 V.S. Vavilov, B.M. Mukashev, A.V. Spitsyn: In *Radiation Damage and Defects in Semiconductors*, ed. by J.E. Whitehouse (Inst. Phys., London 1975) p.191
9.45 G.D. Watkins: Phys. Rev. *155*, 802 (1967)
9.46 G.D. Watkins: In *Radiation Effects in Semiconductors*, ed. by F.L. Vook (Plenum, New York 1968) p.67
9.47 G.D. Watkins, J.W. Corbett: Phys. Rev. *134*, A 1359 (1964)
9.48 B.L. Gregory, C.W. Gwyn: *Radiation Damage in Semiconductor Devices* Sandia Lab. report Sc-M-71 0887 (1971)
9.49 L.C. Kimerling, H.H. De Angelis, J.W. Diebold: Solid State Commun. *16*, 171 (1975)
9.50 K. Matsui, R.R. Hasiguti, T. Shoji, A. Ohkawa: In *Lattice Defects in Semiconductors*, ed. by F.A. Huntley (Inst. Phys. London 1975) Conf. Ser. *23*, p.572
9.51 G.D. Watkins: In *Radiation Effects in Semiconductors*, ed. by J.W. Corbett and G.D. Watkins (Gordon and Breach, New York 1971) p.301
9.52 A.D. Evwaraye, E. Sun: J. Appl. Phys. *47*, 3776 (1976)
9.53 E.L. Elkin, G.D. Watkins: Phys. Rev. *174*, 881 (1968)
9.54 M. Hirata, M. Hirata, H. Saito: J. Phys. Soc. Jpn. *27*, 405 (1969)
9.55 E. Sonder, L.C. Templeton: J. Appl. Phys. *34*, 3295 (1963)
9.56 G.D. Watkins: IEEE Trans. MS-*16*, 13 (1969)
9.57 Y.H. Lee, J.W. Corbett, K.L. Brower: Phys. Stat. Solidi (a) *41*, 637 (1977)
9.58 A.R. Bean, R.C. Newman, R.S. Smith: J. Phys. Chem. Solids *31*, 739 (1970)
9.59 Y.H. Lee, J.W. Corbett: Phys. Rev. B*13*, 2653 (1976)

Subject Index

Applied Physics A Solids and Surfaces

Applied Physics A "Solids and Surfaces" is devoted to concise accounts of experimental and theoretical investigations that contribute new knowledge or understanding of phenomena, principles or methods of applied research.

Emphasis is placed on the following fields:

Solid-State Physics
Semiconductor Physics: **H. J. Queisser,** MPI Stuttgart
Amorphous Semiconductors: **M. H. Brodsky,** IBM Yorktown Heights
Magnetism (Materials, Phenomena): **H. P. J. Wijn,** Philips Eindhoven
Metals and Alloys, Solid-State Electron Microscopy: **S. Amelinckx,** Mol
Positron Annihilation: **P. Hautojärvi,** Espoo
Solid-State Ionics **W. Weppner,** MPI Stuttgart

Surface Science
Surface Analysis: **H. Ibach,** KFA Jülich
Surface Physics: **D. Mills,** UC Irvine
Chemisorption: **R. Gomer,** U. Chicago

Surface Engineering
Ion Implantation and Sputtering: **H. H. Andersen,** U. Aarhus
Laser Annealing: **G. Eckhardt,** Hughes Malibu
Integrated Optics, Fiber Optics, Acoustic Surface Waves: **R. Ulrich,** TU Hamburg

Coordinating Editor: **H. K. V. Lotsch,** Heidelberg

Special Features:
- Rapid publication (3–4 months)
- No page charges for concise reports
- 50 complimentary offprints
- Microform edition available

Subscription information and/or **sample copies** are available from your bookseller or directly from Springer-Verlag, Journal Promotion Dept., P.O.Box 105280, D-6900 Heidelberg, FRG

Springer-Verlag
Berlin
Heidelberg
New York

Thermally Stimulated Relaxation in Solids

Editor: **P. Bräunlich**

1979. 142 figures, 1 table. XXI, 331 pages (Topics in Applied Physics, Volume 37) ISBN 3-540-09595-0

Contents: P. Bräunlich: Introduction and Basic Principles. – P. Bräunlich, P. Kelly, J.-P. Fillard: Thermally Stimulated Luminescence and Conductivity. – D.V. Lang: Space-Charge Spectroscopy in Semiconductors. – J. Vanderschueren, J. Gasiot: Field-Induced Thermally Stimulated Currents. – H. Glaefeke: Exoemission. – L.A. DeWerd: Applications of Thermally Stimulated Luminescence.

Very Large Scale Integration (VLSI)

Fundamentals and Applications
Editor: **D.F. Barbe**
2nd corrected and updated edition

1982. 147 figures. XI, 302 pages (Springer Series in Electrophysics, Volume 5) ISBN 3-540-11368-1

Contents: D.F. Barbe: Introduction. – J.L. Prince: VLSI Device Fundamentals. – R.K. Watts: Advanced Lithography. – P. Losleben: Computer Aided Design for VLSI. – R.C. Eden, B.M. Welch: GaAs Digital Integrated Circuits for Ultra High Speed LSI/VLSI. – E.E. Swartzlander: VLSI Architecture. – B.H. Whalen: VLSI Applications and Testing. – D.F. Barbe, E.C. Urban: VHSIC Technology and Systems. – R.I. Scace: VLSI in Other Countries. – Addenda. – Subject Index.

Springer-Verlag
Berlin
Heidelberg
New York

Insulating Films on Semiconductors

Proceedings of the Second International Conference, INFOS 81, Erlangen, Federal Republic of Germany, April 27–29, 1981
Editors: **M. Schulz, G. Pensl**

1981. 240 figures. X, 316 pages (Springer Series in Electrophysics, Volume 7) ISBN 3-540-11021-6

Contents: Si-SiO_2 Interface. – Thin Insulating Films. – Charge Injection Into Insulators. – Multilayer Structures. – Interface Characterization Techniques. – Breakdown and Instability of the SiO_2-Si System. – Technology. – Laser Processing. – Transport Properties in Inversion Layers. – Films on Compound Semiconductors. – Index of Contributors.

Ion Implantation Techniques

Lectures given at the Ion Implantation School in Connection with the Fourth International Conference on Ion Implantation: Equipment and Techniques
Berchtesgaden, Federal Republic of Germany, September 13–15, 1982
Editors: **H. Ryssel, H. Glawischnig**

1982. 245 figures. XII, 372 pages (Springer Series in Electrophysics, Volume 10)
ISBN 3-540-11878-0

Contents: Machine Aspects of Ion Implantation: Ion Implantation System Concepts. Ion Sources. Faraday Cup Designs for Ion Implantation. Safety and Ion Implanters. – Ion Ranges in Solids: The Stopping and Range of Ions in Solids. The Calculation of Ion Ranges in Solids with Analytic Solutions. Range Distributions. – Measuring Techniques and Annealing: Electrical Measuring Techniques. Wafer Mapping Techniques for Characterization of Ion Implantation Processing. Non-Electrical Measuring Techniques. Annealing and Residual Damage. – Appendix: Modern Ion Implantation Equipment: Evolution and Performance of the Nova NV-10 Predep TM Implanter. Ion Implantation Equipment from Veeco. The Series IIIA and IIIX Ion Implanters. Standard High-Voltage Power Supplies for Ion Implantation. The IONMICROPROBE A-DIDA 3000-30 for Dopant Depth Profiling and Impurity Bulk Analysis. – List of Contributors. – Subject Index.

Springer Series in Solid-State Sciences

Editors: M. Cardona P. Fulde H.-J. Queisser

1 **Principles of Magnetic Resonance** 2nd Edition 2nd Printing By C. P. Slichter

2 **Introduction to Solid-State Theory** 2nd Printing. By O. Madelung

3 **Dynamical Scattering of X-Rays in Crystals** By Z. G. Pinsker

4 **Inelastic Electron Tunneling Spectroscopy** Editor: T. Wolfram

5 **Fundamentals of Crystal Growth I** Macroscopic Equilibrium and Transport Concepts. 2nd Printing By F. Rosenberger

6 **Magnetic Flux Structures in Superconductors** By R. P. Huebener

7 **Green's Functions in Quantum Physics** By E. N. Economou

8 **Solitons and Condensed Matter Physics** 2nd Printing Editors: A. R. Bishop and T. Schneider

9 **Photoferroelectrics** By V. M. Fridkin

10 **Phonom Dispersion Relations in Insulators** By H. Bilz and W. Kress

11 **Electron Transport in Compound Semiconductors** By B. R. Nag

12 **The Physics of Elementary Excitations** By S. Nakajima, Y. Toyozawa, and R. Abe

13 **The Physics of Selenium and Tellurium** Editors: E. Gerlach and P. Grosse

14 **Magnetic Bubble Technology** 2nd Edition By A. H. Eschenfelder

15 **Modern Crystallography I** Symmetry of Crystals, Methods of Structural Crystallography By B. K. Vainshtein

16 **Organic Molecular Crystals** Their Electronic States. By E.A. Silinsh

17 **The Theory of Magnetism I** Statics and Dynamics. By D. C. Mattis

18 **Relaxation of Elementary Excitations** Editors: R. Kubo and E. Hanamura

19 **Solitons,** Mathematical Methods for Physicists. By G. Eilenberger

20 **Theory of Nonlinear Lattices** By M. Toda

21 **Modern Crystallography II** Structure of Crystals. By B. K. Vainshtein, V. M. Fridkin, and V. L. Indenbom

22 **Point Defects in Semiconductors I** Theoretical Aspects By M. Lannoo and J. Bourgoin

23 **Physics in One Dimension** Editors: J. Bernasconi, T. Schneider

24 **Physics in High Magnetic Fields** Editors: S. Chikazumi and N. Miura

25 **Fundamental Physics of Amorphous Semiconductors** Editor: F. Yonezawa

26 **Elastic Media with Microstructure I** One-Dimensional Models. By I.A. Kunin

27 **Superconductivity of Transition Metals** Their Alloys and Compounds. By S.V. Vonsovsky, Yu.A. Izyumov, and E.Z. Kurmaev

28 **The Structure and Properties of Matter** Editor: T.Matsubara

29 **Electron Correlation and Magnetism in Narrow-Band Systems** Editor: T. Moriya

30 **Statistical Physics I** By M. Toda, R. Kubo, N. Saito

31 **Statistical Physics II** By R. Kubo, M. Toda, N. Hashitsume

32 **Quantum Theory of Magnetism** By R. M. White

33 **Mixed Crystals** By A. I. Kitaigorodsky

34 **Phonons: Theory and Experiments I** Lattice Dynamics and Models of Interatomic Forces. By P. Brüesch

35 **Point Defects in Semiconductors II** Experimental Aspects By J. Bourgoin and M. Lannoo

36 **Modern Crystallography III** Crystal Growth By A. A. Chernov et al.

37 **Modern Crystallography IV** Physical Properties of Crystals By L. A. Shuvalov et al.

38 **Physics of Intercalation Compounds** Editors: L. Pietronero and E. Tosatti

39 **Anderson Localization** Editors: Y. Nagaoka and H. Fukuyama

Springer Series in Solid-State Sciences

Editors: M. Cardona P. Fulde H.-J. Queisser

1 **Principles of Magnetic Resonance** 2nd Edition 2nd Printing By C. P. Slichter

2 **Introduction to Solid-State Theory** 2nd Printing. By O. Madelung

3 **Dynamical Scattering of X-Rays in Crystals** By Z. G. Pinsker

4 **Inelastic Electron Tunneling Spectroscopy** Editor: T. Wolfram

5 **Fundamentals of Crystal Growth I** Macroscopic Equilibrium and Transport Concepts. 2nd Printing By F. Rosenberger

6 **Magnetic Flux Structures in Superconductors** By R. P. Huebener

7 **Green's Functions in Quantum Physics** By E. N. Economou

8 **Solitons and Condensed Matter Physics** 2nd Printing Editors: A. R. Bishop and T. Schneider

9 **Photoferroelectrics** By V. M. Fridkin

10 **Phonom Dispersion Relations in Insulators** By H. Bilz and W. Kress

11 **Electron Transport in Compound Semiconductors** By B. R. Nag

12 **The Physics of Elementary Excitations** By S. Nakajima, Y. Toyozawa, and R. Abe

13 **The Physics of Selenium and Tellurium** Editors: E. Gerlach and P. Grosse

14 **Magnetic Bubble Technology** 2nd Edition By A. H. Eschenfelder

15 **Modern Crystallography I** Symmetry of Crystals, Methods of Structural Crystallography By B. K. Vainshtein

16 **Organic Molecular Crystals** Their Electronic States. By E.A. Silinsh

17 **The Theory of Magnetism I** Statics and Dynamics. By D. C. Mattis

18 **Relaxation of Elementary Excitations** Editors: R. Kubo and E. Hanamura

19 **Solitons,** Mathematical Methods for Physicists. By G. Eilenberger

20 **Theory of Nonlinear Lattices** By M. Toda

21 **Modern Crystallography II** Structure of Crystals. By B. K. Vainshtein, V. M. Fridkin, and V. L. Indenbom

22 **Point Defects in Semiconductors I** Theoretical Aspects By M. Lannoo and J. Bourgoin

23 **Physics in One Dimension** Editors: J. Bernasconi, T. Schneider

24 **Physics in High Magnetic Fields** Editors: S. Chikazumi and N. Miura

25 **Fundamental Physics of Amorphous Semiconductors** Editor: F. Yonezawa

26 **Elastic Media with Microstructure I** One-Dimensional Models. By I.A. Kunin

27 **Superconductivity of Transition Metals** Their Alloys and Compounds. By S.V. Vonsovsky, Yu.A. Izyumov, and E.Z. Kurmaev

28 **The Structure and Properties of Matter** Editor: T.Matsubara

29 **Electron Correlation and Magnetism in Narrow-Band Systems** Editor: T. Moriya

30 **Statistical Physics I** By M. Toda, R. Kubo, N. Saito

31 **Statistical Physics II** By R. Kubo, M. Toda, N. Hashitsume

32 **Quantum Theory of Magnetism** By R. M. White

33 **Mixed Crystals** By A. I. Kitaigorodsky

34 **Phonons: Theory and Experiments I** Lattice Dynamics and Models of Interatomic Forces. By P. Brüesch

35 **Point Defects in Semiconductors II** Experimental Aspects By J. Bourgoin and M. Lannoo

36 **Modern Crystallography III** Crystal Growth By A. A. Chernov et al.

37 **Modern Crystallography IV** Physical Properties of Crystals By L. A. Shuvalov et al.

38 **Physics of Intercalation Compounds** Editors: L. Pietronero and E. Tosatti

39 **Anderson Localization** Editors: Y. Nagaoka and H. Fukuyama

Springer Series in Solid-State Sciences

Editors: M. Cardona P. Fulde H.-J. Queisser

MIX
Papier aus verantwortungsvollen Quellen
Paper from responsible sources
FSC® C105338

If you have any concerns about our products,
you can contact us on
ProductSafety@springernature.com

In case Publisher is established outside the EU,
the EU authorized representative is:
Springer Nature Customer Service Center GmbH
Europaplatz 3, 69115 Heidelberg, Germany

Printed by Libri Plureos GmbH
in Hamburg, Germany